Badis Djamaa · Abdelhamid Boudane ·
Oussama Mazari Abdessameud ·
Adil Imad Eddine Hosni
Editors

Advances in Computing Systems and Applications

Proceedings of the 6th Conference on Computing Systems and Applications

Springer

Editors
Badis Djamaa
Computer Science Department
Ecole Militaire Polytechnique
Algiers, Algeria

Abdelhamid Boudane
Computer Science Department
Ecole Militaire Polytechnique
Algiers, Algeria

Oussama Mazari Abdessameud
Computer Science Department
Ecole Militaire Polytechnique
Algiers, Algeria

Adil Imad Eddine Hosni
Computer Science Department
Ecole Militaire Polytechnique
Algiers, Algeria

ISSN 2367-3370 ISSN 2367-3389 (electronic)
Lecture Notes in Networks and Systems
ISBN 978-3-031-71847-2 ISBN 978-3-031-71848-9 (eBook)
https://doi.org/10.1007/978-3-031-71848-9

This Springer imprint is published by the registered company Springer Nature Switzerland AG
The registered company address is: Gewerbestrasse 11, 6330 Cham, Switzerland

Lecture Notes in Networks and Systems 1145

The series "Lecture Notes in Networks and Systems" publishes the latest developments in Networks and Systems—quickly, informally and with high quality. Original research reported in proceedings and post-proceedings represents the core of LNNS.

Volumes published in LNNS embrace all aspects and subfields of, as well as new challenges in, Networks and Systems.

The series contains proceedings and edited volumes in systems and networks, spanning the areas of Cyber-Physical Systems, Autonomous Systems, Sensor Networks, Control Systems, Energy Systems, Automotive Systems, Biological Systems, Vehicular Networking and Connected Vehicles, Aerospace Systems, Automation, Manufacturing, Smart Grids, Nonlinear Systems, Power Systems, Robotics, Social Systems, Economic Systems and other. Of particular value to both the contributors and the readership are the short publication timeframe and the worldwide distribution and exposure which enable both a wide and rapid dissemination of research output.

The series covers the theory, applications, and perspectives on the state of the art and future developments relevant to systems and networks, decision making, control, complex processes and related areas, as embedded in the fields of interdisciplinary and applied sciences, engineering, computer science, physics, economics, social, and life sciences, as well as the paradigms and methodologies behind them.

Indexed by SCOPUS, EI Compendex, INSPEC, WTI Frankfurt eG, zbMATH, SCImago.

All books published in the series are submitted for consideration in Web of Science.

For proposals from Asia please contact Aninda Bose (aninda.bose@springer.com).

Preface

This book is a collection of the articles presented at the sixth conference on Computing Systems and Applications (CSA 2024), held at Ecole Militaire Polytechnique (EMP) on April 23–24, 2024. It explores the latest advances on the rapidly evolving field of computing systems and applications, covering a wide array of modern computer science disciplines from artificial intelligence and data science to computer networks and security and from information systems and software engineering to computer vision and graphics.

For this edition, the organizing committee of the conference received over 182 research articles from more than ten countries worldwide. Each submission underwent a rigorous screening process and was reviewed by at least two competent reviewers from an international technical program committee. Ultimately, 37 articles were accepted, resulting in an acceptance rate of approximately 20%. These articles were divided and presented across eight oral and one poster sessions held over the conference's two-day period. During the closing ceremony, the chairs awarded the best paper award to the presentation that received the highest review score and demonstrated outstanding quality.

Three keynote speeches were presented by leading researchers in the rapidly evolving field of computing systems and applications. The first keynote, delivered by Prof. Elhadj BENKHELIFA from Staffordshire University, UK, explored how to build effective cybersecurity solutions and discussed the common pitfalls. The second keynote was given by Dr. Riyadh BEGHDADI, a research affiliate at MIT, USA, and New York University Abu Dhabi, UAE, who discussed recent advancements in optimizing deep learning models for efficient inference. The final presentation, given by Prof. Hamamache KHEDDOUCI from Claude Bernard Lyon 1 University, France, highlighted the latest developments in bridging the gap between graph-theoretical models and real graph data. The follow-up discussions have underlined the current opportunities and challenges in the field and set up the ground for collaboration.

We are especially delighted to observe the exceptional quality of presentations and research articles featured in the book. We believe that the presented prototypes and obtained results will help advancing research in the field and serve as a valuable resource for researchers, practitioners, and students eager to explore the potential and impact of computing systems across diverse domains and disciplines.

Finally, we extend our heartfelt gratitude to everyone who contributed to making this event a success. In particular, we are thankful to all members of the technical program committee and all authors who contributed their expertise and research. We also express our appreciation to the session chairs and keynote speakers for their valuable insights and engaging presentations. A special thank you goes to the local organizing committee and all the staff at the EMP for their support, hospitality, and dedication, which played a crucial role in ensuring the success of this event.

Kadda Beghdad Bey
Badis Djamaa
Conference General Chairs

Organization

Conference General Chairs

Kadda Beghadad Bey	EMP
Badis Djamaa	EMP

Steering Committee

Abdelhamid Boudane	EMP
Oussama Mazari Abdessameud	EMP
Adil Imad Eddine Hosni	EMP
Omar Zelmati	EMP

Website Chair

M'hamed Mataoui	EMP

Technical Program Committee

Mourad Adnane	ENP, Algeria
Mohamed Ahmed Nacer	USTHB, Algeria
Rachid Ahmed Ouamer	LARI laboratory, Algeria
Amar Aissani	USTHB, Algeria
Yamine Ait Ameur	Univ. of Toulouse, France
Ahcen Aliouat	IMT, Atlantique
Hadjali Allel	ENSMA, France
Abdenour Amamra	EMP, Algeria
Idir Amine Amarouche	EMP, Algeria
Saliha Aouat	USTHB, Algeria
Oualid Araar	EMP, Algeria
Abdelmalik Bachir	ENSIA, Algeria
Miloud Bagaa	UQTR, Canada
Zakaria Bairi	EMP, Algeria
Abdelghani Bakhtouchi	EMP, Algeria
Fairouz Beggas	ASSI, Algeria

Djamal Belazzougui	CERIST, Algeria
Ali Belgacem	M'hamed Bougara Univ., Algeria
Mahmoud Belhocine	CDTA, Algeria
Ladjel Bellatreche	ENSMA, France
Redha Benaissa	EMP, Algeria
Akrem Benatia	EMP, Algeria
Samir Benbelkacem	CDTA, Algeria
Fatima Benbouzid-Si Tayeb	ESI, Algeria
Salem Benferhat	Artois Univ., France
Sid Ahmed Berrani	ENP, Algeria
Salim Bitam	Univ. of Biskra, Algeria
Tassadit Bouadi	Univ. de Rennes 1, France
Louiza Bouallouche	Univ. of Bejaia, Algeria
Boudjemaa Boudaa	Univ. of Tiaret, Algeria
Mourad Boudhar	USTHB, Algeria
Mohamed Amine Boudia	LAMIH, France
Mohand Boughanem	Univ. Paul Sabatier Toulouse, France
Kamel Boukhalfa	USTHB, Algeria
Said Yacine Boulahia	EMP, Algeria
Omar Boussaid	Univ. Lumiere Lyon 2, France
Abderraouf Bouziane	BBA Univ., Algeria
Abbas Bradai	Univ. de Côte d'Azur, France
Talal Ashraf Butt	HCT, UAE
Yacine Challal	UDST, Qatar
Djalel Chefrour	Univ. of Souk Ahras, Algeria
Walid Cherifi	EMP, Algeria
Lounis Chermak	Cranfield Univ., UK
Islam Debicha	EMP, Algeria
Oualid Demigha	EMP, Algeria
Fethi Demim	EMP, Algeria
Adel Djama	EMP, Algeria
Ahmed El Amine Djebri	EMP, Algeria
A. Oualid Djekoune	CDTA, Algeria
Younes Djellouli	ENPEI, Algeria
Yassine Djouadi	Univ. Alger 1, Algeria
Habiba Drias	USTHB, Algeria
Amar El Maadi	DRC, Algeria
Ahmed Fathy	Al Jouf Univ., Saudi Arabia
Mohamed Amine Ferrag	TII, UAE
Krasin Georgiev	Technical Univ., of Sofia, Bulgaria
Abdessamed Reda Ghomari	ESI, Algeria
Ahmed Guessoum	ENSIA, Algeria

Mohamed Guiatni	EMP, Algeria
Karen Gupta	SunPower Corporation, USA
Abdenour Hacine Gharbi	BBA Univ., Algeria
Abderrahim Halimi	Heriot-Watt Univ., UK
Sherief Hashima	Riken-AIP, Japan
Xuanxiang Huang	Toulouse Univ., France
Malika Ioualalen	USTHB, Algeria
Yacine Izza	National Univ. of Singapore
Akhil Kallepalli	Glasgow Univ., Scotland
Okba Kazar	Univ. of Biskra, Algeria
Tayeb Kenaza	EMP, Algeria
Selma Khouri	ESI, Algeria
Assia Kourgli	USTHB, Algeria
Ichraf Lahouli	Apixa, Belgium
Rabah Louali	EMP, Algeria
Khaled Loukhaoukha	CRD, Algeria
Mohamed Maazouz	Univ. of MEDEA, Algeria
Saïd Mahmoudi	Univ. de Mons, Belgium
Mohamed Maiza	EMP, Algeria
Ricardo Marques	Univ. de Barcelona, Spain
Mhamed Mataoui	EMP, Algeria
Fatiha Merazka	USTHB, Algeria
Mahammed Messadi	Univ. of Tlemcen, Algeria
Rahim Messaoud	Univ. of MEDEA, Algeria
Samira Moussaoui	USTHB, Algeria
Abdelkrim Nemra	EMP, Algeria
Farid Nouioua	BBA Univ., Algeria
Boubakr Nour	Ericsson Research, Canada
Abdelkader Ouali	Caen Normandy Univ., France
Ismail Oukid	Snowflake Computing, Germany
Belkacem Ould Bouamama	Lille 1 Univ., France
Badran Raddaoui	Artois Univ., France
Abdenebi Rouigueb	EMP, Algeria
Lamia Sadeg	EMP, Algeria
Said Sadoudi	EMP, Algeria
Zakaria Sahraoui	EMP, Algeria
Lakhdar Sais	Univ. d'Artois, France
Faouzi Sebbak	EMP, Algeria
Mustapha Reda Senouci	EMP, Algeria
Benslimane Sidi Mohamed	ESI Sidi Bel Abbes, Algeria
Sérgio Silva	Univ. Aveiro, Portugal
Karim Tabia	Artois Univ., France

Johan Van Kerckhoven	RMA, Belgium
Amara Yacine	EMP, Algeria
Saïd Yahiaoui	CERIST, Algeria
Khalil Zebbiche	DRC, Algeria
Nadia Zenati-Henda	CDTA, Algeria

Contents

Artificial Intelligence and Data Science

Multimodal Contactless Human Stress Detection Using Deep Learning

Wafa Mellouk(✉) and Wahida Handouzi

Laboratoire d'automatique de tlemcen LAT, Tlemcen University, Chetouane, Algeria
wafa.mellouk@univ-tlemcen.dz

Abstract. It is widely acknowledged that everyone is subjected to some pressure or stress in his daily life. This is a feeling that always accompanies the person's behavior and affects it negatively. For instance, driving a car under extreme stress can generally lead to a fatal accident. Today, the topic of automatic human stress detection is attracting an increasing number of researchers due to its great importance. The present article aims primarily to explore this subject using deep learning techniques with the purpose to detect and classify human stress and non-stress states. It has been revealed that the multimodal approach, which is based on data fusion, is a remarkably successful method. In this study, the fusion between facial expressions and physiological signals is used. The originality of the present study lies in the fact that the PPG signals used are remotely extracted from RGB facial videos. Moreover, the experimentation employs the UBFC-Phys dataset, a publicly available multimodal dataset. Different deep learning architectures, such as 3D-convolutional neural network (CNN) for the facial expressions and 1D-CNN for contactless PPG signals, are proposed in this article. The findings showed that the proposed methods can achieve validation accuracies as high as 83.33%, 75%, and 100% for facial expression, contactless PPG signals, and multimodal systems, respectively.

Keywords: Stress · Facial expressions · iPPG signal · Deep learning · Multimodal

1 Introduction

Stress is one of the negative emotions that an individual often experiences in his daily life. For example, giving a presentation and speaking to an audience, getting through an evaluation exam, waiting for medical analysis results, can be highly stressful. In this context, the results of a research study carried out on this topic indicated that the constant stress-feeling can decrease body immunity and may cause several diseases [1].

Researchers have reported that a person can avoid the risk of being stressed all the time by recognizing his/her stress and doing whatever is necessary to relax. It has also been reported that the transition from a neutral state to a stress state is generally accompanied by several changes in the human body. Some of these changes may be external, i.e. facial expressions, or internal, i.e. physiological signals.

B. Djamaa et al. (Eds.): CSA 2024, LNNS 1145, pp. 3–12, 2025.
https://doi.org/10.1007/978-3-031-71848-9_1

The purpose of the present work is to propose a classification of the human stressed or unstressed states. The experiment conducted in this article used two different types of data, i.e. facial expressions and physiological signals.

It is worth emphasizing that facial expressions are among the first external signs that a person shows when feeling some kind of emotion. This form of non-verbal communication, which conveys interesting information about the emotional state of a person, has been widely utilized in numerous research studies. Almeida et Rodrigues [3] used the pre-trained neural networks VGG16, VGG19 and Inception-ResnetV2 network to select the best model for stress detection under two different classifier. The experiment was conducted using the CK+ and KDFE datasets. In the same context, Zhang et al. [4] proposed a novel deep learning CNN architecture that combines low features with high features. It is worth indicating that high accuracies, i.e. 91.2%, 80.4% and 99.3%, were obtained with the CK+, Oulu-CASIA, and KMU-FED datasets, respectively.

It is worth highlighting that the physiological modality has attracted the interest of many researchers working in the field of emotion classification. It should be mentioned that the self-governing nervous system is automatically activated when a person feels some kind of emotion, which leads to a number of physiological changes in the heart rate, human body temperature, etc. [5].

However, the sensors that are placed on the human body are not usually accepted by people as they cannot stand them. For this, numerous researchers have turned to using physiological signals extracted remotely from facial videos in the human emotion classification [6, 7]. The contactless signals extracted from videos are referred to as photoplethysmography (PPG) signals. It is interesting to note that there have been numerous studies on PPG signals that are based on the extraction of the red, green and blue (RGB) colors from facial videos [8].

Maaoui et al. [9] used the non-contact PPG signals to recognize human stress states and classify them. A high recognition rate of 94.4% was obtained by the SVM classifier and a radial basis function (RBF).

Furthermore, a novel multimodal database, i.e. UBFC-Phys, was created in order to carry out studies on human stress detection [10]. Researchers have used this dataset to show that the Plane-Orthogonal-to-Skin (POS) algorithm for PPG signal extraction achieved higher accuracy in contactless PPG signals extracted from videos. They obtained a recognition rate of 85.48% in stress classification.

It is noteworthy that facial expressions and physiological signals are two different modalities that proved to be of major importance in the field of research about the classification of human emotions. Ouzar et al. [6] combined this two modalities for the purpose of classifying stress or non-stress state. These researchers used the MTTS-CAN algorithm to extract PPG signals from videos, and the VGG16 network for facial expressions. They obtained 91.07% of accuracy. In this regard, Fig. 1 clearly illustrates the method proposed to be used in this study.

Through this rich contribution, different types of deep learning (DL) architectures for stress classification are introduced, using two different modalities, such as the 3D-convolutional neural network (3D-CNN) for facial expression recognition and the 1D-convolutional neural network (1D-CNN) for contactless PPG signals. The early fusion of multimodal approach was then applied.

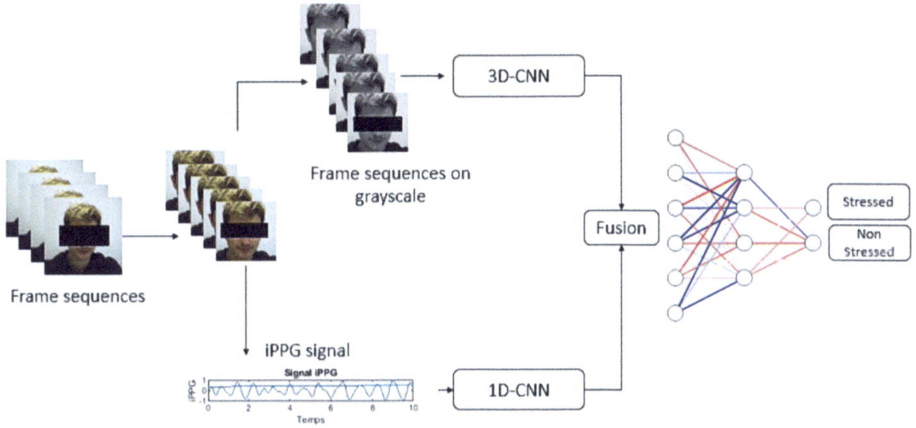

Fig. 1. Multimodal human stress status classification using facial expressions and contactless PPG signals from facial videos. The participant image used is from the publicly available dataset called UBFC-Phy [10].

The remainder of this paper is organized as follows. In section 2, it presents the steps used to preprocess facial expression data, as well as those used to extract PPG signals for later use, while section 3 describes the experiments carried out using deep learning techniques along with the results obtained. The fourth section examines and discusses these results. Finally, a conclusion to the article is drawn in the fifth and last section.

2 Data Preparation and Collection

In order to achieve the objective of this study, it was deemed interesting to use the multimodal UBFC-Phys dataset that is primarily dedicated to the recognition of human stress [10].

2.1 Data Preparation for Facial Expression

As mentioned above, the study conducted here is based on deep learning techniques. Data preprocessing is an important step that helps to clarify and retrain important features for learning and eliminate unimportant features for better training. It should be pointed out that higher quality RGB facial videos were obtained with a frame rate of 35fps and 1024 × 1024 pixels, using the UBFC-Phys dataset.

The videos of 12 participants from the UBFC-Phys database (12 videos for the stress state and 12 for the unstressed state) were used in this experiment. Moreover, we used10 seconds per video, which obtained 350 frames per video.

The cropping technique was applied to each frame. Here, the focus was put on the face in order to detect only the face images while eliminating the background [11].

The images, which were initially in RGB color, were converted to black and white (grayscale). The images used as input to the deep learning network were resized to 75 × 75 pixels with gray color (Fig. 2).

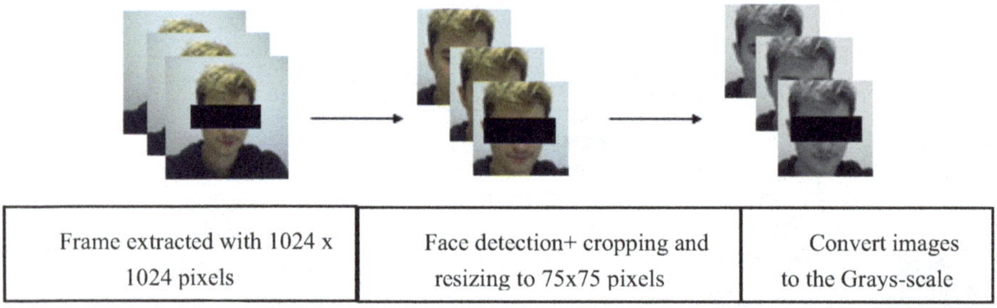

Fig. 2. Preprocessing steps proposed for facial expression recognition. The participant image used is from the publicly available dataset called UBFC-Phys [10].

2.2 Collection of Contactless PPG Signals

This paper seeks to present a human stress classification that is based on contactless physiological signals. It is worth reminding that, in our previous work, a comparative study was conducted on four contactless PPG signal extraction methods, under two different face regions [12]. It was thus found that the independent component analysis (ICA) method [8] with good full-face detection could achieve accurate extraction of PPG signals.

This step of the study was carried out using the iphys Toolbox MATLAB code that was originally designed and presented by McDuff and Blackford [13]. It should be noted that the cropping technique was applied so as to concentrate on the face in order to detect the features of the face only.

3 Experimentation and Results

This section aims primarily to present the proposed deep learning architectures for human stress classification along with the results obtained. For this, it was considered more interesting to begin our study with the unimodal approach separately, and then move on to the multimodal approach.

3.1 Stress Classification Using Facial Expressions

Being aware of the great success that the deep learning networks have previously witnessed in the computer vision area, it was deemed appropriate to propose a 3D-CNN to classify human stress and non-stress states.

It has been reported that the most widely used deep learning architecture for video classification is the 3D-CNN [14]. It was found that this 3D-CNN architecture is the most adequate, as it allows analyzing a set of sequential frames at the same time, which can help to preserve important features over time. The following table presents the proposed 3D-CNN for human stress detection (Table 1).

Table 1. 3D-CNN proposed for human stress classification.

Layer	Filter size	Output size	Parameters
Conv-3D	(16, (3,3,50))	(None, 16, 73, 73, 301)	7216
Max-Pooling	(3,3,3)	(None, 16, 24, 24, 100)	0
Dropout	0.5	(None, 16, 24, 24, 100)	0
Conv-3D	(32, (3,3,25))	(None, 32, 22, 22, 76)	115232
Max-Pooling	(3,3,3)	(None, 32, 7, 7, 36)	0
Dropout	0.5	(None, 32, 7, 7, 25)	0
Conv-3D	(64, (3,3,10))	(None, 64, 5, 5, 16)	184384
Max-Pooling	(2,2,2)	(None, 64, 2, 2, 5)	0
Dropout	0.5	(None, 64, 2, 2, 5)	0
Flatten	/	(None, 320)	0
Fully connected	1024	(None, 1024)	328704
Dropout	0.5	(None, 1024)	0
Fully connected	512	(None,512)	524800
Dropout	0.5	(None, 512)	0
Fully connected	256	(None, 256)	131328
Dropout	0.5	(None, 256)	0
Fully connected	128	(None, 128)	32896
Dropout	0.5	(None, 128)	0
Output	2	(None, 2)	258

The proposed 3D-CNN architecture was trained for 100 epochs, using the Adam optimizer, with a learning rate of 0.001 and the categorical cross-entropy loss function. Figure 3 presents the confusion matrix obtained in our experiment.

The model achieved a high recognition rate, with 80.33% of validation accuracy and 23% of validation loss. The next section deals with another step of the study, which consists of classifying the stressed or non-stressed state using contactless PPG.

3.2 Stress Classification Using Contactless PPG Signals

The next step, following the collection of all the required data presented in Sect. 2, was to focus on stress classification based on the extracted. Based on the idea presented in our previous work [7], it was considered more appropriate to use the 1D-CNN model for the classification of signals after their normalization and segmentation. It should be emphasized that the contactless PPG signals extracted before have a size of 1054×1. In the segmentation step, the PPG signal was divided into two signals with a size of 527×1 each corresponding to five seconds. These two smaller signals (527×1) were then used at the input of the deep learning network. The 1D-CNN architecture, which

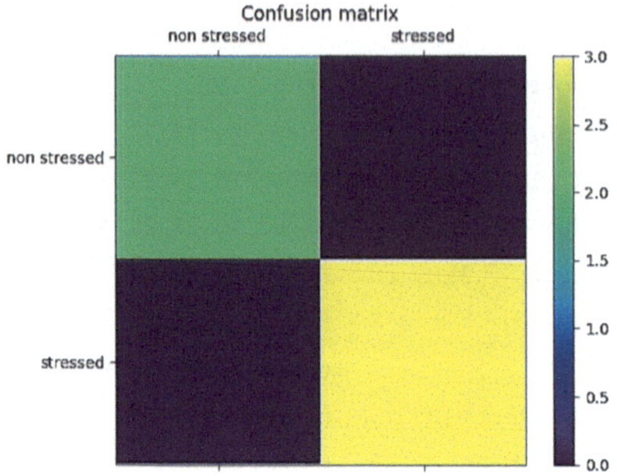

Fig. 3. The confusion matrix obtained using facial expressions.

contains one-dimensional filters, turned out to be highly suitable for one-dimensional signals. Table 2 presents the proposed 1D-CNN architecture for stress detection using contactless PPG signals.

Table 2. 1D-CNN proposed for stress detection using contactless PPG signals.

Layer	Filter size	Output size	Parameters
Conv-1D	32.527	(None, 527.32)	16896
Max-Pooling	1.2	(None, 263.32)	0
Batch normalization	/	(None, 263.32)	128
Dropout	0.5	(None, 263.32)	0
Conv-1D	64.250	(None, 263.32)	256032
Max-Pooling	1.2	(None, 131.32)	0
Batch normalization	/	(None, 131.32)	128
Dropout	0.5	(None, 131.32)	0
Flatten	/	(None, 4192)	0
Fully connected	1024	(None, 1024)	4293632
Fully connected	256	(None, 256)	262400
Fully connected	128	(None, 128)	32896
Fully connected	64	(None, 64)	8256
Output	1	(None, 1)	65

A binary classification of either stressed state or non-stressed state was carried out in this experiment. It is important to note that the proposed model was trained for 50 epochs, using the Adam optimizer, with 0.01 learning rate. In order to evaluate the model proposed here, four essential parameters, namely Accuracy (AC), Score (F1), Recall (RE), and Precision (PR), were calculated. Once the training phase completed, the following results, i.e. 75% of accuracy, 75% of precision, 50% of recall, and 66% of F1 score, were obtained. Figures 4 and 5 depict the resulting confusion matrix and validation and training loss, respectively.

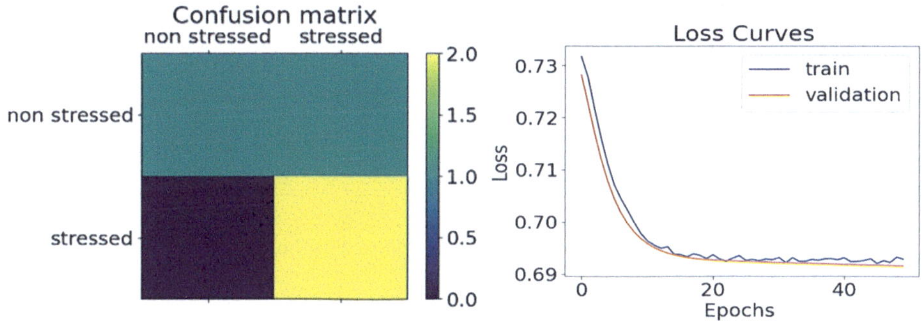

Fig. 4. Confusion matrix and loss curves obtained using contactless PPG signals.

The next step after the unimodal stress classification was to create a multimodal system by merging the two modalities mentioned before, namely facial expressions and physiological signals.

3.3 Multimodal Stress Classification

This section completes our study by creating a multimodal system. Figure 5 illustrates the proposed multimodal system for the classification of human stress states using deep learning techniques. The findings of a large number of studies have confirmed that the multimodal approach achieves a higher recognition rate than the unimodal approach [6]. As indicated in Fig. 5, the proposed model is based on the approaches presented in the previous sections, which are data processing and the deep learning model. The feature extraction step for each modality was first carried out. Next, after the parameter fusion stage, there comes the classification stage in which fully connected layers were used to classify the stressed or unstressed human states. The proposed model was trained for 50 epochs using the Adam optimizer. A higher recognition rate was achieved with 100% validation accuracy. Figure 6 presents the confusion matrix obtained with a multimodal system.

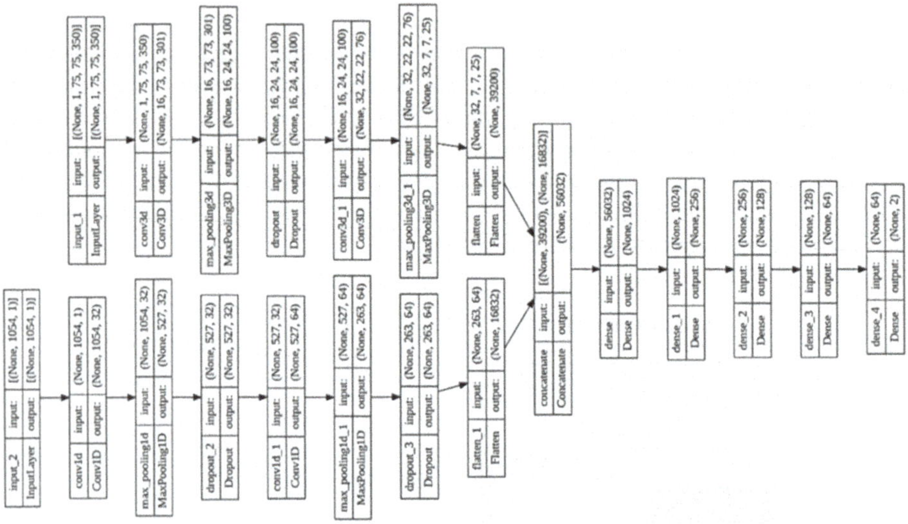

Fig. 5. Multimodal deep learning model proposed for human stress detect.

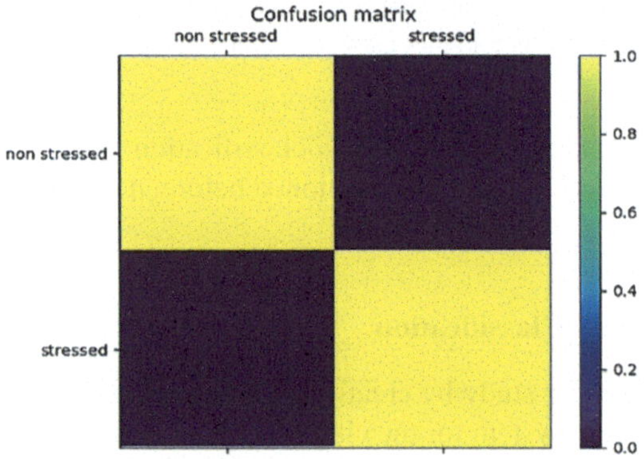

Fig. 6. Confusion matrix obtained using a contactless multimodal system.

4 Discussion and the Comparative Study with the State of the Art Methods

The multimodal system proposed in this study is based on the merging of facial expressions and contactless physiological signals PPG. The table below summarizes the results obtained in this study.

The comparison results presented in Table 3 suggest that our methods are similar if not better than those reported in other works, whether unimodal or multimodal. It should be highlighted that the deep learning models used in this study have the potential to execute classification tasks and identify patterns in signals and images. Furthermore, it was also found that the multimodal system improves the unimodal system performance.

Table 3. Comparative results for human stress detection using offline physiological signals, facial expressions, as well as a multimodal approach using the UBFC-Phys dataset.

Proposals	Modality	Methods	Accuracy
Ouzar et al. [6]	Contactless PPG signal	RF	62.4%
	Facial expressions	VGG16	82.48%
	Multimodal	MLP	91.07%
Our method	Contactless PPG signal	1D-CNN	75%
	Facial expressions	3D-CNN	83.33%
	Multimodal	MLP	100%

5 Conclusion

The classification model proposed in this study is highly efficient. It helps researchers in the field of automatic human stress recognition, to further develop and confirm these findings. The first step of the model consisted of investigating stress recognition using the facial expressions modality, by applying deep learning techniques such as 3D-CNN. The second step of the study focused on extracting PPG signals from RGB facial videos and the use of these signals as a second modality for the classification of stress and non-stress states, using the 1D-CNN network. For the purpose of achieving more accurate results in this study, it was deemed more adequate to apply the concept of multimodality by merging two types of data, i.e. facial expressions and physiological signals data. The originality of this work lies in the fact that it deals with a multimodal study that considers single input data, i.e. facial videos. In addition, this study, which was based on the public UBFC-Phys database, allowed achieving 100% accuracy.

In future research, it would be highly recommended to include all participants from the UBFC-Phys database, provided that powerful computers are available.

Acknowledgments. This work is supported by the Directorate General of Scientific Research and Technological Development DGRSDT.

References

1. Lagraauw, H.M., Kuiper, J., Bot, I.: Acute and chronic psychological stress as risk factors for cardiovascular disease: Insights gained from epidemiological, clinical and experimental studies. Brain Behav. Immun. **50**, 18–30 (2015). https://doi.org/10.1016/j.bbi.2015.08.007
2. Mahesh, B., Hassan, T., Prassler, E., Garbas, J.-U.: Requirements for a reference dataset for multimodal human stress detection. In: Proceedings of the 2019 IEEE International Conference on Pervasive Computing and Communications Workshops (PerCom Workshops), pp. 492–498 (2019). https://doi.org/10.1109/PERCOMW.2019.8730884
3. Almeida, Rodrigues, F.: Facial expression recognition system for stress detection with deep learning. In: Proceedings of the 23rd International Conference on Enterprise Information Systems, Online Streaming, pp. 256–263. SCITEPRESS - Science and Technology Publications (2021). https://doi.org/10.5220/0010474202560263

4. Zhang, J., Mei, X., Liu, H., Yuan, S., Qian, T.: Detecting negative emotional stress based on facial expression in real time. In: Proceedings of the 2019 IEEE 4th International Conference on Signal and Image Processing (ICSIP), pp. 430–434 (2019). https://doi.org/10.1109/SIPROCESS.2019.8868735
5. Shu, L., et al.: A review of emotion recognition using physiological signals. Sensors **18**(7), Art. no 7 (2018). https://doi.org/10.3390/s18072074
6. Ouzar, Y., Lagha, L., Bousefsaf, F., Maaoui, C.: Multimodal stress state detection from facial videos using physiological signals and facial features. In: Rousseau, J.J., Kapralos, B. (eds.) Pattern Recognition, Computer Vision, and Image Processing. ICPR 2022 International Workshops and Challenges. ICPR 2022. Lecture Notes in Computer Science, vol. 13646, pp. 139–150. Springer, Cham (2023). https://doi.org/10.1007/978-3-031-37745-7_10
7. Mellouk, W., Handouzi, W.: CNN-LSTM for automatic emotion recognition using contactless photoplythesmographic signals. Biomed. Signal Process. Control **85**, 104907 (2023). https://doi.org/10.1016/j.bspc.2023.104907
8. Poh, M.-Z., McDuff, D.J., Picard, R.W.: Non-contact, automated cardiac pulse measurements using video imaging and blind source separation. Opt. Express OE **18**(10), 10762–10774 (2010). https://doi.org/10.1364/OE.18.010762
9. Maaoui, C., Bousefsaf, F., Pruski, A.: Automatic human stress detection based on webcam photoplethysmographic signals. J. Mech. Med. Biol. 16(04), 1650039 (2016). https://doi.org/10.1142/S0219519416500391
10. Meziati Sabour, R., Benezeth, Y., De Oliveira, P., Chappe, J., Yang, F.: UBFC-phys: a multimodal database for psychophysiological studies of social stress. IEEE Trans. Affect. Comput., 1 (2021). https://doi.org/10.1109/TAFFC.2021.3056960
11. Rössler, A., Cozzolino, D., Verdoliva, L., Riess, C., Thies, J., Nießner, M.: FaceForensics: a large-scale video dataset for forgery detection in human faces. arXiv:1803.09179, 24 March 2018
12. Mellouk, W., Handouzi, W.: Comparison and evaluation of IPPG methods for HR estimation under different face regions. In: Proceedings of the 2022 19th International Multi-Conference on Systems, Signals & Devices (SSD), pp. 1956–1961 (2022). https://doi.org/10.1109/SSD54932.2022.9955726
13. McDuff, D., Blackford, E.: iPhys: an open non-contact imaging-based physiological measurement toolbox. arXiv:1901.04366 [cs]
14. Haddad, J., Lezoray, O., Hamel, P.: 3D-CNN for facial emotion recognition in videos. In: Bebis, G., et al. Advances in Visual Computing. ISVC 2020. Lecture Notes in Computer Science, vol. 12510, pp. 298–309. Springer, Cham (2020). https://doi.org/10.1007/978-3-030-64559-5_23

Arabic Lipreading Using YOLO and CNN Models

Ali Baaloul[1(✉)], Nadjia Benblidia[1], Abdelkader Ouared[2], and Fatma Zohra Reguieg[1]

[1] LRDSI Laboratory, Faculty of Sciences, Blida1 University, Blida, Algeria
baaloulali@gmail.com, {benblidia_nadjia,reguieg_fzohra}@univ-blida.dz
[2] Tiaret University, Tiaret, Algeria
a_ouared@esi.dz

Abstract. Lipreading is a vital aspect of human communication and requires effective computational methods. Lip movements, integral to this process, present challenges such as variability and context dependence. Recent developments in deep learning show potential for enhancing Arabic visual speech recognition (VSR) systems. This paper focuses on leveraging deep learning to assist Arabian individuals with hearing impairments, reduce their communication barriers, and enhance their quality of life. We employed our Arabic-created dataset, including YOLO version V7 as a frontend for mouth detection and CNN models (i) InceptionV3, and (ii) custom CNN model for speech classification. Our approach aims to address the complexities of lipreading. Our results show promise, with an impressive 90% speech recognition accuracy. These results underscore the capacity of deep learning to improve visual speech recognition, Facilitating the development of more effective and precise methods for detection and recognition.

Keywords: Lipreading · YOLO · Visual Speech Recognition VSR · CNN

1 Introduction

Lipreading, a vital and crucial communication mode for those with hearing impairments, relies on the visual interpretation of lip movements for effective language comprehension, promoting communication effectively. Recent deep learning advances hold to comprehension of spoken language. This visual method fosters inclusively and serves as an alternative means of effective communication. With applications ranging from aiding individuals with hearing challenges to surveillance and human-computer interaction, recent advances in deep learning show promise in automating lip movement extraction, improving and enhancing VSR accuracy, and enhancing accessibility for applications in various fields.

Analyzing lip movement patterns and behaviors during verbal interactions presents significant challenges, demanding careful oversight of each stage and

B. Djamaa et al. (Eds.): CSA 2024, LNNS 1145, pp. 13–23, 2025.
https://doi.org/10.1007/978-3-031-71848-9_2

comprehension of how lip movements differ, particularly in sequences of intermediate steps transitioning into finer, sequential actions. Ensuring the model's capacity to generalize across different speakers and accurately interpret various lip movements necessitates the identification and extraction of relevant visual features. The research inquiry guiding this study is: "How can advanced analysis methods be utilized to proficiently extract subtle expressions in lip movements?"

Answering this research question RQ is highly valuable for various fields such as accessibility, education, healthcare, technology, cultural preservation, and advancements in linguistic and cognitive sciences, especially for the Arabic-speaking world and beyond.

- Our main goal is to enhance Arabic VSR through advancements in Deep Learning and computer vision.
- We integrate the YOLO (You Only Look Once) model [18–20] for precise detection of regions of interest (ROI) in our dataset.
- We leverage the strengths of our proposed CNNs in extracting and understanding intricate features within the localized regions.

The paper is divided into several sections: Sect. 2 reviews existing literature, Sect. 3 outlines the proposed approach, Sect. 4 presents results and discussions, and Sect. 5 concludes by summarizing findings and suggesting future research directions.

2 Related Work

In advancing VSR systems, researchers have pursued two primary tracks: building comprehensive databases and proposing innovative approaches.

2.1 VSR Datasets

To build VSR systems, researchers stress the need for extensive databases, facing challenges like language diversity, speaker variation, vocabulary size, lighting conditions, and cost models [14,15]. Diverse datasets are categorized into three sections: Individual Letters and Digits, Words, and Sentences and Phrases. The foundational category focuses on recognizing isolated letters or digits with datasets like AVletters [11], AV@CAR [13], and CUAVE [17]. Transitioning to word-level recognition, the intermediate category includes datasets like MIRACL-VC [21] and MODALITY [6]. The advanced category deals with complexities in recognizing sentences and phrases, utilizing datasets like VIDTIMIT [23], OuluVS [28], and LRS [4] for comprehensive studies.

2.2 VSR Approaches

The second research track in VSR proposes novel strategies to improve automatic recognition systems, specifically emphasizing feature extraction from the mouth region in images.

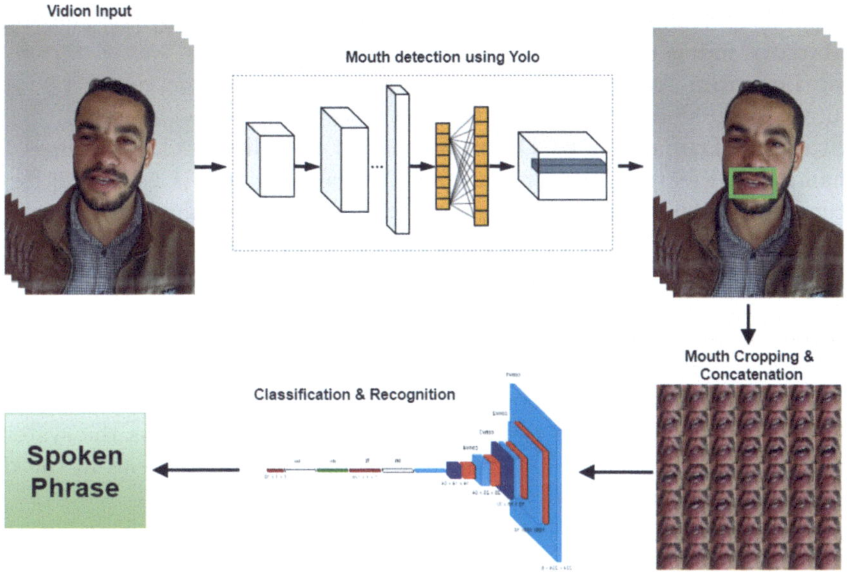

Fig. 1. Overview of the model we propose.

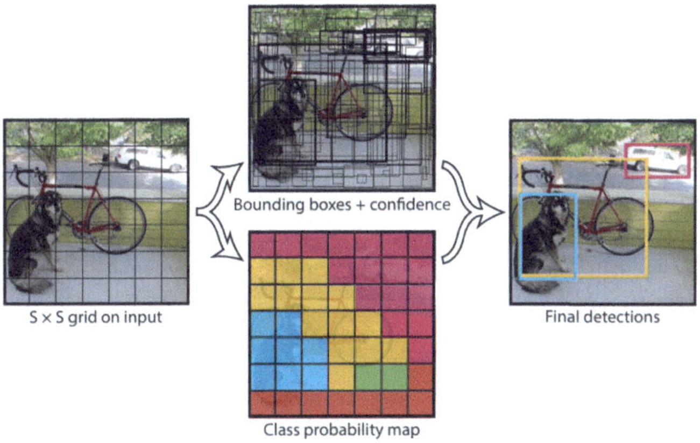

Fig. 2. The YOLO model treats detection as a regression task, dividing the image into an S×S grid. Each grid cell predicts B bounding boxes, their confidences, and C class probabilities, encoded as an S×S×(Bx5+C) tensor.

Traditional Approaches: These approaches utilize domain-expert knowledge to guide feature extraction through methods like Discrete Cosine Transform (DCT) [10,24], Active Appearance Models (AAM) [7,25], and Histogram of Oriented Gradients (HOG) [21,22]. Subsequently, extracted features are classified using techniques like Support Vector Machines (SVM) [29] and Hidden Markov Models (HMM) [9].

Deep Learning Approaches: The DL approaches have emerged as a powerful alternative, evolving through three distinct stages:

Stage 1: Transition to Deep Learning: Initially, the focus shifted from handcrafted features to Convolutional Neural Networks (CNNs) [12] for feature extraction, demonstrating the efficacy of deep learning in VSR.

Stage 2: Embracing Deep Learning: This stage saw the full potential of CNNs being harnessed, as they were employed for both feature extraction and classification, further solidifying the advantages of deep learning [3].

Stage 3: Advanced Deep Learning Architecture: Our work falls under this stage, which leverages the combined power of CNNs and recurrent neural networks (RNNs) like Long Short-Term Memory (LSTM) [5] and Gated Recurrent Unit (GRU) networks [27]. This integration allows the system to not only extract features but also learn temporal dependencies in the data, potentially leading to superior performance.

In our position, we believe the integration of YOLO with proposed CNN models emerges as a promising pathway for advancing VSR, The amalgamation of YOLO and CNN architectures consistently exhibits superior performance across diverse studies, highlighting its potential to enhance both real-time mouth detection and recognition tasks.

3 Our Proposal

This section delineates our proposed methodology for the comprehensive VSR system, rooted in deep learning principles. Focusing on the intricate analysis of lip movements, our approach combines a mouth region detection phase utilizing the YOLO architecture with a subsequent recognition phase employing CNN architectures. By integrating these stages (see Fig. 1), our approach seamlessly merges various aspects of deep learning to establish a more thorough and potentially more efficient diagnostic system for individuals with hearing impairment.

3.1 Mouth Detection Based on YOLO Model

In 2016, Joseph Redmon et al. developed YOLO, an acronym for "You Only Look Once," a real-time object detection system. It introduced an end-to-end approach for object detection, performing the task in a single pass (Fig. 2).

YOLO's distinct feature is its capability to achieve real-time, high-accuracy object detection by dividing the input image into a grid. It predicts bounding boxes and class probabilities for each grid cell, allowing for the detection of multiple objects in a single pass, making it renowned for its speed and efficiency in real-time applications.

With versions like YOLOv1 to YOLOv8 and variants like YOLO-NAS and YOLO with Transformers, YOLO has expanded its capabilities, finding applications in robotics, driverless cars, video monitoring, action recognition, sports

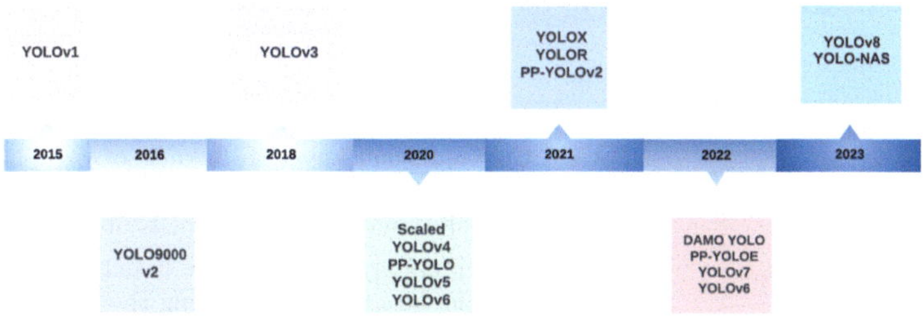

Fig. 3. A timeline of YOLO versions.

analysis, and human-computer interaction. The YOLO object detection framework has undergone significant evolution with each version, introducing improvements in accuracy, speed, and adaptability. A timeline of key YOLO versions is depicted in (Fig. 3).

3.2 Lipreading Based on CNN Architectures

LeCun's CNNs, a deep learning algorithm [8], are widely used in computer vision for tasks like image classification and facial recognition. Their significance lies in superior image recognition and effective handling of high-dimensional input data. CNNs are applied in diverse domains, eliminating the need for manual engineering by automatically extracting relevant features through convolutional filters.

Our CNN Architecture. The proposed CNN architecture, shown in Fig. 4, is designed for the accurate classification of complex images, comprising 16 layers with various components to enhance performance. Key features of this architecture as shown in the Table 1:

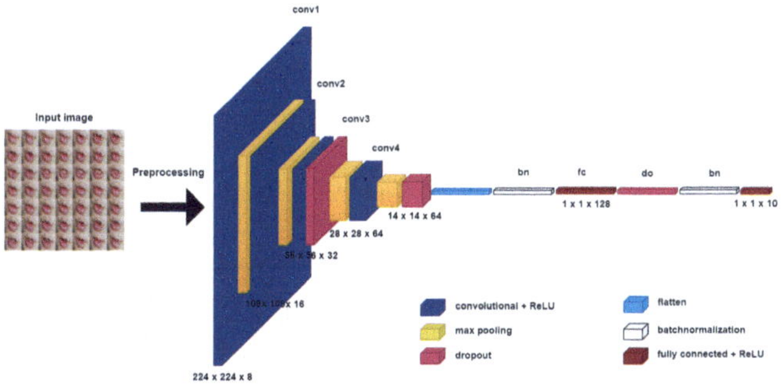

Fig. 4. Our CNN architecture

Table 1. Details of the Proposed CNN Architecture

Component	Details
Convolutional Layers	– First Convolutional Layer: Eight kernels of size (3,3). – Second Convolutional Layer: 16 kernels of size (3,3). – Third Convolutional Layer: 32 kernels of size (4,4). – Fourth Convolutional Layer: 64 kernels of size (5,5).
Feature Extraction	Every convolutional layer plays a role in extracting unique features from the input image, allowing the model to acquire hierarchical representations.
MaxPooling Layers	Six MaxPooling layers with kernel sizes (2,2) reduce feature map dimensionality by a factor of two.
Dropout Layers	Three dropout layers mitigate overfitting by deactivating neurons during training.
BatchNormalization Layers	The 2 BatchNormalization layers are utilized to standardize activations within batches, expediting training and improving generalization capabilities.
Dense Layers	The final output is generated by dense layers. The first layer with 128 neurons and ReLU activation, followed by a second layer with 10 neurons for classification. The output layer employs Softmax activation for class probability interpretation.

InceptionV3. Introduced by Google researchers in 2015 [26], InceptionV3 stands out as a highly efficient and scalable Convolutional Neural Network (CNN) tailored for image classification and object recognition tasks. Its architecture, illustrated in Fig. 5, integrates inception modules, strategically combining parallel convolutional operations with varying filter sizes and stride lengths to effectively capture features across different scales and resolutions. Noteworthy performance improvements, such as batch normalization, residual connections, and global average pooling, significantly contribute to its outstanding performance on benchmark datasets like ImageNet and COCO. With its remarkable computational efficiency,

4 Experimentations And Results

4.1 Dataset Description

The scarcity of work and limited development in the field of VSR systems in Arabic has made it extremely challenging to obtain readily available datasets in the Arabic language. This difficulty prompted us to create our dataset [1], comprising 1382 videos obtained through requests from 9 individuals pronouncing 10 everyday sentences in Modern Standard Arabic (refer to Table 2).

4.2 YOLO Model Training

During YOLOv7 training for mouth region detection, the *LabelImg* tool played a pivotal role in annotating dataset images and ensuring precise labels. Optimizing

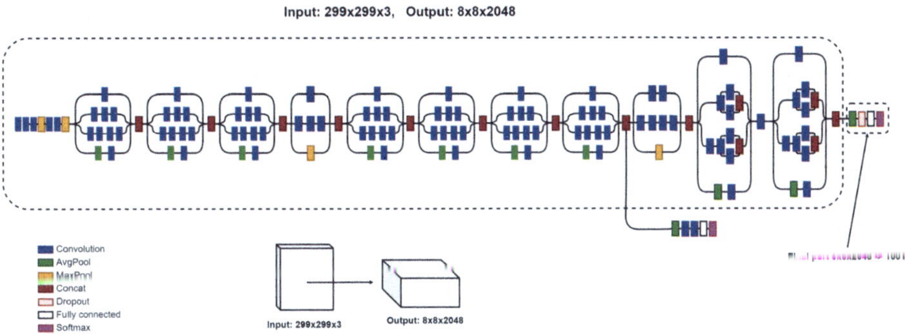

Fig. 5. InceptionV3 architecture diagram, Reproduced from "A Convolutional Neural Network for Automatic Identification and Classification of Fall Army Worm Moth" by Francis Chulu et al. [2], licensed under CC BY 4.0

Table 2. The ten daily Arabic phrases selected.

Phrase in Arabic	Phrase Syllables	Phrase in English
السلام عليكم	Assalamu alaikum	Hello
صباح الخير	Sabah al-khair	Good morning
متأسف جدا	Mataasifun jiddan	I am so sorry
ماذا تفعل	Maadha taf'al	What do you do
هل انت بخير	Hal anta bi-khair	Are you fine
تعال معي	Ta'al ma'i	Come with me
كيف حالك	Kayf halak	How are you
كن شجاعا	Kun shuja'an	Be brave
أريد الدخول	'Ureed al-dkhul	I want to come in
شكرا لك	Shukran lak	Thank you

model parameters enhanced accuracy in identifying and localizing mouth regions. Iterative refinement fine-tuned the YOLOv7 model for proficient mouth area detection. Precise annotations from *LabelImg* played a crucial role in improving model efficacy and ensuring robust generalization across diverse datasets. This section highlights the successful integration of advanced object detection techniques tailored for mouth region identification, achieved through YOLOv7 training with *LabelImg*.

Next, employing *OpenCV*, we cropped the mouth area in each video, resizing it to 32 by 32 pixels, resulting in 60 images per video and a total of 83,160 images. For optimization, we concatenated frames into a single 224×224-pixel image. This consolidated image, illustrated in Fig. 1, serves as input for our proposed models.

4.3 Results

In this phase, we showcase the outcomes of our dataset experimentation. Table 3 compiles the performance metrics of various models utilized, encompassing their training and validation accuracy percentages, Figs. 6 and 7 show the accuracy results. These findings offer valuable insights into the efficacy of each model within the scope of our research.

Due to our inability to access an Arabic dataset previously tested for comparison purposes, we experimented with the VGG16 model. This model has been previously employed in VSR on English datasets [16].

Table 3. The outcomes from employing various models in our study.

Methods	Train (%)	Validation (%)
OurCNN with colors	90	85
OurCNN grayscale	96	90
InceptionV3	83	79
VGG16	80	80

In terms of model performance, particularly with OurCNN Grayscale, high accuracy is observed in both training and validation 6b, indicating effective learning and generalization from the dataset. When comparing models, OurCNN Grayscale stands out, surpassing others in both training and validation. This highlights its robustness and effectiveness in capturing features for lipreading tasks.

Choosing a model involves considering the trade-off between complexity and accuracy. For resource-conscious situations, custom models like OurCNN might be more practical than big models like VGG16 or InceptionV3 7. The comparison between OurCNN with Colors 6a and OurCNN Grayscale implies that the grayscale model achieves comparable accuracy while potentially being more computationally efficient.

Discussion

The results emphasize the significance of choosing an appropriate model architecture. OurCNN Grayscale, with its superior performance, stands out as a promising candidate for lipreading tasks. Further investigation into regularization techniques or data augmentation methods may help alleviate overfitting concerns and enhance model generalization. The high accuracy across models showcases the potential of deep learning in VSR, opening avenues for refining existing models and exploring novel architectures.

Overall, the findings indicate positive outcomes, laying the groundwork for continued research and optimization in the field of VSR.

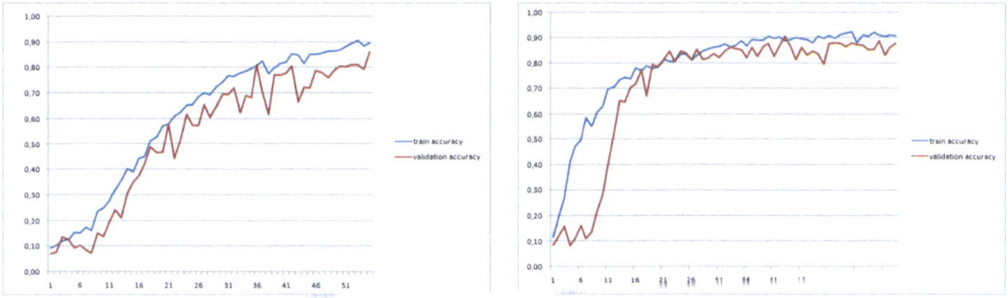

(a) OurCNN with colors accuracy result (b) OurCNN with grayscale accuracy result

Fig. 6. Accuracy results obtained for our proposed CNN methods

Fig. 7. InceptionV3 accuracy result

5 Conclusion

We highlight the potential of object detection and recognition to advance Arabic VSR. Through our exploration of deep learning, we achieved successful mouth detection using YOLO version 7 and attained exceptional lipreading performance with our proposed CNN architectures, achieving an impressive 90% accuracy. These methods demonstrate effectiveness in enhancing VSR, offering promising pathways to enhance the accuracy of this crucial technology. However, it's important to note that our experiments were conducted with a limited dataset. To improve the performance and applicability of our models, we propose several avenues for future research and development. These include expanding our dataset to cover a broader range of cases and refining model architectures. These steps are vital for ensuring that the benefits of deep learning in VSR can be more widely realized, benefiting both the scientific community and individuals with hearing impairments.

Declarations

Conflict of Interest. The authors declare that they have no conflict of interest.

References

1. Baaloul, A., Benblidia, N., Reguieg, F.Z., Bouakkaz, M., Felouat, H.: An arabic visual speech recognition framework with cnn and vision transformers for lipreading. Multimed. Tools Appl., 1–35 (2024)
2. Chulu, F., Phiri, J., Nkunika, P.O., Nyirenda, M., Kabemba, M.M., Sohati, P.H.: A convolutional neural network for automatic identification and classification of fall army worm moth. Int. J. Adv. Comput. Sci. Appl. **10**(7) (2019)
3. Chung, J., Gulcehre, C., Cho, K., Bengio, Y.: Empirical evaluation of gated recurrent neural networks on sequence modeling. arXiv preprint arXiv:1412.3555 (2014)
4. Chung, J.S., Senior, A., Vinyals, O., Zisserman, A.: Lip reading sentences in the wild. In: 2017 IEEE Conference on Computer Vision and Pattern Recognition (CVPR), pp. 3444–3453. IEEE (2017)
5. Chung, J.S., Zisserman, A.: Out of time: automated lip sync in the wild. In: Asian Conference on Computer Vision, pp. 251–263. Springer (2016)
6. Czyzewski, A., Kostek, B., Bratoszewski, P., Kotus, J., Szykulski, M.: An audiovisual corpus for multimodal automatic speech recognition. J. Intell. Inf. Syst. **49**(2), 167–192 (2017)
7. Hilder, S., Harvey, R.W., Theobald, B.J.: Comparison of human and machine-based lip-reading. In: AVSP, pp. 86–89 (2009)
8. LeCun, Y., Bottou, L.: Yb, Haffner, P.: Gradientbased learning applied to document recognition. Proc. IEEE (1998)
9. Lee, D., Lee, J., Kim, K.E.: Multi-view automatic lip-reading using neural network. In: Asian Conference on Computer Vision, pp. 290–302. Springer (2016)
10. Lucey, P., Sridharan, S., Dean, D.: Continuous pose-invariant lipreading. In: Proceedings of the 9th Annual Conference of the International Speech Communication Association (Interspeech 2008) incorporating the 12th Australasian International Conference on Speech Science and Technology (SST 2008), pp. 2679–2682. International Speech Communication Association (2008)
11. Matthews, I., Cootes, T.F., Bangham, J.A., Cox, S., Harvey, R.: Extraction of visual features for lipreading. IEEE Trans. Pattern Anal. Mach. Intell. **24**(2), 198–213 (2002)
12. Noda, K., Yamaguchi, Y., Nakadai, K., Okuno, H.G., Ogata, T.: Lipreading using convolutional neural network. In: Fifteenth Annual Conference of the International Speech communication Association (2014)
13. Ortega, A., Sukno, F., Lleida, E., Frangi, A.F., Miguel, A., Buera, L., Zacur, E.: Av@ car: A spanish multichannel multimodal corpus for in-vehicle automatic audiovisual speech recognition. In: LREC (2004)
14. Ouared, A., Kharroubi, F.Z.: Moving database cost models from darkness to light. In: Smart Applications and Data Analysis: Third International Conference, SADASC 2020, Marrakesh, Morocco, June 25–26, 2020, Proceedings 3, pp. 17–32. Springer (2020)
15. Ouared, A., Ouhammou, Y.: Capitalizing the database cost models process through a service-based pipeline. Concurrency Comput. Practice Exp. **35**(11), e6463 (2023)
16. Patilkulkarni, S., Murthy, N.S., et al.: Visual speech recognition using vgg16 convolutional neural network (2021)
17. Patterson, E.K., Gurbuz, S., Tufekci, Z., Gowdy, J.N.: Cuave: A new audio-visual database for multimodal human-computer interface research. In: 2002 IEEE International Conference on Acoustics, Speech, and Signal Processing, vol. 2, pp. II–2017. IEEE (2002)

18. Redmon, J., Divvala, S., Girshick, R., Farhadi, A.: You only look once: Unified, real-time object detection. In: Proceedings of the IEEE Conference on Computer Vision and Pattern Recognition, pp. 779–788 (2016)
19. Redmon, J., Farhadi, A.: Yolo9000: better, faster, stronger. In: Proceedings of the IEEE Conference on Computer Vision and Pattern recognition, pp. 7263–7271 (2017)
20. Redmon, J., Farhadi, A.: Yolov3: An incremental improvement. arXiv preprint arXiv:1804.02767 (2018)
21. Rekik, A., Ben-Hamadou, A., Mahdi, W.: A new visual speech recognition approach for rgb-d cameras. In: International Conference Image Analysis and Recognition, pp. 21–28. Springer (2014)
22. Saeed, V.A.: A framework for recognition of facial expression using hog features. Int. J. Math. Stat. Comput. Sci. **2**, 1–8 (2024)
23. Sanderson, C.: The vidtimit database. Tech. rep, IDIAP (2002)
24. Seymour, R., Stewart, D., Ming, J.: Comparison of image transform-based features for visual speech recognition in clean and corrupted videos. EURASIP J. Image Video Process. **2008**, 1–9 (2007)
25. Sterpu, G., Harte, N.: Towards lipreading sentences with active appearance models. arXiv preprint arXiv:1805.11688 (2018)
26. Szegedy, C., et al.: Going deeper with convolutions. In: Proceedings of the IEEE Conference on Computer Vision and Pattern Recognition, pp. 1–9 (2015)
27. Xu, K., Li, D., Cassimatis, N., Wang, X.: Lcanet: End-to-end lipreading with cascaded attention-ctc. In: 2018 13th IEEE International Conference on Automatic Face & Gesture Recognition (FG 2018), pp. 548–555. IEEE (2018)
28. Zhao, G., Barnard, M., Pietikainen, M.: Lipreading with local spatiotemporal descriptors. IEEE Trans. Multimedia **11**(7), 1254–1265 (2009)
29. Zhou, Z., Zhao, G., Pietikäinen, M.: Towards a practical lipreading system. In: CVPR 2011, pp. 137–144. IEEE (2011)

Investigating Person Attribute Recognition in Challenging Environments

Ilyes Oukid(✉), Said Yacine Boulahia, and Abdenour Amamra

École Militaire Polytechnique, BP 17, 16046 Bordj el Bahri, Algiers, Algeria
ilyes.oukid1@gmail.com

Abstract. This study delves into attribute recognition challenges within computer vision and deep learning methodologies, aiming to address real-world complexities encountered in the identification of nuanced traits like gender, age, clothing styles, and accessories. Despite remarkable advancements in this field, obstacles persist due to occlusion, lighting variations, and image noise. Conventional approaches exhibit limitations in handling these disruptions, leading to the introduction of an innovative approach augmenting traditional deep learning techniques. Through experiments utilizing perturbed datasets and a filtering mechanism, this study demonstrates enhanced model resilience against disruptions, surpassing traditional models in accuracy amidst visual noise and reduced resolutions. These findings underscore the practical potential and robustness of the augmented approach in overcoming real-world complexities in attribute recognition, suggesting avenues for broader applicability across diverse domains.

Keywords: Attribute recognition · Image filtering · Augmented training · Deep learning · Convolutional neural network

1 Introduction

Advancements in attribute recognition through computer vision and deep learning have significantly improved the accuracy of identifying traits such as gender, age, clothing styles, and accessories using Convolutional Neural Networks (CNNs). However, challenges persist, including variations in lighting, occlusion, diverse poses, low resolution, and image noise, which hinder precise recognition in real-world scenarios.

Conventional methodologies in attribute recognition excel in controlled environments but struggle with real-world complexities, such as occlusion and varying lighting conditions, leading to reduced accuracy. To address these challenges, innovative methodologies are necessary. In this paper, we propose an approach that augments traditional deep learning techniques by introducing a filtering mechanism to perturb the dataset, simulating real-world disruptive conditions. By retraining models using this perturbed dataset while leveraging previous iterations' weights, our approach aims to enhance model robustness and mitigate challenges posed by occlusion and other disruptive elements.

B. Djamaa et al. (Eds.): CSA 2024, LNNS 1145, pp. 24–35, 2025.
https://doi.org/10.1007/978-3-031-71848-9_3

This paper emphasizes the importance of dataset selection, model architecture, and iterative training in deep learning. It details how our innovative approach enhances model resilience against disruptive elements and discusses the potential implications of this augmentation in improving attribute recognition under adverse conditions.

The remainder of this paper is structured as follows. Section 2 reviews previous works on attributes recognition techniques. In Sect. 3, we present our proposed approach which introduces an additional step to the traditional deep learning process. Section 4 discusses the achieved performances. Concluding insights and a forward-looking perspective into the future of attribute recognition are encapsulated in Sect. 5.

2 Related Work

Recent strides in attribute recognition have been largely powered by innovations in computer vision and deep learning. Convolutional Neural Networks (CNNs) have particularly revolutionized the field, significantly improving precision and effectiveness in identifying individual traits. Leveraging pre-trained CNNs, along with sophisticated architectures like Residual Networks (ResNet) [2] and Neural Networks (OSNet) [9], has notably boosted recognition accuracy.

Various strategies have been explored to tackle the complexities of attribute recognition. Multi-label learning [10] techniques enable models to predict multiple attributes simultaneously, capitalizing on interrelationships between attributes to enhance accuracy. Multi-task learning [8] paradigms have also played a vital role by enabling models to learn and execute multiple tasks concurrently, thereby amplifying their ability to discern nuanced attributes effectively.

Recent advancements have introduced deep models like DeepMAR [3], HydraPlusNet [6], RCRA [11], and PGDM [4] into attribute recognition. These models integrate multi-attribute learning strategies, attention mechanisms, cascaded refining and classifying frameworks, and progressive growth approaches, respectively, to tackle complex attribute recognition tasks.

Despite these advancements, challenges persist. Variations in lighting conditions, occlusion, diversity of poses, low image resolution, and noisy images pose significant hurdles to precise attribute recognition. In response, our proposed solution aims to address these challenges, as detailed in the subsequent sections.

3 Proposed Approach

Deep learning relies on training models with datasets, involving dataset curation, model architecture design, and iterative training. Our approach introduces a distinctive step to the conventional deep learning paradigm by integrating a filtering mechanism into the dataset, simulating real-world disruptions such as occlusion or blur. The model is then retrained using prior weights, aiming to enhance resilience and performance in extreme conditions. Figure 1 illustrates

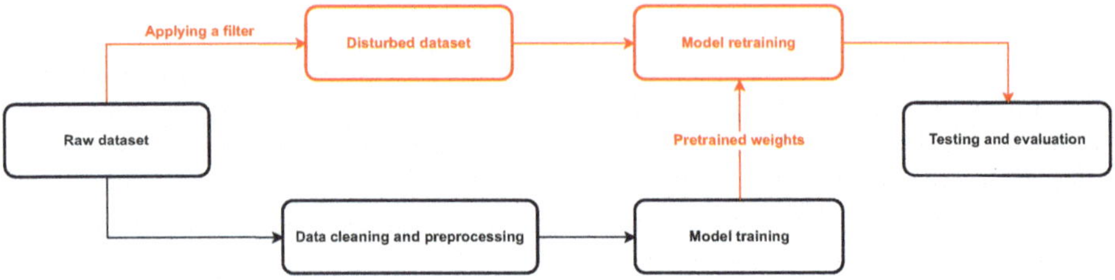

Fig. 1. Proposed approach pipeline

the detailed pipeline of the proposed approach, showcasing the various stages from processing the raw dataset to evaluating the model.

The rationale behind retraining with preexisting weights lies in leveraging previously acquired knowledge. By capitalizing on prior information, the model adapts and fine-tunes its parameters, aligning them more effectively to handle perturbations introduced by occlusion or similar disruptive elements in the dataset. This approach holds promise for advancing attribute recognition in challenging environments.

Additionally, our proposed approach aims to investigate the impacts of perturbed environments on attribute recognition. These environments encompass various scenarios, including occlusion, blurring, grayscale images, low resolution, and low brightness. Each of these factors can significantly alter image quality and, consequently, the accuracy of attribute recognition. In the following subsections, we explore the effects of these disturbances and highlight how our approach effectively addresses these challenges.

3.1 Occlusion

The impact of occlusion on person attribute recognition is notable, as it obstructs crucial visual information needed for accurate identification. To simulate this disruption, we apply filters to images, randomly adding black circles to them, as illustrated in Fig. 2. This simulated occlusion challenges the model's ability to correctly identify attributes associated with obscured or masked regions.

Additionally, we retrain the model using the perturbed dataset, starting from pre-trained weights on the raw dataset. This refinement of parameters aims to enhance the model's capacity, rendering it more robust and yielding superior results in handling occlusion-related challenges.

3.2 Blurring

The impact of blurring on attribute recognition accuracy is substantial, especially when fine details become less discernible. To replicate this situation, we introduce blurring filters to images, reducing their sharpness and creating a blurred effect, as depicted in Fig. 3. This deliberate blurring simulates real-world scenarios where images may lack clarity due to various factors.

Furthermore, we perform model retraining using the dataset affected by blurring, leveraging pretrained weights from the original dataset. This retraining process aims to refine the model's parameters, enhancing its adaptability and resilience against blurring effects. By fine-tuning the model on the perturbed dataset, we strive to improve its ability to accurately identify attributes even in the presence of blurred image inputs.

Fig. 2. Example of occlusion application

Fig. 3. Example of blurring application

3.3 Grayscale

The impact of grayscale images on attribute recognition accuracy is notable, particularly regarding color-specific attributes. To simulate this limitation, we convert images to grayscale, eliminating color information while retaining luminance, as exemplified in Fig. 4. This conversion challenges the model's ability to discern attributes reliant on color cues.

Additionally, we conduct model retraining using the grayscale-affected dataset while utilizing pretrained weights from the original dataset. This retraining process aims to adapt the model's parameters to grayscale inputs, thereby enhancing its capability to identify attributes in images lacking color information. By fine-tuning the model on the perturbed dataset, we endeavor to improve its accuracy in recognizing attributes, specifically in grayscale images.

3.4 Low Resolution

The impact of low-resolution images on attribute recognition accuracy is substantial, primarily due to the limited information available for feature extraction. To simulate this constraint, we downscale the images, reducing their resolution while maintaining the same aspect ratio, as depicted in Fig. 5. This reduction in image resolution imposes challenges on the model's ability to extract detailed attributes accurately.

Furthermore, we retrain the model using the low-resolution-affected dataset, starting from pre-trained weights derived from the original dataset. This retraining process aims to refine the model's parameters and adapt its features to handle low-resolution inputs. By reiteratively training the model with low-resolution images, our goal is to improve its capacity to recognize attributes even in images with reduced clarity. Ultimately, this process seeks to enhance the model's accuracy in low-resolution scenarios, improving its attribute recognition capabilities.

Fig. 4. Example of converting to grayscale

Fig. 5. Example of resolution reduction

3.5 Low Light

The significant impact of low-light images on attribute recognition accuracy lies in the obscured or shadowed areas that limit the extraction of essential features. To simulate these conditions, we adjust the lighting conditions in images, creating darker areas to mimic low-light settings, as demonstrated in Fig. 6. These modifications pose challenges to the model's ability to accurately recognize attributes in such environments.

Our method retrains the model using a low-light dataset with pre-existing weights. This enhances its ability to recognize attributes in darker areas, improving accuracy under low-light conditions. Ultimately, our approach aims to enhance model performance in reduced illumination, ensuring more reliable attribute recognition.

4 Experimental Results

In this section, we aim to provide a comprehensive overview of our experimental evaluation, starting with a description of the dataset used, the metrics employed for evaluation, and a detailed analysis of the results obtained.

Fig. 6. Example of brightness reduction

4.1 Dataset

We chose to use the PETA dataset [1] due to its significant size, allowing us to extensively explore various person attributes and their possible variations. It is widely recognized and adopted within the research community, serving as a reference in numerous previous studies.

The PETA dataset [1] comprises 19,000 images with resolutions ranging from 17×39 to 169×365 pixels. These images are annotated with a total of 105 person-specific attributes.

4.2 Evaluation Metrics

We evaluate our approach using multiple criteria. Firstly, we employ the *mean accuracy* as an overall measure of the model's performance:

$$\text{Mean Accuracy} = \frac{1}{N} \sum_{i=1}^{N} \frac{\text{Number of correct predictions for class } i}{\text{Total number of predictions for class } i} \quad (1)$$

Next, we consider *accuracy*, which measures the proportion of correct predictions:

$$\text{Accuracy} = \frac{\text{Number of correct predictions}}{\text{Total number of predictions}} \quad (2)$$

Additionally, we assess the model's performance using *precision*, representing the ratio of true positive predictions to total predicted positives:

$$\text{Precision} = \frac{\text{True Positives}}{\text{True Positives} + \text{False Positives}} \quad (3)$$

The *recall*, or sensitivity, measures the ratio of true positive predictions to total actual positives:

$$\text{Recall} = \frac{\text{True Positives}}{\text{True Positives} + \text{False Negatives}} \quad (4)$$

Finally, we use the *F1-score*, which balances precision and recall:

$$\text{F1-score} = 2 \times \frac{\text{Precision} \times \text{Recall}}{\text{Precision} + \text{Recall}} \tag{5}$$

This metric is particularly useful for imbalanced class distributions.

4.3 Findings and Analysis

Initially, we conducted a comparative study of recognition models to analyze and select the most suitable ones for the task of recognizing person attributes. We trained several base deep neural networks on the PETA dataset [1], adjusting various parts of the neural networks, such as loss functions, heads, and batch normalization (BN). The training results are presented in Table 1.

Table 1. Training results of deep neural networks

Network	BN	Head	Loss Func.	mA (%)	Acc. (%)	Prec. (%)	Recall (%)	F1 (%)
ResNet50 [2]	Aft.	BNH	CEL_Sig.	80.60	79.89	88.49	85.12	86.77
ResNet50 [2]	Aft.	BNH	Sin_BCE	**85.95**	71.60	76.41	**87.26**	80.92
ResNet50 [2]	Bef.	BNH	CEL_Sig.	79.95	79.12	88.50	84.71	86.22
ResNet50_nl [5]	Aft.	BNH	CEL_Sig.	81.11	**80.62**	**89.27**	86.01	**87.31**
ResNet50_ibn_a [7]	Aft.	BNH	CEL_Sig.	81.02	80.32	89.23	85.77	87.14
ResNet50_ibn_a_nl [7]	Aft.	BNH	CEL_Sig.	80.59	80.05	89.09	85.44	86.90
ResNet101 [2]	Aft.	BNH	CEL_Sig.	80.55	79.79	88.57	85.47	86.70
ResNet101 [2]	Aft.	RedH	CEL_Sig.	80.39	79.36	88.79	84.88	86.38
ResNet101_nl [5]	Aft.	BNH	CEL_Sig.	81.12	80.36	89.07	85.78	87.09
ResNet101_ibn_a [7]	Aft.	BNH	CEL_Sig.	81.05	80.21	88.78	85.89	85.89
ResNet101_ibn_a_nl [7]	Aft.	BNH	CEL_Sig.	81.68	79.58	87.84	85.83	86.56
OSNet [9]	Aft.	BNH	CEL_Sig.	76.07	73.67	79.34	84.45	82.11

Loss functions, head configurations, and batch normalization impact model performance. CEL_Sigmoid outperforms Singular_BCE, likely due to its suitability for object detection tasks. Batch normalization after linear layers improves stability and gradient propagation. Additionally, the BNHead (BNH) configuration yields better results compared to the ReductionHead (RedH) configuration, which performs dimension reduction. This may be because BNHead captures finer and layer-specific information. Notably, the ResNet50_nl [5] model with BNHead and CEL_Sigmoid emerged as the top performer across most evaluated metrics, demonstrating enhanced attribute recognition capabilities.

After analyzing different models and their configurations, we examine the impact of the disturbed environment on recognition performance. Furthermore, we assess the effects of our proposed approach on addressing these issues. We conduct tests in different environments, including occlusion, blurring, grayscale, low resolution, and varying brightness levels.

For our experiments, we employ the DeepMAR [3] deep neural network as it utilizes a Residual Network (ResNet) [2] as its backbone. Leveraging ResNet offers several advantages, such as improved gradient flow, efficient training even in deeper architectures, and a residual learning framework that mitigates vanishing gradient issues during training. Its ability to retain learned features effectively contributes to robustness in recognizing complex attributes within various perturbed environments.

Occlusion. We applied two occlusion filters to the PETA dataset [1], randomly adding black circles to the images. The first filter occludes **20%** of the image, while the second occludes **30%** of the image. We subsequently retrained the DeepMAR [3] model on the disturbed dataset using pre-trained weights from the undisturbed dataset. The results from the various tests are presented in Table 2.

Table 2. Results of occlusion tests in percentage (%)

Dataset	mA	Accuracy	Precision	Recall	F1-score
Raw before retraining	85.34	83.06	89.87	88.41	89.13
Occlusion 20% before retraining	68.84	60.54	76.21	69.22	72.55
Occlusion 30% before retraining	58.69	40.24	57.47	50.99	54.03
Occlusion 20% after retraining	82.94	80.60	88.40	86.41	87.40
Occlusion 30% after retraining	82.58	79.61	87.80	85.62	86.70
Raw after retraining	84.33	82.84	90.00	87.92	88.95

Introducing **20%** and **30%** occlusion led to a significant decrease in model performance. Accuracy dropped by **22.52%** and **42.82%** respectively, compared to the initial accuracy, indicating a substantial loss in the model's ability to recognize person attributes based on the introduced occlusion percentage. The mA, precision, recall, and F1-score also decreased, highlighting the negative impact of occlusion on the overall model performance. After retraining on the disturbed data, the model's performance showed a marked improvement, successfully overcoming the challenge posed by occlusion. During testing of the retrained model on the raw dataset, high performance is observed, even surpassing the initial precision by **0.13%**. This improvement demonstrates the effectiveness of retraining in strengthening the recognition capacity of person attributes.

Blurring. We applied a Gaussian filter to the PETA dataset [1] to make the images blurry. We then retrained the DeepMAR [3] model on the perturbed dataset using pre-trained weights from the unperturbed dataset. The results of different tests are presented in Table 3.

Table 3. Results of blurring tests in percentage (%)

Dataset	mA	Accuracy	Precision	Recall	F1-score
Raw before retraining	85.34	83.06	89.87	88.41	89.13
Blurring before retraining	80.94	78.92	87.49	84.86	86.16
Blurring after retraining	84.63	82.61	89.92	87.69	88.79
Raw after retraining	82.21	79.97	87.55	86.34	86.94

The introduction of blurring in the data also had a negative impact on the model's performance. We observed that performance measures decreased compared to the initial values of the unperturbed dataset. This decrease is relatively modest, with only a **4.14%** decrease in accuracy compared to the initial accuracy. However, after retraining on the perturbed data, we observe an improvement in the model's performance. The mA, accuracy, precision, recall, and F1-score have increased compared to the results obtained before retraining. This improvement demonstrates the effectiveness of retraining in overcoming the challenges posed by blurring and strengthening the model's capability to recognize people's attributes.

Grayscale. We converted the images into grayscale to test the impact of colors on recognizing person attributes. Subsequently, we retrained the DeepMAR [3] model on the altered dataset using pre-trained weights from the unaltered dataset. The results of the various tests are presented in Table 4.

Table 4. Results of tests of conversion into grayscale in percentage (%)

Dataset	mA	Accuracy	Precision	Recall	F1-score
Raw before retraining	85.34	83.06	89.87	88.41	89.13
Grayscale before retraining	68.71	68.55	80.81	77.05	78.89
Grayscale after retraining	78.52	78.34	87.28	84.20	85.71
Raw after retraining	79.61	79.00	87.79	84.69	86.21

The conversion of images into grayscale led to a significant decrease in the model's performance. Accuracy dropped by **14.51%** compared to the initial accuracy, indicating a considerable loss in the recognition ability of person attributes. The mA, precision, recall, and F1-score also decreased, highlighting the importance of colors in the overall model performance. After retraining on the altered data, the model's performance showed a marked improvement, although it did not reach the initial scores due to difficulties in recognizing attributes related to clothing colors.

Low Resolution. We applied a low-resolution filter to test the model's attribute extraction performance under reduced image resolution. Subsequently, we retrained the DeepMAR [3] model on the altered dataset using pretrained weights from the original dataset. Results are presented in Table 5.

Table 5. Results of resolution reduction tests in percentage (%)

Dataset	mA	Accuracy	Precision	Recall	F1-score
Raw before retraining	85.34	83.06	89.87	88.41	89.13
Low-resolution before retraining	81.87	79.85	88.09	85.61	86.83
Low-resolution after retraining	85.17	82.87	89.96	88.09	89.02
Raw after retraining	83.33	80.96	88.53	86.81	87.66

Reducing image resolution resulted in a slight decline in model performance. Accuracy decreased by **3.21%** compared to the initial accuracy. Additionally, mA, precision, recall, and F1-score experienced a slight decrease, indicating that deep learning models have some capability to handle low-resolution issues. After retraining on altered data, the model's performance significantly improved, even surpassing the initial precision by **0.09%**. This improvement demonstrates the efficacy of retraining in overcoming challenges posed by low-dimensional data.

Low Light. We decreased the brightness of the images by **50%** to test the impact of brightness on the recognition of people's attributes. Subsequently, we retrained the DeepMAR [3] model on the altered dataset using the pre-trained weights of the unaltered dataset. The results of various tests are presented in Table 6.

Table 6. Results of brightness reduction tests in percentage (%)

Dataset	mA	Accuracy	Precision	Recall	F1-score
Raw before retraining	85.34	83.06	89.87	88.41	89.13
Low brightness before retraining	76.85	75.36	85.02	82.73	83.86
Low brightness after retraining	85.04	82.81	89.76	88.13	88.94
Raw after retraining	84.93	82.91	89.77	88.25	89.00

Reducing the brightness resulted in a significant decrease in the model's performance. Accuracy decreased by **7.70%** compared to the initial accuracy, indicating a considerable loss in the recognition ability of people's attributes. Similarly, mA, precision, recall, and F1-score also decreased, emphasizing the influence of image lighting on the overall model performance. After retraining on the altered data, the model's performance showed significant improvement, almost reaching the initial performance. This improvement demonstrates the effectiveness of retraining in overcoming the challenges posed by low image brightness.

4.4 Discussion

The impact of disturbed environments on attribute recognition models revealed significant performance declines under various conditions such as occlusion, blurring, reduced brightness, and low resolution. These distortions notably affected the model's accuracy in extracting attributes from images. Retraining the model on disturbed data showed promising results, enhancing performance and improving the model's robustness against environmental disruptions. Despite challenges, the retraining methodology effectively countered adversities, enhancing attribute recognition accuracy even in challenging conditions.

5 Conclusion

In conclusion, this study tackled challenges in attribute recognition using computer vision and deep learning. Despite remarkable advancements, real-world complexities like occlusion and lighting variations hinder accurate recognition. Our proposed approach, augmenting traditional deep learning with perturbed datasets, aimed to enhance model resilience.

Experiments demonstrated that models trained on perturbed datasets showed increased resilience, outperforming traditional models in accuracy, even in challenging conditions. These findings suggest practical applicability and highlight the adaptability of our approach. Future research can further optimize these techniques for broader use across diverse domains.

References

1. Deng, Y., Luo, P., Loy, C.C., Tang, X.: Pedestrian attribute recognition at far distance. In: Proceedings of the 22nd ACM International Conference on Multimedia, pp. 789–792 (2014)
2. He, K., Zhang, X., Ren, S., Sun, J.: Deep residual learning for image recognition. In: Proceedings of the IEEE Conference on Computer Vision and Pattern Recognition, pp. 770–778 (2016)
3. Li, D., Chen, X., Huang, K.: Multi-attribute learning for pedestrian attribute recognition in surveillance scenarios. In: 2015 3rd IAPR Asian Conference on Pattern Recognition (ACPR), pp. 111–115. IEEE (2015)
4. Li, D., Chen, X., Zhang, Z., Huang, K.: Pose guided deep model for pedestrian attribute recognition in surveillance scenarios. In: 2018 IEEE International Conference on Multimedia and Expo (ICME), pp. 1–6. IEEE (2018)

5. Lin, M., Chen, Q., Yan, S.: Network in network. arXiv preprint arXiv:1312.4400 (2013)
6. Liu, X., Zhao, H., Tian, M., Sheng, L., Shao, J., Yi, S., Yan, J., Wang, X.: Hydraplus-net: attentive deep features for pedestrian analysis. In: Proceedings of the IEEE International Conference on Computer Vision, pp. 350–359 (2017)
7. Pan, X., Luo, P., Shi, J., Tang, X.: Two at once: Enhancing learning and generalization capacities via ibn-net. In: Proceedings of the European Conference on Computer Vision (ECCV), pp. 464–479 (2018)
8. Su, C., Yang, F., Zhang, S., Tian, Q., Davis, L.S., Gao, W.: Multi-task learning with low rank attribute embedding for multi-camera person re-identification. IEEE Trans. Pattern Anal. Mach. Intell. **40**(5), 1167–1181 (2017)
9. Xie, B., Wu, X., Zhang, S., Zhao, S., Li, M.: Learning diverse features with part-level resolution for person re-identification. In: Pattern Recognition and Computer Vision: Third Chinese Conference, PRCV 2020, Nanjing, China, October 16–18, 2020, Proceedings, Part III 3, pp. 16–28. Springer (2020)
10. Zhang, M.L., Zhou, Z.H.: A review on multi-label learning algorithms. IEEE Trans. Knowl. Data Eng. **26**(8), 1819–1837 (2013)
11. Zhao, X., Sang, L., Ding, G., Han, J., Di, N., Yan, C.: Recurrent attention model for pedestrian attribute recognition. In: Proceedings of the AAAI Conference on Artificial Intelligence, vol. 33, pp. 9275–9282 (2019)

Layer-Wise Compression Analysis of CNN Architectures: Insights from VGG16

Fethi Ourghi(✉), Abdenour Amamra, and Youcef Azri

Ecole Militaire Polytechnique, Bordj El-Bahri BP 17,, Algiers, Algeria
fethiourghi@gmail.com

Abstract. The state-of-the-art performance of most visual tasks is accomplished by deep convolutional neural networks, which have millions to billions of parameters, resulting in large neural networks. This introduced more redundant and oversized layers, which pose challenges in terms of memory requirements and computational efficiency.

To address these issues, this paper investigates the compressibility of CNN layers, with the aim of reducing the size of the model. In this paper, we propose a method based on an autoencoder, specifically CSNet (Compressive Sensing Network) for layer-wise compression analysis. Our study reveals that CNN models can be effectively compressed using autoencoder-based techniques. We assess the performance drop by compressing individual layers and combinations thereof, providing insight into the layers that contribute the most to the inference result. Our contributions include training and evaluating an autoencoder compression network (CSNet), to compress layer-per-layer a CNN model (VGG16), in order to conduct a comprehensive analysis of CNN compressibility and evaluating the performance impact of layer-wise compression.

Keywords: Convolutional Neural Networks · Model Compression · Compressive Sensing Network (CSNet) · Layer-wise Compression · VGG16 · Computational Efficiency

1 Introduction

Convolutional neural networks have shown remarkable success in various tasks such as classification and detection. This is evidenced by the high accuracy and precision that it provides in many applications. To achieve this performance, many deep architectures were proposed; These architectures have a large number of parameters, reaching millions or even billions of parameters, such as VGG16 [1], which have around 138 million parameters, and represent a good example of the level of complexity that is now common in modern CNNs.

However, the increasing number of parameters in these deep CNN architectures presents various challenges, including the increase in the required memory and computation power, affecting the time required to perform the inference and

B. Djamaa et al. (Eds.): CSA 2024, LNNS 1145, pp. 36–45, 2025.
https://doi.org/10.1007/978-3-031-71848-9_4

the memory needed to process and store the model, making the model inefficient in both real-time scenarios and storage cases.

Many CNN architectures include a larger number of parameters compared to the task they perform, leading to the following hypothesis. CNN models contain degenerated and redundant filters that do not contribute to inference results, which subsequently affects the performance of the model in general. In this paper, we investigate the compressibility of the CNN layers in order to reduce the model size, and we support the above hypothesis by conducting an analysis study of the layers of CNN models, based on layer-wise compression and evaluation of the drop in the classification accuracy of the chosen models. This study allows us to identify the layers that contribute the most to the inference result. Our contributions in this paper include the following:

1. Conduct an analysis to show that CNN models can be compressed using Auto-Encoder based techniques.
2. Analysis of the performance drop when we compress each layer apart.

The rest of this paper is structured as follows. Section 2, provides a background on the techniques used to compress deep CNN models. Section 3 details the proposed approach to analyze CCN architectures and the compressibility investigation. Section 4 describes the experimental setup in addition to the results obtained with their evaluation and discussion. Finally, Sect. 5 concludes our work and highlights some future research directions.

2 Background and Related Work

As the size and complexity of deep convolutional networks such as VGG16 [1], is increasing, many challenges hinder their deployment, particularly in terms of computational resources and storage efficiency. To overcome these issues, many approaches were proposed. Model compression techniques can be divided into two principal categories: The first pertains to modifications in the network's structural and topological characteristics, and the second is related to the temporal aspect in which the compression is applied within the network.

The first category includes two methods, structure-preserving compression and structure-altering methods [2]. The structure-preserving does not modify the topology of the networks, where the second changes the network structure (i.e. filters, connections, neurons, number of layers, etc.). Among number of approaches based on structure-preserving compression, the most well-known techniques are pruning, weight quantization, and low-rank matrix decomposition. On the other hand, non-preserving methods include: knowledge distillation [3] and skeletonization [4], those methods change the structure of the network, e.g., distilling the knowledge acquired by a teacher model in order to transfer it to a lighter student model, or even using more than one teacher to extract the learned and soft outputs to learn smaller model, while skeletonization focuses on removing layers, channels, or filters, based on certain criteria or heuristics.

The second category focuses on when we apply compression, that is, pre-training, in-training, or post-training compression. Generally non-preserving compression methods such as knowledge distillation use pre-training compression techniques, where the model is resized before the training process. Skeletonization techniques use an in-training compression approach that means including the process of refinement and fine-tuning of the network in a loop with the process of compression or generally pruning, in order to adapt to changes in structure like filters, channels, connections, or even layers. Structure-preserving approaches use both in-training and post-training techniques. In-training, compression in this class preserves the network structure while sparsifying the weights (i.e., filters) or channels and then keeps fine-tuning the remaining weights. Post-training techniques are usually used to compress pre-trained models. Figure 1 shows the categories of the existing methods.

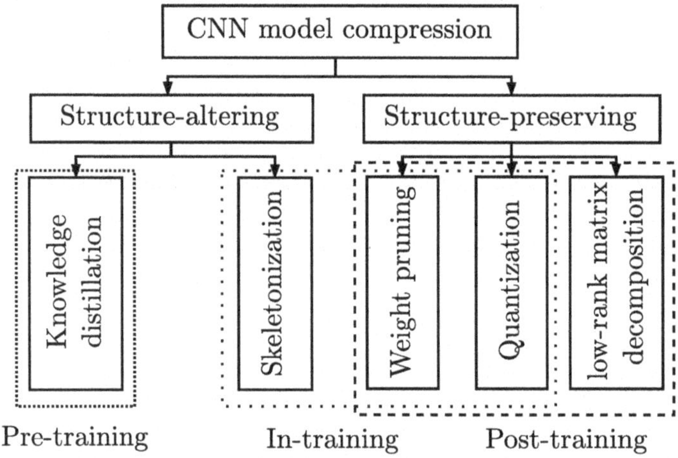

Fig. 1. Model compression techniques

Hinton et al. [3] proposed a method to distill the knowledge of a larger model into a smaller one. The goal of this method is to align the class probability distributions of the teacher and student networks. The method uses a distillation loss between the output logits of both the teacher and the student models, combined with the loss of the student model. Gou et al. [5] proposed comparing the weights of the teacher and student models. Other methods [6,7] proposed to match the intermediate features generated by the layers of each model, minimizing the distance between them. Zagoruyko et al. [8] proposed to transfer the attention of the teacher model to the student model; this method ensures that the student model will look at the region of interest in the image. Heo et al. [9] method consists of matching the sparsity patterns of the model weights of both teacher and student.

Han et al. [10] in their paper discussed a combination of multiple methods, including pruning, trained quantization, and Huffman coding. This three-stage pipeline consists of pruning redundant connections between neurons and then quantizing the network parameters to a lower floating point representation (i.e., reducing the precision), which may result in collisions among the new values, thereby leading to the creation of shared weights. These stages are followed by a retraining to fine-tune the remaining weights. The final stage of the pipeline consists of the application of Huffman coding in order to achieve additional compression by encoding redundant weights with shorter codes, thus reducing the overall size of the network. This pipeline was able to reduce the size of the neural network by a factor of 35× applied to AlexNet to 49× when applied to VGG16.

Jeong et al. [11] proposed a method to reduce the size of neural networks for transfer and storage purposes, such as transferring the network over WiFi. This method is derived from the low-rank matrix approximation approach, where the generative basis is smaller in size compared to the original. The method consists of generating and not training CNN filters using a combination of trained coefficients and a set of input basis filters (IBF). The input basis filters for the first part of the network is generated using heuristic methods such as the Sobel edge filter basis and line detection filter basis. The second and third parts consist of generating random IBFs based on well-determined statistical distributions such as uniform, normal, and exponential distributions. This method shows considerable gains in model size in both VGG16 [1] and ResNet-18 [12].

3 Layer-Wise Analysis of VGG16 Compression Using CSNet Autoencoder

In this section, we describe the method used to investigate the compressibility of a Convolutional Neural Network. Specifically, we examine the convolutional layers of VGG16[1]. We hypothesize that the network can be compressed without a high drop in accuracy because the layers have varying levels of redundancy.

3.1 Neural Networks Selection

We investigated a state-of-the-art convolutional neural network VGG16 [1] pre-trained on ImageNet-1k dataset [13], which is known for its consistent architecture with the 3 by 3 convolutional filters over 12 convolutional layers, allowing us to explore how various structural designs affect the compressibility of the layer.

3.2 Compression Technique

The compression technique used in this work involves an unsupervised reconstruction autoencoder called CSNet (Compressive sensing Network), which takes a single channel tensor as input and reconstructs an output that is very similar to the input of the network. The CSNet is defined by two convolutional layers; the first layer is used to compress the input to a low-dimensional tensor, called the latent-space representation. The second layer is designed to reconstruct a tensor similar to the input.

CSNet is characterized not only by its architecture, but also by a compression rate, which determines the dimensionality of the latent space. In this study, various compression rates were used to assess the degree of redundancy present within the layers of the convolutional neural network. CSNet was trained on a filter dataset presented by the work of P. Gavrikov et al. [14]. The latter work consists of collecting and analyzing a large collection of convolutional models from different tasks such as segmentation, detection, and classification, seeking for redundant patterns in layers' filters, thus motivating the current work.

3.3 Layer-Wise Compression Approach

In this study, our goal is to evaluate the individual effect of each layer on network performance after compression. An analysis was conducted that took into account our initial observation of the different levels of compressibility of the initial and subsequent layers. This process included isolating, compressing, and then recombining each layer and then evaluating the performance of the network to determine the effect of the compressed layer.

4 Performance Evaluation

In this section, we present the experiments carried out to evaluate the performance of our proposed compression method. A series of tests were carried out, in which each layer is compressed and its impact on the overall accuracy and efficiency of the network is investigated.

4.1 Metrics and Experimental Setup

In this study, our focus is solely on measuring the accuracy of the model and comparing it with the baseline, which represents its accuracy prior to compression. Using a convolutional auto-encoder (CSNet) to perform the compression, we measured the performance by comparing the results of the reconstructions of the compressed layers with the input ones; thus, we used a metric called Structural Similarity (SSIM) [15], which is mainly used in images to measure their quality.

This evaluation is crucial for understanding the level to which we can compress a CNN model and to determine the feasibility of these approaches in real-world applications. The experiments were carried out on an NVIDIA Tesla V100 GPU with 32 GB of VRAM and 16 GB of RAM.

4.2 Datasets

In this work, we used two datasets, each of which is used to perform different tasks. The first dataset is used to train the CSNet model to reconstruct the filter map for each layer. The second is used to evaluate the VGG16 [1] model after each compression of its layers.

CNN Filter DB. [14] is a collection of 3×3 filters extracted from 647 publicly available pre-trained CNN models, in addition to other models trained on various 2D visual tasks. The dataset consists of 1, 464, 797, 156 filters from 21, 436 layers of the 647 pre-trained models trained to perform various tasks ranging from classification, segmentation, detection, and many more visual tasks.

Due to its large size, a subset of 20% of the classification models presented in the dataset is used mainly to train, validate, and test the CSNet autoencoder for reconstruction purposes. The 3 by 3 filters presented in the dataset were used to construct a filter map of size 192×192 that is used to train the CSNet model on how to reconstruct the filter map and find redundancy patterns.

ImageNet. [13] is a well-known benchmark in object category classification and detection. The dataset consists of three subsets: training, validation, and test sets. It contains 1000 object classes, with the subsets comprising 1, 281, 167, 50, 000, and 100, 000 images, respectively.

Due to the unavailability of the test set labels, we have used the validation set as a benchmark for the tests. This dataset is used to evaluate the models after compression of each layer. The selection of this dataset was motivated primarily by its widespread use in the training of various models, such as VGG16 [1], AlexNet [16], making it a standard benchmark dataset.

4.3 Results and Discussion

This section is divided into two parts. The first discusses the results achieved through CSNet training, while the second part provides an in-depth analysis of the compression results obtained from the selected CNN architecture VGG16 [1].

CSNet Training Results. The training of the autoencoder was performed over a series of epochs (10 epochs), with the loss curves for each compression rate shown in Fig. 2. All loss curves, regardless of compression rate, exhibit a consistent decrease over training epochs. This indicates that the autoencoder is effectively learning and adapting during the training process.

Each curve represents a different compression rate, ranging from 10% to 90%. The increase in compression rate is observed to lead to a model that becomes incapable of reconstructing the original input, thus increasing the loss between the input and the output, as shown in Fig. 2b. The compression rate defines the latent space size that influences the reconstruction of the input image.

The previous observed results are demonstrated by the results of the SSIM metric shown in Table 1, which measures the structural similarity between the input and output of the CSNet compression model. SSIM values decrease as the compression rate increases.

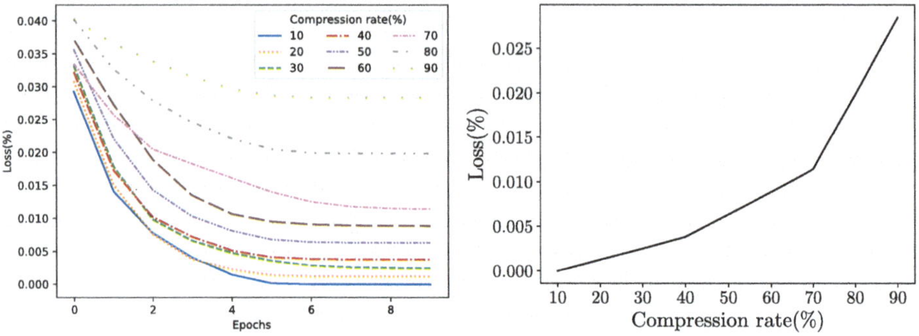

(a) CSNet training loss on the CNN Filter DB dataset

(b) Compression rate influence on the loss values

Fig. 2. Performance measurements for training CSNet, where each compression rate represent a CSNet model. Figure 2a shows the different loss curves for each CSNet model. Figure 2b shows the variation in the loss values in relation to the compression rate, where it increases whenever the compression rate increases, due to the loss in data caused by the decrease in the latent space size.

Table 1. Structural Similarity for each compression rate

Compression rate (%)	10	20	30	40	50	60	70	80	90
SSIM (%)	100.0	85.08	78.19	73.71	63.91	55.56	48.59	34.57	21.09

Model Compression Results. Figure 3 shows the relationship between compression rate and classification accuracy between the layers of the VGG16[1] convolutional neural network. As the compression rate increases, the earlier layers (layers 0–3) show a decrease in accuracy. This downward trend is amplified in the middle layers (layers 4–7), which exhibit substantial degradation in accuracy under increasing compression. Interestingly, later layers (layers 8–12) demonstrate considerable robustness to high compression ratios, maintaining stable accuracy despite extreme parameter reduction. This divergence in accuracy loss in the earlier and middle layers, in contrast to preservation of performance in the latter layers, reveals a variable susceptibility to compression throughout the VGG16 model. Therefore, our results highlight the importance of considering layer-specific compression to determine the contribution of each layer in the overall classification. Figure 3a illustrates the impact of compression on the initial layers, revealing a higher sensitivity. Figure 3b explores the middle layers, demonstrating moderate sensitivity to compression. The last figure, Fig. 3c focuses on the latter layers, highlighting their lower sensitivity to the compression process. This analysis sheds light on the varying degrees of compressibility between different layers, providing information on the overall performance of the network.

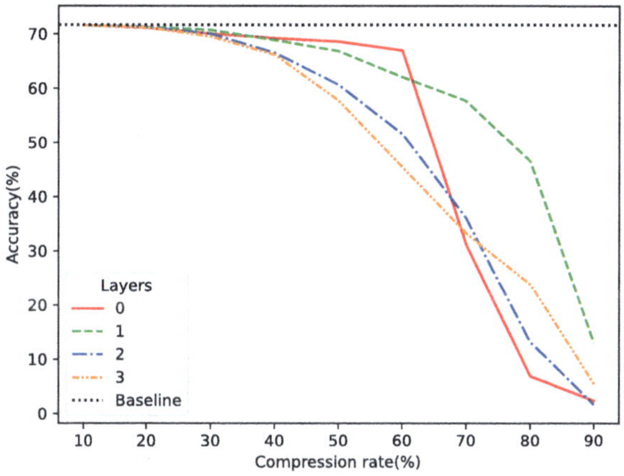

(a) First layers(0, 1, 2, and 3)

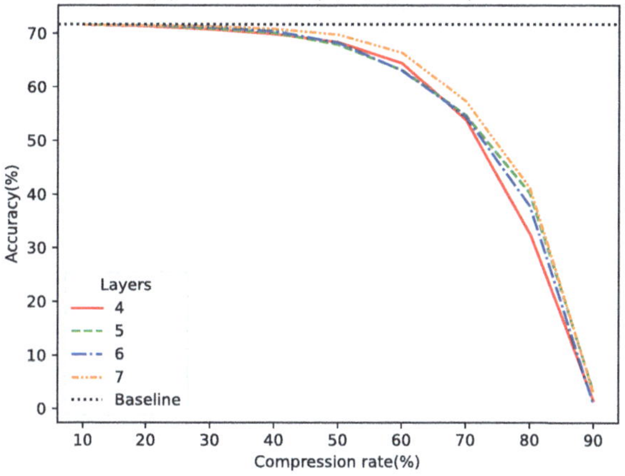

(b) Middle layers(4, 5, 6, and 7)

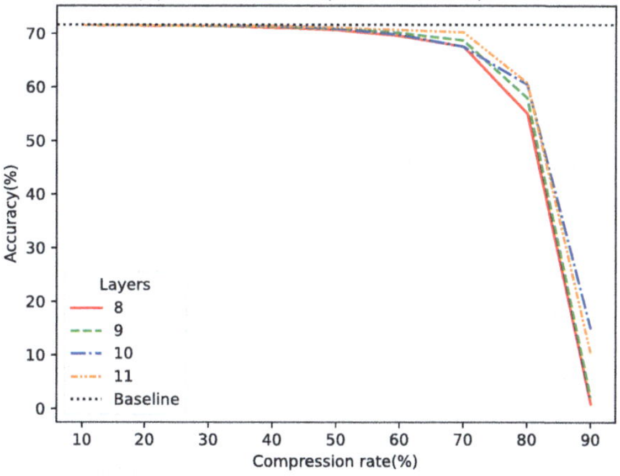

(c) Last layers(8, 9, 10, 11, and 12)

Fig. 3. Performance obtained by compressing the VGG16 layer per layer compared to the baseline accuracy of 71.65%.

5 Conclusion

In this paper, we have presented the investigation method that allows us to determine the impact of the layer on the overall accuracy of the CNN model. Through our study, which focuses on compressing the CNN model layer by layer, we present significant variations in the accuracy sensitivity between layers. In the case studied of VGG16, the early layers are shown to result in weak performance under medium and high compression rates, whereas the middle layers show a degradation of performance at high rates. In contrast, the last layers demonstrated a high level of compressibility compared to the first and middle layers, thus maintaining the classification accuracy.

One promising direction for future work is developing layer-adaptive compression techniques that tune compression rates based on layer sensitivity measurements. This could enable aggressive compression focused on resilient layers while minimizing distortion of vulnerable layers. Examining the effects of techniques such as knowledge distillation, pruning, and quantization through a layer-wise lens could drive the creation of customized pipelines that integrate multiple methods.

References

1. Simonyan, K., Zisserman, A.: Very deep convolutional networks for large-scale image recognition. In: Y. Bengio, Y. LeCun (eds.) 3rd International Conference on Learning Representations, ICLR 2015, San Diego, CA, USA, May 7-9, 2015, Conference Track Proceedings (2015)
2. Marinó, G.C., Petrini, A., Malchiodi, D., Frasca, M.: Deep neural networks compression: a comparative survey and choice recommendations. Neurocomputing **520**, 152–170 (2023). https://doi.org/10.1109/10.1016/j.neucom.2022.11.072
3. Hinton, G., Vinyals, O., Dean, J.: Distilling the knowledge in a neural network. Stat **1050**, 9 (2015)
4. Mozer, M.C., Smolensky, P.: Skeletonization: A technique for trimming the fat from a network via relevance assessment. Advances in neural information processing systems **1** (1988)
5. Gou, J., Yu, B., Maybank, S.J., Tao, D.: Knowledge distillation: a survey. Int. J. Comput. Vision **129**(6), 1789–1819 (2021). https://doi.org/10.1109/10.1007/s11263-021-01453-z
6. Huang, Z., Wang, N.: Like what you like: knowledge distill via neuron selectivity transfer. CoRR abs/1707.01219 (2017)
7. Kim, J., Park, S., Kwak, N.: Paraphrasing complex network: network compression via factor transfer. In: Proceedings of the 32nd International Conference on Neural Information Processing Systems, NIPS 2018, pp. 2765–2774. Curran Associates Inc., Red Hook (2018)
8. Zagoruyko, S., Komodakis, N.: Paying more attention to attention: Improving the performance of convolutional neural networks via attention transfer. In: 5th International Conference on Learning Representations, ICLR 2017, Toulon, France, April 24–26, 2017, Conference Track Proceedings (2017)

9. Heo, B., Lee, M., Yun, S., Choi, J.Y.: Knowledge transfer via distillation of activation boundaries formed by hidden neurons. In: Proceedings of the Thirty-Third AAAI Conference on Artificial Intelligence and Thirty-First Innovative Applications of Artificial Intelligence Conference and Ninth AAAI Symposium on Educational Advances in Artificial Intelligence. AAAI Press (2019). https://doi.org/10.1109/10.1609/aaai.v33i01.33013779
10. Han, S., Mao, H., Dally, W.J.: Deep compression: Compressing deep neural network with pruning, trained quantization and huffman coding. In: Y. Bengio, Y. LeCun (eds.) 4th International Conference on Learning Representations, ICLR 2016, San Juan, Puerto Rico, May 2–4, 2016, Conference Track Proceedings (2016)
11. Jeong, J., Cho, J.H., Lee, J.G.: Filter combination learning for CNN model compression. ICT Express **7**(1), 5–9 (2021). https://doi.org/10.1109/10.1016/j.icte.2021.01.001
12. He, K., Zhang, X., Ren, S., Sun, J.: Deep residual learning for image recognition. In: 2016 IEEE Conference on Computer Vision and Pattern Recognition (CVPR), pp. 770–778 (2015)
13. Russakovsky, O., Deng, J., Su, H., Krause, J., Satheesh, S., Ma, S., Huang, Z., Karpathy, A., Khosla, A., Bernstein, M., Berg, A.C., Fei-Fei, L.: ImageNet large scale visual recognition challenge. Int. J. Comput. Vis. (IJCV) **115**(3), 211–252 (2015). https://doi.org/10.1109/10.1007/s11263-015-0816-y
14. Gavrikov, P., Keuper, J.: Cnn filter db: an empirical investigation of trained convolutional filters. In: Proceedings of the IEEE/CVF Conference on Computer Vision and Pattern Recognition (CVPR), pp. 19,066–19,076 (2022)
15. Wang, Z., Bovik, A., Sheikh, H., Simoncelli, E.: Image quality assessment: From error visibility to structural similarity **13**(4), 600–612. https://doi.org/10.1109/10.1109/TIP.2003.819861
16. Krizhevsky, A., Sutskever, I., Hinton, G.E.: ImageNet classification with deep convolutional neural networks. In: Advances in Neural Information Processing Systems, vol. 25. Curran Associates, Inc. (2012)

Enhanced ECG Signals Classification with Image-Driven Ensemble Deep Transfer Learning

Hadjer Bechinia[1(✉)], Djamel Benmerzoug[1], and Nawres Khlifa[2]

[1] LIRE Laboratory, University Abdelhamid Mehri Constantine 2, Constantine, Algeria
{Hadjer.Bechinia,Djamel.Benmerzoug}@univ-constantine2.dz

[2] Higher Institute of Medical Technologies of Tunis, Research Laboratory of Biophysics and Medical Technologies, University of Tunis el Manar, Tunis, Tunisia
nawres.khlifa@istmt.utm.tn

Abstract. Cardiovascular diseases (CVDs) stand as a significant global health concern, characterized by a rising number of cases and fatalities. Within this context, arrhythmias emerge as critical manifestations that require advanced diagnostic tools. In this study, we focus on ECG signals analysis plotted and saved as images employing deep learning methods to address the challenges associated with limited data in medical datasets. Through transfer learning with a relatively small ECG dataset, our research explores the effectiveness of three pre-trained models, inception-V3, DenseNet121, and Xception, in developing robust ECG image classification models. Furthermore, we utilize ensemble learning techniques, specifically averaging, to enhance classification accuracy. Our evaluation results present individual model accuracies and an enhanced weighted average ensemble accuracy of 97.82%, providing a comprehensive and effective approach for arrhythmia detection.

Keywords: ECG signals · Arrhythmia Classification · Deep Learning · Ensemble Learning · Transfer Learning

1 Introduction

Cardiovascular diseases (CVDs) are designated by the World Health Organization as the foremost single cause of mortality. In recent times, there has been a continual increase in both the prevalence and fatality rates of these diseases, a trend that is notably evident, extending even to the younger population [1]. Arrhythmia manifests as irregularities in heart rhythm, stemming from abnormal electrical conduction within the heart. It encompasses a range of irregular heart tremors to tachycardia (resting heart rate exceeding 100 beats per minute) and bradycardia (resting heart rate falling below 60 beats per minute) [2]. An electrocardiogram (ECG)is a diagnostic test that captures electrical signals generated by the heart's activity. This examination provides crucial insights into

B. Djamaa et al. (Eds.): CSA 2024, LNNS 1145, pp. 46–56, 2025.
https://doi.org/10.1007/978-3-031-71848-9_5

the heart's condition and serves as a valuable indicator for various cardiovascular diseases (CVDs). ECG is a non-invasive and cost-effective tool, adept at diagnosing cardiac disorders, including arrhythmias, through continuous monitoring of the heart's electrical activity. The signals extracted from the ECG offer valuable information for analyzing and comprehending an individual's cardiac activity, including aspects such as heart rate, rhythm, and morphology [3]. To detect and to classify ECG arrhythmia, Deep Learning, known for its robust feature extraction capability, has experienced remarkable success over the past few years [1]. Among various deep learning models, the convolutional neural network (CNN) frequently emerges. Its effectiveness extends to diverse applications such as medical imaging [4,5], gene recognition, speech processing, sleep apnea detection, and other domains, showcasing its versatility and robust performance [1]. Nevertheless, Deep learning models, including CNNs, typically demand a substantial volume of data for effective training, thereby giving rise to the challenge of limited data under optimal conditions. This issue is particularly pronounced in the context of medical datasets, where the constrained availability of expert-annotated datasets poses well-acknowledged challenges. Various research initiatives have actively addressed this challenge through the application of transfer learning (TL) models [6].

Therefore, this research endeavors to harness the power of transfer learning with a modest ECG image dataset by proposing three different pre-trained models: Inception-V3, DenseNet121, and Xception. It showcases the benefits of this approach in achieving optimal results in the classification of ECG signals, ultimately contributing to more effective and accessible cardiovascular health diagnostics. Furthermore, the incorporation of ensemble learning techniques aims to enhance the classification accuracy by leveraging the strengths of combining multiple models. The rest of the paper is organized as follows, in Sect. 2 we navigate through a review of current literature and previous research efforts using deep learning methodologies for ECG arrhythmia diagnosis, while in Sect. 3 we present our proposed approach in detail. Our experimental results and corresponding discussions are presented in Sect. 4. Finally, Sect. 5 concludes the paper while offering insights into future directions.

2 Related Works

In this section, we review the literature pertaining to the detection and classification of ECG arrhythmia, placing particular emphasis on the integration of ensemble and deep transfer learning models in these studies. In [7], the authors introduced an automated myocardial infarction (MI) detection system utilizing ECG signals through a CNN model, a hybrid CNN-long short-term memory network (LSTM), and an ensemble technique called a post-processing method to determine the optimal performing model. Trained on two datasets, the experiments unfolded in achieving the highest classification accuracy of 99.89% using the ensemble technique utilizing a balanced dataset generated via the synthetic minority oversampling technique (SMOTE). In [8], an innovative end-to-end approach for automating cardiac arrhythmia classification utilizing transfer learning was proposed. Three pre-trained CNNs, namely, AlexNet, Resnet18, and GoogleNet were explored and crafted by block-wise fine-tuning each layer with varying model training parameters. For the evaluation, impressively, the proposed fine-tuned ResNet18 attained the highest accuracy at 99.56% using RMSprop optimizer, surpassing the other two models. The work of [9] introduced an ensemble approach for heart arrhythmia classification using ECG signals. The ensemble comprised two deep learning models, CNN-LSTM and RRHOS-LSTM, addressing local features, temporal dynamics, and class imbalances. Through a meta-classifier and subsequent verification with a CNN-LSTM model, the ensemble achieved an impressive 95.81% overall accuracy in patient-independent evaluation on MIT-BIH arrhythmia data. Moreover, the authors in [10] proposed a deep-learning approach for detecting arrhythmia utilizing the continuous wavelet transform (CWT) to obtain scalograms, which are then classified by a 2 dimensional CNN model based on arrhythmia signatures. The model achieved an average accuracy and 99.65%. In [11], they leveraged the capabilities of CNNs in conjunction with transfer learning techniques, specifically ResNet50, VGG16, VGG19 and Inception, to enable the analysis of both image representations and one-dimensional heartbeat signal. Through the application of the ensemble weighted average method to consolidate the predictions from these diverse models, they achieved an accuracy of 94% for ECG signal classification and 93% for ECG image data. Furthermore, in [12] they introduced an ensemble methodology for classifying 2D scalogram ECG images. The approach combined two adapted VGG-16 models and one InceptionResNetV2 model, featuring added feature-extracting layers and ImageNet weights, to generate predictions through a fusion of their outputs opting for the averaging technique, where each model's prediction is equally weighted. Through a rigorous 5-fold cross-validation using the modified Dataset Combining three different datasets, the ensemble model achieved an impressive accuracy of 99.98%. By the same token, the work of [13], also proposed an innovative ensemble-based classification model that combines the effectiveness of CNN using the pre-trained VGG16 model with linear and bio-inspired classifiers such as Random Forests (RF), Support Vector Classifier (SVC), and k-Nearest Neighbor (KNN). The classifier was applied to arrhythmia datasets sourced from the MIT-BIH databases, achieving the same remarkable accuracy of 99.98%.

3 Materials and Methods

In this work, our focus lies in the analysis of ECG signals extracted from the MIT-BIH database as these signals undergo a series of preprocessing steps to facilitate their classification. Afterward, we harness the power of three distinct pre-trained deep learning models, inception-V3, DenseNet121, and Xception to develop a robust system that classifies the ECG signals converted as image representations. These models, each with its unique architecture and strengths, were utilized to capture intricate features from the input ECG image data. The outcomes from these three models are combined into an ensemble learning model using the averaging technique. Figure 1 presents the overall workflow of the proposed approach.

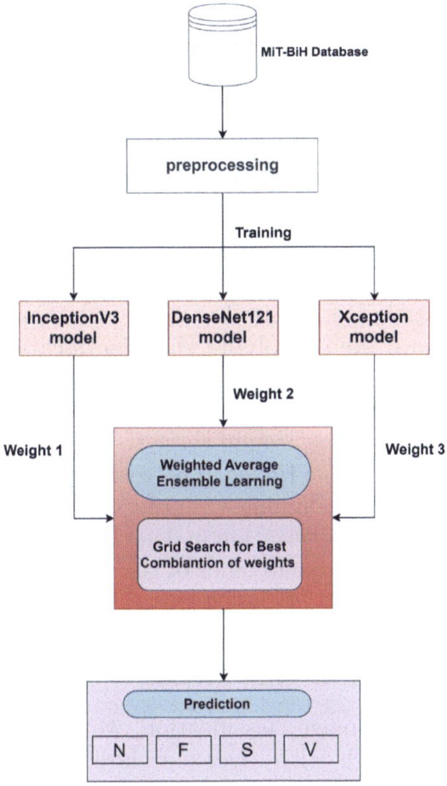

Fig. 1. Flowchart of the Proposed Approach

3.1 Dataset

Our work relies on using the MIT-BIH Arrhythmia Database[1], a widely recognized repository of ECG signal data conveniently accessible through the PhysioNet platform. This database comprises 48 half-hour two-lead ECG recordings sampled at 360 Hz, originating from 47 patients. Notably, the dataset offers diversity in subjects and provides annotated beat-by-beat labels, encompassing normal heartbeats, premature ventricular contractions (V), supraventricular premature beats (S), and fusion beats (F).

Given the imbalanced nature of the MIT-BIH dataset, we strategically selected only 20,000 ECG signals for our research. Out of these, 8,983 represent abnormal beats, while the remaining signals correspond to normal beats.

[1] https://archive.physionet.org/physiobank/database/mitdb/.

This approach allows us to work on a more manageable dataset, focusing on the four distinct categories within lead MLII. This dataset significantly contributes to the understanding and diagnosis of cardiac arrhythmias.

3.2 Preprocessing

In our research, we undertake a thorough preprocessing of ECG signals sourced from the MIT-BIH Arrhythmia Database to establish a foundation for robust analysis. To optimize data quality, the Butterworth filter is used, which effectively eliminates undesirable noise. ECG signals are particularly susceptible to two primary categories of noise: low-frequency noise, arising from baseline fluctuations due to bodily motions and respiration, and high-frequency noise, stemming from interference such as power lines and the digitization process during analog electrical potentials [14]. Following noise reduction, we proceed to beat segmentation, an integral step involving the identification of R-peaks within the signals. For their precise detection, we employ established techniques like the Pan-Tompkins algorithm. Once identified, individual beats are encapsulated using the fixed window technique around these R-peaks, ensuring a rigorous segmentation of each cardiac event. Hence, The segmented beats are transformed into grayscale 2D image plots, aligning with our objective to leverage computer vision techniques for enhanced utilization in deep learning models for arrhythmia classification and detection. The outcomes of the preprocessing step are showed in Fig. 2.

3.3 Ensemble Transfer Learning Model

In our exploration of ECG image classification, we embraced the utilization of three transfer learning models with diverse architectural strengths and feature extraction capabilities. Transfer learning originates from cognitive research and is grounded in the concept that knowledge can be transferred across related tasks to enhance performance on a new task [6]. This approach significantly mitigates the limited data challenge by transferring learned features, patterns, and representations, accelerating the convergence and enhancing the generalization capability of the deep learning model. First, the inception-V3 architecture, a powerful pre-trained deep learning model, renowned for its efficient use of inception modules, allowing the network to capture multi-scale features efficiently, providing a sophisticated understanding of intricate patterns present in ECG signals [15]. DenseNet121, another pivotal pre-trained model in our study, is known for its dense connections between layers. These connections enhance feature reuse, allowing for a more profound understanding of the hierarchical representations in ECG images [16]. Our third chosen model, Xception, integrates depth-wise separable convolutions to enhance computational efficiency. This model architecture allows for effective extraction of features from the ECG images [17]. In our adaptation, we retained the pre-trained weights of the three models and extended the architectures with the same custom classification layers. These model architectures serve as the foundation for our comprehensive exploration of ECG image classification, capturing diverse features crucial for accurate and robust predictions.

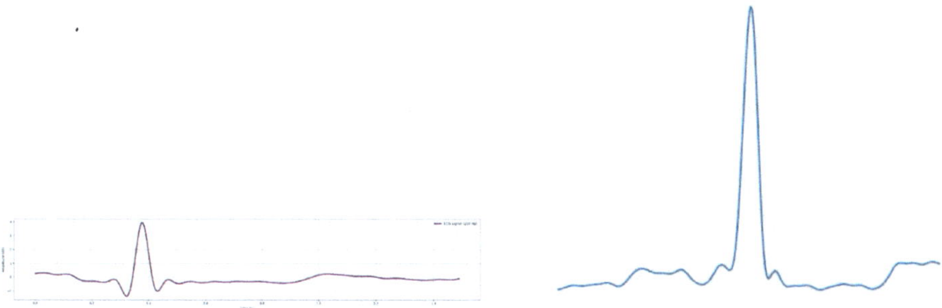

(a) Denoised ECG Signal

(b) Saved ECG Beat Image

Fig. 2. Preprocessing Results

To enhance the accuracy and robustness of our ECG classification, we employed an ensemble learning approach. After independently training each pre-trained model, we adopted a sophisticated ensemble technique to combine their predictions, thereby enhancing the overall classification accuracy. This ensemble strategy involves the application of a weighted average technique, where careful consideration is given to the influence of each individual model in the final classification. The weighted average ($\hat{y}$) for a given sample is mathematically expressed as follows:

$$\hat{y} = \frac{\sum_{i=1}^{M} w_i \cdot y_i}{\sum_{i=1}^{M} w_i} \tag{1}$$

Here, M denotes the number of pre-trained models, and y_i signifies the prediction of the ith model. The weights (w_i) assigned to each model's output serve as a critical factor in fine-tuning their contribution to the ensemble. This ensemble approach is designed to capitalize on the synergies and unique strengths of the diverse pre-trained models. By adjusting the weights, we strategically blend the predictions of each model, compensating for their individual limitations and creating a more robust and accurate overall prediction. To determine the optimal combination of weights for our ensemble models, we conducted a thorough grid search. This involved systematically exploring a range of weight values ($W = (w_1, w_2, \ldots, w_M)$) for each model. The search is driven by an objective function (Objective(W)) representing the overall accuracy of our ensemble model, as outlined below:

$$W^* = \arg\max_W \text{Objective}(W) \tag{2}$$

The grid search process ensures the identification of the most effective combination of weights, further elevating the predictive power of our ensemble models.

This meticulous approach not only strengthens the reliability of our predictions but also underscores the adaptability and versatility of the ensemble strategy in optimizing model performance.

4 Results and Discussion

In this section, we evaluate the effectiveness of our ensemble transfer model, explaining the details of our experiment and conducting a comprehensive analysis of the achieved results. During the training process, inception-V3 emerges as the most successful in terms of validation accuracy, reaching a peak of 96.88% with an early stop of the training after 27 epochs. In contrast, Xception and DenseNet, demonstrated commendable performance, stabilizing at slightly lower validation accuracies of 95.61% and 96.27% respectively. Despite minor differences in the validation accuracies, all three models exhibit consistent learning trends with diminishing training losses and increasing accuracies, showcasing their capabilities in capturing intricate patterns in the training data inception-V3 with its superior validation accuracy, stands out as particularly effective in generalizing well to unseen data, while Xception and DenseNet demonstrate robust and stable performances in their respective training processes. Figure 3 shows the training and validation accuracies and losses curves for the three models.

In the evaluation phase, individual model accuracies were recorded at 97.10% for inception-V3, 95.50% for Xception, and 96.60% for DenseNet. Upon employing a weighted average ensemble with random weights (0.4, 0.3, 0.3) for the trained models, the combined accuracy ranged between 97.50% and 97.70%. Further refinement through a grid search led to the identification of optimal weights (0.2, 0.4, 0.3), resulting in an improved ensemble accuracy of 97.82%. This accurate optimization process showcases the effectiveness of ensemble learning in enhancing overall model performance. The classification report and the confusion matrix depicted in Fig. 4 provide detailed insights into the model's predictive capabilities across four classes, revealing high precision, recall, and F1-score values. The weighted average ensemble, especially with the optimized weights, showed effectiveness in accurately classifying a diverse set of images, achieving an impressive overall accuracy of 97.82%.

As depicted in Table 1, our comprehensive evaluation on the MiT-BiH arrhythmia dataset involved training individual deep learning models, including inception-V3, DenseNet121, and Xception, resulting in accuracies of 97.10%, 96.60%, and 95.50%, respectively. Notably, our ensemble learning approach, utilizing a weighted average method, yielded an exceptional accuracy of 97.82%, surpassing state of the art's outcomes of 1 Dimensional signal and 2D image ensembles (94% and 93%, respectively) on the same dataset. This underscores the effectiveness of our ensemble strategy in achieving superior classification performance for ECG arrhythmia, emphasizing the potential significance of combining deep learning models for enhanced diagnostic capabilities.

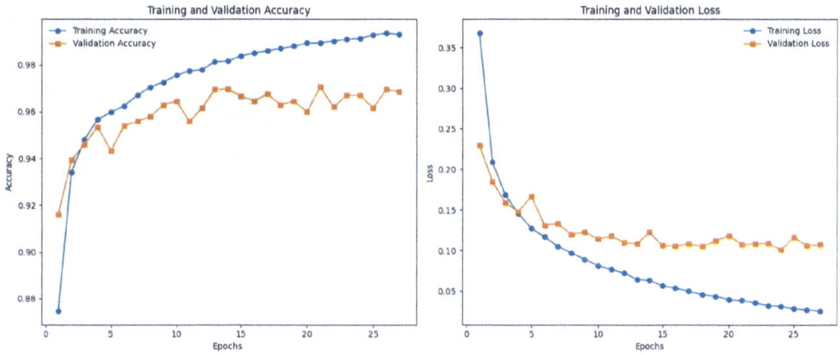

(a) Inception-V3 Training and Validation Curves

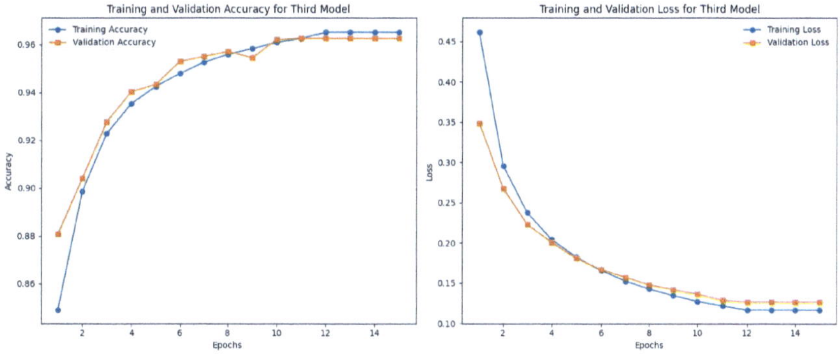

(b) DenseNet121 Training and Validation Curves

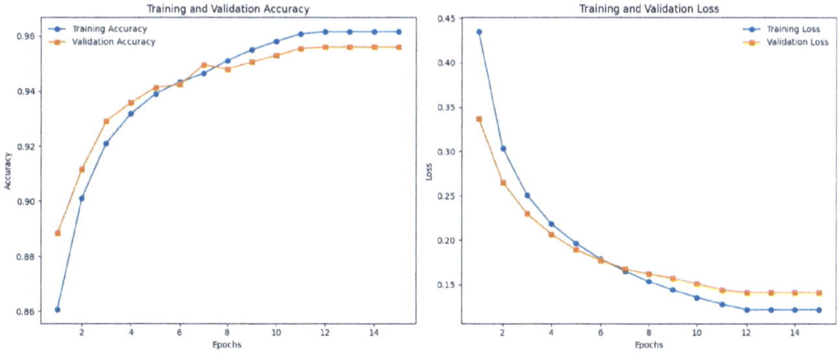

(c) Xception Training and Validation Curves

Fig. 3. Proposed Models' Accuracies and Losses Curves

```
Best Weights: (0.2, 0.4, 0.30000000000000004)
Best Accuracy: 97.82%
Accuracy (Best Weighted Average): 97.82%
              precision    recall  f1-score   support

           F       0.91      0.77      0.84       139
           N       0.98      0.99      0.98      2241
           S       0.99      0.94      0.97       305
           V       0.98      0.98      0.98      1315

    accuracy                           0.98      4000
   macro avg       0.97      0.92      0.94      4000
weighted avg       0.98      0.98      0.98      4000
```

(a) Classification Report

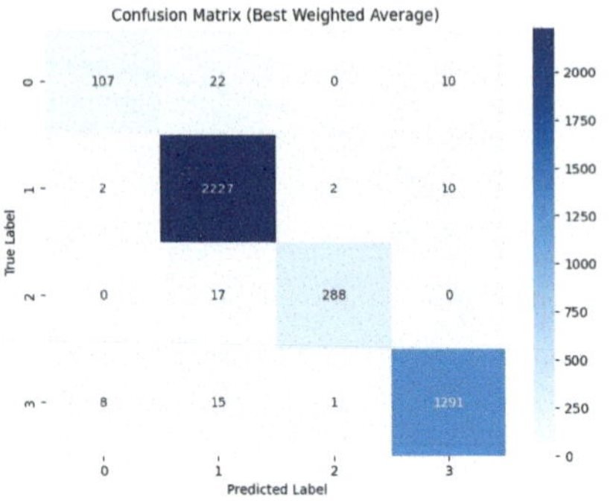

(b) Confusion Matrix

Fig. 4. Experimentation Results of the Proposed Ensemble Learning

Table 1. Comparison With Related Work

Reference	Dataset	Deep Learning Model	Accuracy
[11]	MiT-BiH arrhythmia	1D ensemble(VGG16, CNN, Inception)	94%
		2D ensemble(VGG16,VGG19, ResNet50)	93%
our work	MiT-BiH arrhythmia	inception-V3	97.10%
		DenseNet121	96.60%
		Xception	95.50%
		Ensemble Learning	**97.82%**

5 Conclusion

This research has addressed critical challenges in cardiovascular diseases (CVDs) and, specifically, arrhythmia detection using ECG signals. Leveraging the power of deep learning, our study utilized transfer learning with a dataset comprising approximately 20,000 ECG signal images, demonstrating the efficacy of inception-V3, DenseNet121, and Xception pre-trained models. Through an ensemble learning technique, we achieved enhanced classification accuracy, with a accurately optimized weighted average ensemble accuracy of 97.82%. Looking forward, in the future work we would explore alternative ensemble learning techniques and incorporate additional deep learning models to further enhance the robustness of the classification system. Additionally, assessing the effectiveness of the model on a larger dataset would provide valuable insights into its scalability and generalization.

References

1. Chen, S.W., Wang, S.L., Qi, X.Z., Samuri, S.M., Yang, C.: Review of ecg detection and classification based on deep learning: coherent taxonomy, motivation, open challenges and recommendations. Biomed. Signal Process. Control **74**, 103493 (2022)
2. Ma, S., Cui, J., Chen, C.-L., Chen, X., Ma, Y.: An effective data enhancement method for classification of ecg arrhythmia. Measurement **203**, 111978 (2022)
3. Musa, N., et al.: A systematic review and meta-data analysis on the applications of deep learning in electrocardiogram. J. Ambient Intell. Humanized Comput. **14**(7), 9677–9750 (2023)
4. Khachnaoui, H., Chikhaoui, B., Khlifa, N., Mabrouk, R.: Enhanced Parkinson's disease diagnosis through convolutional neural network models applied to spect datscan images. IEEE Access,(2023)
5. Rahmany, I., Guetari, R., Khlifa, N.: A fully automatic based deep learning approach for aneurysm detection in dsa images. In: 2018 IEEE International Conference on Image Processing, Applications and Systems (IPAS), pp. 303–307. IEEE (2018)
6. Kim, H.E., Cosa-Linan, A., Santhanam, N., Jannesari, M., Maros, M.E., Ganslandt, T.: Transfer learning for medical image classification: a literature review. BMC Med. Imaging **22**(1), 69 (2022)
7. Hari Mohan Rai and Kalyan Chatterjee: Hybrid cnn-lstm deep learning model and ensemble technique for automatic detection of myocardial infarction using big ecg data. Appl. Intell. **52**(5), 5366–5384 (2022)
8. Mohebbanaaz, Rajani Kumar, L.V., Padma Sai, Y.: A new transfer learning approach to detect cardiac arrhythmia from ecg signals. Signal, Image Video Process. **16**(7), 1945–1953 (2022)
9. Essa, E., Xie, X.: An ensemble of deep learning-based multi-model for ecg heartbeats arrhythmia classification. IEEE Access **9**, 103452–103464 (2021)
10. Mohonta, S.C., Motin, M.A., Kumar, D.K.: Electrocardiogram based arrhythmia classification using wavelet transform with deep learning model. Sensing Bio-Sensing Res. **37**, 100502 (2022)
11. Mahmud, T., et al.: Ensemble deep learning approach for ecg-based cardiac disease detection: Signal and image analysis. In: 2023 International Conference on Information and Communication Technology for Sustainable Development (ICICT4SD), pp. 70–74. IEEE (2023)
12. Ovi, T.B., Naba, S.S., Chanda, D., Onim, M.S.H.: A transfer-learning based ensemble architecture for ecg signal classification. In: 2022 IEEE Region 10 Symposium (TENSYMP), pp. 1–6. IEEE (2022)
13. Goswami, A.D., Bhavekar, G.S., Chafle, P.V.: Electrocardiogram signal classification using vggnet: a neural network based classification model. Int. J. Inf. Technol. **15**(1), 119–128 (2023)
14. Merdjanovska, E., Rashkovska, A.: Comprehensive survey of computational ecg analysis: Databases, methods and applications. Expert Syst. Appl. **203**, 117206 (2022)

15. Christian Szegedy, Vincent Vanhoucke, Sergey Ioffe, Jonathon Shlens, and Zbigniew Wojna. Rethinking the inception architecture for computer vision. 2015. *arXiv preprint* arXiv:1512.00567, 2015
16. Huang, G., Liu, Z., Van Der Maaten, L., Weinberger, K.Q.: Densely connected convolutional networks. In: Proceedings of the IEEE Conference on Computer Vision and Pattern Recognition, pp. 4700–4708 (2017)
17. Chollet, F.: xception: deep learning with depthwise separable convolutions. arxiv preprint. arXiv preprint arXiv:1610.02357 (2016)

Learning $N{:}M$ Structured Sparse Neural Networks from Scratch: A Comparative Study

Dehbia Ahmed Zaid(✉), Badis Djamaa, and Mohamed Akrem Benatia

Computer Science Departement, Ecole Militaire Polytechnique, BP 17, 16046 Bordj El-Bahri Algiers, Algeria
{d_AHMEDZAID.Dehbia,badis.djamaa,mohamedakrem.benatia}@emp.mdn.dz

Abstract. Sparsity in deep neural networks (DNNs) has attracted significant attention in recent years due to its ability to compress DNN models and speed up their execution in resource-constrained environments. Among different approaches, $N{:}M$ structured sparsity has taken the lead thanks to its efficient support in modern hardware accelerators. While numerous training methods for $N{:}M$ sparse DNNs have been proposed, there is a lack of comparisons among these methods, which prevents their fair and consistent evaluations. To address such a gap, this paper presents a comparative study of some state-of-the-art methods for training $N{:}M$ structured sparse neural networks. We first start by proposing a taxonomy of $N{:}M$ structured pruning configuration settings to systematically categorize and organize these methods. Building upon this taxonomy, we conduct several experiments to, qualitatively and quantitatively, compare the prominent works in the field under a uniform set of conditions. Our findings highlight opportunities for improving sparse neural network performance through approaches involving layer-wise $N{:}M$ sparse patterns, pretrained weight initialization, suitable pruning ratios, and the selection of effective sparse training strategies.

Keywords: Deep Learning · Sparse Neural Networks · $N{:}M$ Structured Sparsity

1 Introduction

Deep neural networks (DNNs) have made significant performance and generalization advancements, impacting fields like computer vision, natural language processing, robotics, and healthcare [10,19,21,25]. To achieve such performance, deep learning models have significantly increased in size over the last few years. Indeed, going deeper into neural network architectures resulted in very huge models with millions or even billions of parameters to train and use for inference. These huge models, despite their promising performances, are not practical for real-world applications that are deployed on mobile and embedded systems, which are known for their limitations in memory and computational capabilities.

B. Djamaa et al. (Eds.): CSA 2024, LNNS 1145, pp. 57–69, 2025.
https://doi.org/10.1007/978-3-031-71848-9_6

In response, efficient model compression techniques have emerged as a pressing research area. These techniques aim to reduce the size and computational complexity of neural networks while retaining their predictive capabilities [9].

Sparsity, a key model compression method for (DNNs) [13], simplifies models by zeroing out certain weights via pruning techniques [8,9]. There are two primary methods for integrating sparsity in DNNs training [30]: (i) a two-phase process involving pruning an already trained dense network followed by retraining the sparse model; and (ii) a single-phase method where a dense model is gradually pruned from the early stages [3,5]. The latter, while efficient in reducing training time and costs, often necessitates an extra fine-tuning phase due to the lingering effects of redundant structures on outputs [8,22]. Current research focuses on obviating the need for fine-tuning during pruning [30]. A notable development in this area is the SR-STE method [30], which explores training sparse networks from scratch without sacrificing performance.

Furthermore, the sparsification of DNNs utilizes two main pruning techniques: structured and unstructured. Unstructured pruning eliminates single weights at any location, achieving a high compression ratio, yet may reduce hardware performance [28]. Structured pruning, in contrast, targets groups of parameters, enhancing efficiency but potentially affecting precision. Recently, the N:M fine-grained structured pruning method has gained popularity. This approach merges the benefits of both structured and unstructured methods. It imposes limits, permitting only N active values in every M consecutive weights, enhancing processing speed on contemporary hardware [26].

Indeed, the N:M sparse DNNs represent a notable advancement in neural network optimization, as evidenced by various training methods elucidated in academic literature [1,2,6,7,14,23,24,26,30]. These methods, varying in their approaches to achieving sparsity, aim to balance the trade-off between model performance and computational efficiency. Nevertheless, the challenge persists in determining the most suitable method tailored to a particular application. this predicament is primarily due to the absence of comparative studies that critically evaluate these methods against each other. Such studies are essential for a holistic understanding of how different pruning techniques perform under a uniform set of conditions. The necessity for these comparative investigations extends beyond academic curiosity to practical significance, offering practitioners invaluable insights to make informed decisions aligning with the specific requirements of their applications, whether emphasizing speed, accuracy, or resource constraints.

In this paper, we attempt to address the above issue by proposing an in-depth comparative study of methods for training N:M structured sparse neural networks. Thus, our contributions can be summarized as follows: First, we review the latest state-of-the-art methods, delineating the principal parameters that we subsequently present as a taxonomy encompassing N:M structured pruning configuration settings. Second, we carry out, based on this taxonomy, a qualitative comparison of recent methods proposed for training N:M structured sparsity.

Finally, we conduct experiments to consolidate our study by a quantitative comparison of these methods.

The remainder of this paper is organized as follows: Sect. 2 reviews related work and presents our taxonomy of configuration setting of methods for N:M structured sparsity, whereas Sect. 3 presents our qualitative comparative study. Section 4 presents quantitative comparisons of prominent approaches. Section 5 concludes the paper and gives some perspectives.

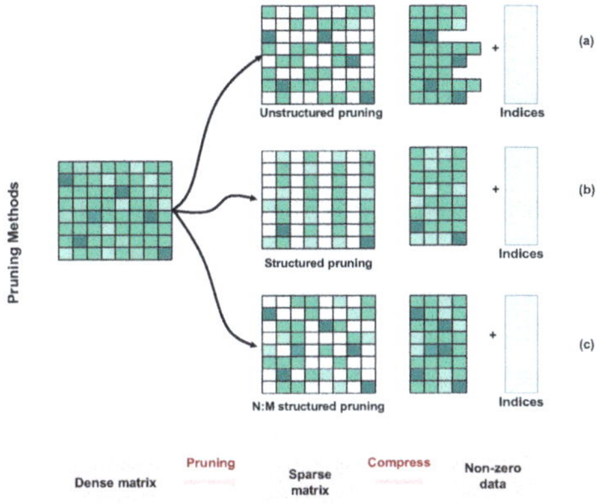

Fig. 1. (a) Unstructured pruning eliminates individual parameters at any location. The result is a sparse matrix with irregular, non-zero data. **(b)** Structured pruning removes parameters in groups. The result can be a regular, sparse matrix. **(c)** Structured N:M pruning satisfies that, in each group of M consecutive parameters of the matrix, there are at most N parameters with non-zero values (here $M = 8$ and $N = 4$).

2 N:M Structured Sparsity in DNNs

In this section, we give a summary of several methods for learning N:M structured sparse neural networks. Subsequently, we introduce a taxonomy of N:M structured pruning configuration settings, serving as the foundation for our comparative study.

2.1 Background on Sparsity

Sparsity in DNNs can be structured or unstructured. Unstructured sparsity is attainable through pruning weights at any location resulting in a notably high sparsity ratio without significant accuracy drop [11,16,17,28]. Nevertheless, it introduces a large execution overhead on modern hardware architectures mainly due to the costs associated with managing irregular non-zero values in the weight tensors. On the other hand, structured sparsity removes predetermined structures, such as blocks or channels [18,29], which can make better use of hardware resources but cannot maintain accuracy [22]. To take advantage of both unstructured and structured sparsity, N:M structured sparsity was proposed as a more advantageous alternative [14,22,30]. In this context, N:M sparsity denotes that, for each consecutive set of M weights, a maximum of N weights are allowed to remain non-zero. Figure 1 illustrates the two traditional types (unstructured and structured) and the N:M structured sparsity.

2.2 State of N:M Structured Sparsity

Authors in [22] proposed the 2:4 structured sparsity, requiring that every group of consecutive four weights must incorporate a minimum of two zero elements. which reduces the model's parameter count by half. The regular structure of the resulting weight matrix facilitates efficient compression, leading to efficient utilization of memory storage by exclusively manipulating non-zero weights. In this study, the authors proposed a workflow for network pruning that encompasses three steps: (1) pretraining a dense model; (2) pruning the dense model to a sparse one by applying a magnitude-based single-shot pruning method to achieve the 2:4 sparsity; and (3) finetuning the sparse model to regain performance.

Although this workflow aims at enhancing inference time acceleration, it remains computationally expensive since it requires twice the training time. To address this challenge, the authors of [30] proposed SR-STE, a framework designed to train sparse neural networks from scratch. Subsequently, [26] extends SR-STE by incorporating dynamic N:M ratios across network layers. Experimental results in [30] have shown that sparse models present a negligible drop in performance and, in certain cases, can even exceed the dense models. Nevertheless, SR-STE only tackles the problem of how to train an N:M sparse neural network in a uniform fashion (i.e., each layer has the same N:M sparsity). Also, as illustrated in [26], the layer-wise N:M sparse networks can achieve comparable accuracy with their unstructured sparse counterparts. In the same line of research, the authors of [23] present a pruning method for image restoration that determines, for every layer, the pruning ratio for N:M structured sparsity. This approach aims to enhance the trade-off between the efficiency and the restoration accuracy.

Despite being hardware-friendly, the underexploited potential to accelerate N:M structured sparse DNNs training highlights a deficiency in tailored efficient hardware [26]. Addressing these challenges involves a growing interest in innovative approaches, such as [7], introducing an algorithm-hardware co-optimized framework to accelerate transformers. Additionally, the authors in [6] propose Computation-Efficient N:M Sparse Training (CEST) as an algorithm-hardware co-design for DNNs, featuring bidirectional weight pruning to significantly reduce computational costs while preserving model accuracy and a sparse accelerator supporting regular dense and N:M sparse operations.

2.3 Taxonomy of N:M Structured Sparsity

Recent N:M structured pruning methods, summarized in table 1, vary primarily in their choices regarding the dynamicity of the N:M sparse pattern, pruning ratio, pruning criterion, model initialization, task dependency, and data dependency. This is illustrated in Fig. 2 and described as follows:

- **Sparse training**: The initial contributions that implemented the N:M structured sparse pattern proceeded with its static variant, meaning that the mask value remains fixed during the training process, such as methods proposed in [20,22,30]. However, opting for a uniform mask during training may adversely affect the solution's performance because all layers of the network are treated equally [4]. On the other hand, modern DNNs tend to have a non-uniform distribution of redundant parameters across layers, requiring the application of appropriate masks for each layer of the network [23,26].
- **Pruning ratio**: Pruning ratios indicate the proportion or percentage of weights to be removed during a neural network pruning process. In general, two ways can be distinguished for determining the pruning ratio, namely, static and dynamic methods. Pre-defining pruning ratios mean that one knows exactly how many parameters will be removed before running the algorithm without adaptation during training. In this case, the pruning ratio can be global [6,20,22,30] (i.e., the number of weights that will be removed from the whole network) or layer-wise (i.e., each network layer has its pruning ratio). Conversely, dynamically defined pruning ratios are typically assessed by implementing regularization-based pruning methods. These methods achieve weight removal by reducing them to zero and utilizing penalty terms during the training process [7,23].
- **Pruning criterion**: The fundamental operation of any neural network sparsification process is to select candidate elements for removal. In practice, the magnitude of weights is the most popular pruning criterion [9,27]. Common methods within this approach include L1 norm pruning and L2 norm pruning [26], both targeting weights with small absolute or squared values, respectively. The process results in the elimination of less influential connections with magnitudes below a certain threshold, fostering model sparsity. Other pruning criteria are illustrated in [13].
- **Model initialization**: Random initialization and pretrained models represent two distinct approaches to initializing neural network parameters, each with its own set of advantages and use cases. Random initialization involves setting the weights to small random values before training [30]. On the other hand, pretrained models leverage knowledge gained from training on a related task or dataset [26]. This initialization approach often proves highly effective for transfer learning, where a model pretrained on a large dataset is fine-tuned on a smaller, task-specific dataset.

- **Task dependency**: The methods based on N:M structured pruning can be categorized into two distinct approaches, according to task dependency: the generic methods and the task-dependent methods. Generic N:M pruning approaches apply the same pruning strategy to all tasks, regardless of their specific characteristics [30]. In contrast, task-dependent N:M pruning approaches tailor the pruning strategy to the specific requirements of each task [23,26].
- **Data dependency**: The methods based on N:M structured pruning can be categorized into two distinct approaches, depending on whether data is used during the sparse training process: the data-free methods and the data-driven methods. Data-free pruning involves pruning the neural network without using any specific training data. The decision to prune parameters is made solely based on the structure of the neural network itself. Common techniques include magnitude-based pruning and are often simpler to implement since they don't require access to training data. In contrast, data-driven methods involve training the model from the data and iteratively performing a pruning operation [23,26,30]. Furthermore, data-driven pruning can potentially lead to better performance preservation because it takes into account the specific characteristics of the data to find good subnetworks.

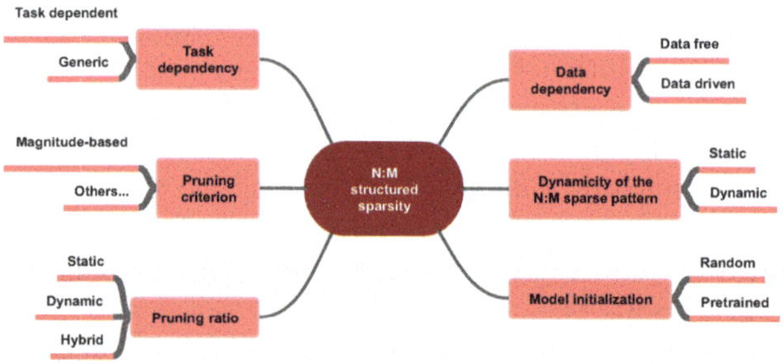

Fig. 2. Taxonomy of N:M structured pruning configuration settings

Based on the table 1, this paper presents a comparison of methods employing structured N:M sparsity for DNNs optimization during the training process, namely SR-STE [30] and DominoSearch (DS) [26] considered, to the best of our knowledge, to be the current state of the art in this field. This choice is also justified by the availability of their respective source codes.

Table 1. Comparative table of solutions, considering the taxonomy of N:M structured pruning configuration settings presented in the Sect. 2.3.

Method	Year	Sparse training	Pruning ratio	Pruning criterion	Model init.	Src. code	Task dependency	Data dependency
SR-STE [30]	2021	Static	Global pre-defined	Magnitude of weights	Random	✓	Generic	Data driven
DS [26]	2021	Dynamic	Hybrid	Magnitude of weights	Pretrained	✓	Image classification	Data driven
SLS [23]	2022	Dynamic	Dynamic layer-wise	Magnitude of weights	Pretrained	✓	Image retauration	Data driven
IDP [7]	2022	Static	Dynamic	Magnitude of weights	Pretrained	X	NLP tasks	Data driven
CEST [6]	2023	Static	Global pre-defined	Magnitude of weights	–	X	Tasks on CNN	Data driven

3 Quantitative Comparative Study

This section provides an overview of the experimental setup, encompassing a brief description of the methods studied, the technical environment, datasets, and model configurations. Furthermore, it presents the experimental protocol.

3.1 Environment and Settings

The experiments shown in this section were conducted on Nvidia Tesla V100 GPUs and based on the PyTorch 1.8.1 framework. Table 2 shows the hyperparameters settings.

Table 2. Hyperparameters settings.

Optimizer	SGD
Batch size	64
Epochs	120
Loss function	cross-entropy
Base LR	0.01
LR policy	- with DominoSearch: 0.01(fixed). - with SR-STE: increases linearly from 0 to 0.01 in the first 5 training epochs, and then is decayed using the cosine annealing

3.2 Datasets and Models

To investigate the performance of different N:M sparse patterns, we exploit the popular ResNet50 [12] model, which has shown to be highly efficient in a wide range of vision tasks, including image classification. We have also used the CIFAR-100 dataset [15]. This is a widely used benchmark dataset in the field of computer vision, namely in the image classification task. It consists of 60000 32x32 color images in 100 classes, each containing 600 images. There are 500 training images and 100 test images per class.

3.3 Metrics and Varied Parameters

Metrics. The performance of the sparse training methods studied in this paper was evaluated using two key metrics: top-1 and top-5 classification accuracies. They are metrics commonly used to assess the performance of classification models, particularly in image recognition tasks. Top-1 classification accuracy measures the percentage of instances where the model's highest predicted class aligns with the true class label. In contrast, Top-5 classification accuracy considers a prediction as correct if the true class label is among the top five predicted classes by the model.

Varied parameters. The selection of parameters for our experimental investigation aligns with the taxonomy proposed in Sect. 2.3, which delineates key dimensions that are instrumental in characterizing the behavior and performance of $N{:}M$ structured sparse neural networks. Specifically, our focus revolves around three pivotal axes:

- The dynamicity of the $N{:}M$ sparse structure across the model layers, which captures the adaptability and variation of sparsity patterns;
- The pruning ratio, a crucial parameter influencing the extent of weight elimination in the model;
- The initialization of the learning model, plays a foundational role in shaping the network's subsequent training dynamics.

Together, these dimensions facilitate a thorough assessment of $N{:}M$ structured sparse neural networks, providing insights into their behaviors and efficiencies. Furthermore, our study is limited to the task of image classification, as it is the sole task investigated in [26]. In our experiments, we use magnitude-based model pruning. This method, as demonstrated by previous research [9], has proven to perform as well as more complex and computationally intensive techniques such as variational dropout and L0 regularization.

3.4 Experimental Protocol

We propose an experimental protocol for evaluating and comparing the performance of $N{:}M$ structured sparse neural network training methods under different training settings: SR-STE, DS, and DS combined with SR-STE (DS+SR-STE) methods. These methods are assessed under a pruning ratio of 75% and an N:8 sparse pattern configuration, as it demonstrated competitive performance in terms of accuracy and adaptability.

Our investigation focuses on assessing the effectiveness of dynamic adaptation through a comparison between SR-STE, featuring a static N:M sparse pattern across all model layers, and DS's adaptive adjustments. Additionally, we analyze the performance of the chosen methods across varying pruning ratios (50%, 75%, and 94%) to comprehend their robustness and adaptability to increased sparsity. Moreover, we explore the influence of distinct weight initialization strategies, such as "Random" and "Pretrained," on the performance outcomes of SR-STE, DS, and DS+SR-STE.

4 Results and Discussion

4.1 Impact of Dynamicity of N:M Pattern

The experimental results presented in the table 3 present a comparative analysis of different training methods, with a pruning ratio of 75%, including SR-STE with 2:8 sparse pattern, DS, and DS+SR-STE, all evaluated under the N:8 sparse pattern configuration. SR-STE, with a 2:8 sparse pattern, achieves a Top-1 accuracy of 60.91% over 120 epochs without dynamic adjustments. DS, employing an N:8 sparse pattern, attains higher accuracy levels of 78.18% (Top-1) with dynamic adjustments over 32 epochs. Notably, the DS+SR-STE method outperforms other approaches, showcasing superior Top-1 and Top-5 accuracies over 120 epochs. The dynamic adaptation further highlights its effectiveness. This comparison underscores the importance of method combination, demonstrating the potential of DominoSearch coupled with SR-STE for enhanced performance in similar contexts.

Table 3. Experimental results of SR-STE, DS, and DS combined with SR-STE training method with different N:M sparse patterns and a pruning ratio of 75%.

Method	N:M	Top-1 Acc.	Top-5 Acc.	epochs	Dynamic
SR-STE	2:8	60.91%	86.27%	120	No
DS	N:8	78.18%	95.51%	32	Yes
DS+SR-STE	**N:8**	**83.04%**	**96.71%**	**120**	**Yes**

4.2 Impact of Different Pruning Ratio

Table 4 provides experimental results for SR-STE, DS, and DS combined with SR-STE under varying pruning ratios and N:M sparse patterns. In the 50% pruning ratio scenario, DS+SR-STE consistently outperforms individual methods, achieving the highest Top-1 and Top-5 accuracies across diverse sparse patterns. Notably, as the pruning ratio increases to 75% and 94%, DS+SR-STE consistently demonstrates superior accuracy, showcasing its effectiveness in mitigating the impact of increased sparsity. This comparative analysis highlights the robustness of the combined DS+SR-STE approach, emphasizing its potential for achieving enhanced classification accuracy under different pruning conditions and sparse patterns. Furthermore, it can be shown that DS presents, in all experiments, a top-1 accuracy superior to that of SR-STE despite a number of epochs significantly lower than that of SR-STE, enabling it to reach the specified sparsity ratio.

Table 4. Experimental results of SR-STE, DS, and DS combined with SR-STE with different pruning ratios.

Method	N:M	Pruning ratio	Top-1 Acc.	Top-5 Acc.	epochs
SR-STE	2:4	50%	59.58%	85.13%	120
DS	N:4	50%	71.98%	93.57%	12
DS+SR-STE	**N:4**	50%	**82.55%**	**96.68%**	**120**
SR-STE	2:8	75%	60.91%	86.27%	120
DS	N:8	75%	78.18%	95.51%	32
DS+SR-STE	**N:8**	75%	**83.04%**	**96.71%**	**120**
SR-STE	1:16	94%	61.38%	86.30%	120
DS	N:16	94%	73.44%	90.62%	50
DS+SR-STE	**N:16**	**94%**	**83%**	**96.83%**	**120**

4.3 Impact of Model Initialization

Table 5 shows the experimental results for SR-STE and DS combined with SR-STE under diverse weights initialization strategies. In the "Random" initialization setting, SR-STE achieves a Top-1 accuracy of 60.91%, while DS+SR-STE reaches a Top-1 accuracy of 61.83% over 120 epochs. However, under the same "Random" initialization, DS demonstrates a significant drop in accuracy, stemming from an interruption in the training process since the early epochs. In the "Pretrained" initialization scenario, SR-STE attains a Top-1 accuracy of 82.94%, DS achieves 78.18%, and DS+SR-STE excels with a Top-1 accuracy of 83.04% over 120 epochs. These findings underscore the impact of weight initialization strategies on the performance of the considered methods, with DS+SR-STE demonstrating competitive results across diverse initialization conditions. This can be illustrated in Fig. 3 showing the learning curves of these methods under different weights initialization configurations. Notably, the accuracy curve of randomly initialized methods is consistently below their pretrained counterparts. Furthermore, "pretrained" initialization outperforms "random" one due to the use of previously trained weights. Indeed, these pretrained weights capture general features and valuable representations that can be transferred to the new model, providing a more effective starting point for subsequent training iterations.

Table 5. Experimental results of SR-STE and DominoSearch combined with SR-STE with different weights initializations.

Method	Weights init.	N:M	Pruning ratio	Top-1 Acc.	Top-5 Acc.	epochs
SR-STE	Random	2:8	75%	60.91%	86.27%	120
DS+SR-STE	**Random**	**N:8**	**75%**	**61.83%**	**86.29%**	**120**
SR-STE	Pretrained	2:8	75%	82.94%	96.55%	120
DS	Pretrained	N:8	75%	78.18%	95.51%	32
DS+SR-STE	**Pretrained**	**N:8**	**75%**	**83.04%**	**96.71%**	**120**

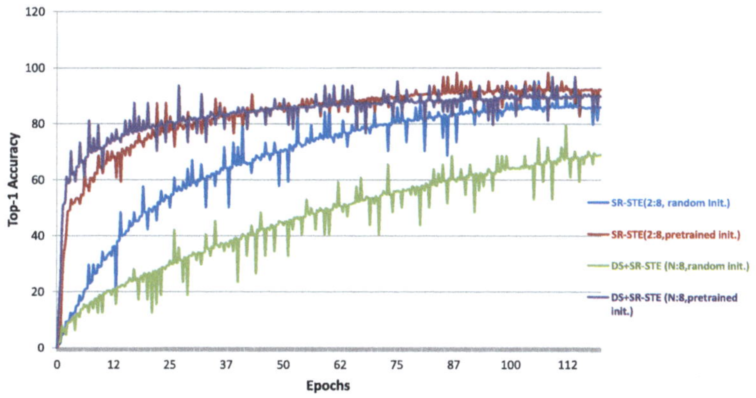

Fig. 3. This figure compares the top-1 accuracy curves of sparse net trained with SR-STE and DominoSearch combined with SR-STE with different N:M patterns and different initialization settings.

5 Conclusion and Future Work

In this work, we have presented a comparison of methods proposed for learning N:M structured sparsity. The experimental results documented shed light on the fact that there is room for enhancing the performance of sparse neural networks. This improvement can be achieved through a combination of key factors, such as the use of dynamic N:M sparse patterns, the use of pretrained weight initialization, the use of a suitable pruning ratio, and the careful selection of an effective sparse training methodology. These observations underscore the significance of well-thought-out technical choices in achieving efficient outcomes in the context of sparse neural networks.

Combining and comparing state-of-the-art methods allows us to identify advancements and potential areas for improvement and contributes to the overall understanding of the evolving landscape of N:M structured pruning. As a future work, we will take advantage of the conclusions drawn from this comparative study to build our framework for learning N:M structured sparsity.

References

1. Chen, T., et al.: Only train once: a one-shot neural network training and pruning framework. Adv. Neural. Inf. Process. Syst. **34**, 19637–19651 (2021)
2. Chen, Z., Qu, Z., Quan, Y., Liu, L., Ding, Y., Xie, Y.: Dynamic n: M fine-grained structured sparse attention mechanism. In: Proceedings of the 28th ACM SIGPLAN Annual Symposium on Principles and Practice of Parallel Programming, pp. 369–379 (2023)
3. Dettmers, T., Zettlemoyer, L.: Sparse networks from scratch: Faster training without losing performance. ArXiv abs/1907.04840 (2019). https://api.semanticscholar.org/CorpusID:195873969
4. Evci, U., Gale, T., Menick, J., Castro, P.S., Elsen, E.: Rigging the lottery: making all tickets winners. In: International Conference on Machine Learning, pp. 2943–2952. PMLR (2020)
5. Evci, U., Pedregosa, F., Gomez, A.N., Elsen, E.: The difficulty of training sparse neural networks. In: Identifying and Understanding Deep Learning Phenomena Workshop, 36 th International Conference on Machine Learning (2019)
6. Fang, C., Sun, W., Zhou, A., Wang, Z.: CEST: computation-efficient N: M sparse training for deep neural networks. In: 2023 Design, Automation and Test in Europe Conference and Exhibition (DATE), pp. 1–2. IEEE (2023)
7. Fang, C., Zhou, A., Wang, Z.: An algorithm–hardware co-optimized framework for accelerating n: M sparse transformers. IEEE Trans. Very Large Scale Integr. (VLSI) Syst. **30**(11), 1573–1586 (2022)
8. Frankle, J., Carbin, M.: The lottery ticket hypothesis: finding sparse, trainable neural networks. In: International Conference on Learning Representations (2018)
9. Gale, T., Elsen, E., Hooker, S.: The state of sparsity in deep neural networks. In: International Conference on Machine Learning (2019)
10. Han, D.K., Mulyana, B., Stanković, V., Cheng, S.: A survey on deep reinforcement learning algorithms for robotic manipulation. Sensors (Basel, Switzerland) **23**(7), 3762 (2023). https://api.semanticscholar.org/CorpusID:258020497
11. Han, S., Pool, J., Tran, J., Dally, W.: Learning both weights and connections for efficient neural network. Adv. Neural Inf. Process. Syst. 28 (2015)
12. He, K., Zhang, X., Ren, S., Sun, J.: Deep residual learning for image recognition. In: 2016 IEEE Conference on Computer Vision and Pattern Recognition (CVPR), pp. 770–778 (2016)
13. Hoefler, T., Alistarh, D., Ben-Nun, T., Dryden, N., Peste, A.: Sparsity in deep learning: pruning and growth for efficient inference and training in neural networks. J. Mach. Learn. Res. **22**(1), 10882–11005 (2021)
14. Hubara, I., Chmiel, B., Island, M., Banner, R., Naor, S., Soudry, D.: Accelerated sparse neural training: A provable and efficient method to find N: M transposable masks. CoRR abs/2102.08124 (2021). arXiv:2102.08124
15. Krizhevsky, A., Hinton, G., et al.: Learning multiple layers of features from tiny images (2009)
16. Kusupati, A., et al.: Soft threshold weight reparameterization for learnable sparsity. In: International Conference on Machine Learning, pp. 5544–5555. PMLR (2020)
17. Lee, J., Park, S., Mo, S., Ahn, S., Shin, J.: Layer-adaptive sparsity for the magnitude-based pruning. In: International Conference on Learning Representations (2021). https://openreview.net/forum?id=H6ATjJ0TKdf
18. Li, H., Kadav, A., Durdanovic, I., Samet, H., Graf, H.P.: Pruning filters for efficient convnets. In: International Conference on Learning Representations (2017). https://openreview.net/forum?id=rJqFGTslg

19. Liu, J., Jin, Y.: A comprehensive survey of robust deep learning in computer vision. J. Autom. Intell. (2023). https://api.semanticscholar.org/CorpusID:265324661
20. Lu, Y., Agrawal, S., Subramanian, S., Rybakov, O., De Sa, C., Yazdanbakhsh, A.: STEP: learning N:M structured sparsity masks from scratch with precondition. In: Proceedings of the 40th International Conference on Machine Learning, ICML 2023, JMLR.org (2023)
21. Michalowski, M., Abidi, S.S.R., Abidi, S.: Artificial Intelligence in Medicine: 20th International Conference on Artificial Intelligence in Medicine, AIME 2022, Halifax, NS, Canada, June 14–17, 2022, Proceedings, vol. 13263. Springer Nature (2022)
22. Mishra, A., et al.: Accelerating sparse deep neural networks. arXiv preprint arXiv:2104.08378 (2021)
23. Oh, J., Kim, H., Nah, S., Hong, C., Choi, J., Lee, K.M.: Attentive fine-grained structured sparsity for image restoration. In: Proceedings of the IEEE/CVF Conference on Computer Vision and Pattern Recognition (CVPR), pp. 17673–17682 (2022)
24. Pool, J., Yu, C.: Channel permutations for n: m sparsity. Adv. Neural. Inf. Process. Syst. **34**, 13316–13327 (2021)
25. Sarker, I.H.: Deep learning: a comprehensive overview on techniques, taxonomy, applications and research directions. SN Comput. Sci. **2**(6), 420 (2021)
26. Sun, W., et al.: Dominosearch: find layer-wise fine-grained N:M sparse schemes from dense neural networks. In: Ranzato, M., Beygelzimer, A., Dauphin, Y., Liang, P., Vaughan, J.W. (eds.) Adv. Neural Inf. Process. Syst. vol. 34, pp. 20721–20732. Curran Associates, Inc. (2021)
27. Wang, H., Qin, C., Bai, Y., Zhang, Y., Fu, Y.: Recent advances on neural network pruning at initialization. arXiv e-prints pp. arXiv–2103 (2022)
28. Wang, Z.: Sparsert: accelerating unstructured sparsity on gpus for deep learning inference. In: Proceedings of the ACM International Conference on Parallel Architectures and Compilation Techniques, p. 31-42, PACT 2020, Association for Computing Machinery, New York, NY, USA (2020). https://doi.org/10.1145/3410463.3414654
29. Wen, W., Wu, C., Wang, Y., Chen, Y., Li, H.: Learning structured sparsity in deep neural networks. Adv. Neural Inf. Process. Syst. **29** (2016)
30. Zhou, A., et al.: Learning n:m fine-grained structured sparse neural networks from scratch. In: International Conference on Learning Representations (2021)

Breast Density Prediction from Mammograms: A Comprehensive Review

Nassima Dif[1(✉)], Mohamed Amine Abdelali[2], Mohamed El Amine Boudinar[2], Jesia Asma Benchouk[3], and Sidi Mohammed Benslimane[1]

[1] LabRI-SBA Lab, Ecole Superieure en Informatique, Sidi Bel Abbes, Algeria
{n.dif,s.benslimane}@esi-sba.dz

[2] Ecole Superieure en Informatique, Sidi Bel Abbes, Algeria
{ma.abdelali,m.boudinar}@esi-sba.dz

[3] Breast and Pelvic Surgery Department, Cancer Center Tidjini Haddam, Sidi Bel Abbes, Algeria

Abstract. Predictive medicine aims to improve the stratification and management of individuals according to their probability of developing a pathology in the future. The risk of breast cancer is influenced by multiple factors, such as lifestyle, age, breast density, and family history. Nevertheless, radiologists face challenges in accurately classifying breast density from mammogram images due to the inter-observer subjectivity problem. To address this issue, artificial intelligence systems, especially deep learning techniques, have been exploited. The purpose of this research paper is to review the current state of research on the use of machine learning and deep learning based approaches for breast density classification.

Keywords: Mammograms · Risk factors · Breast density · BI-RADS · Multi-view CNN

1 Introduction

Breast cancer is currently the most prevalent type of cancer among women, with approximately 2.2 million new cases occurring worldwide each year, as reported by the world health organization (WHO) [1]. Unfortunately, this number is expected to continue rising over time. In Algeria, breast cancer starts to increase at the age of 20 and reaches its peak among individuals aged 50 to 64, then decreases for those over 65 [2–5]. Early detection is considered the most effective solution to combat this disease. The risk of developing breast cancer in women can be computed by assessing various factors, including age, family history, lifestyle, and breast density [6].

Breast density plays a crucial role in determining the risk of breast cancer [7]. For assessing breast density, mammography tests are the recommended methods

B. Djamaa et al. (Eds.): CSA 2024, LNNS 1145, pp. 70–79, 2025.
https://doi.org/10.1007/978-3-031-71848-9_7

that are visually evaluated by radiologists. However, it is essential to note that the diagnosis can vary among radiologists, leading to inter-observer variability. This variability can be attributed to differences in expertise levels among radiologists. Additionally, radiologists often experience significant pressure to provide accurate diagnoses, which can affect the level of reassurance that women receive while awaiting examination results. To address these challenges, the development of a tool to assist radiologists in decision-making becomes imperative. Particularly with the recent revolution in artificial intelligence techniques over the last decade, driven by the remarkable success of deep learning in the medical field. This advancement holds significant promise for overcoming obstacles faced by radiologists and enhancing the accuracy of breast cancer diagnoses.

This paper aims to provide readers with medical knowledge on breast density assessment and review the new proposed techniques for breast density classification from mammography images. This paper proceeds as follows: Sect. 2 provides an overview of breast density and mammography. Section 3 presents, compares, and discusses the related works on deep learning for breast density classification. Finally, the last part concludes this work.

2 Generalities on Breast Density and Mammography

2.1 Breast Anatomy

The female breast is composed of various types of tissue, including glandular, fibrous, and fatty tissue. Glandular tissue contains the breast lobes and ducts [8], which are thin tubes responsible for transporting milk to the nipple, situated in the center of the darker area known as the areola. Fibrous tissue is similar to the tissue found in ligaments and scar tissue, while fatty tissue influences breast size. Each patient has both a left and a right breast, and each breast is examined using two distinct projections: the MLO (mediolateral oblique) projection and the CC (craniocaudal) projection.

2.2 Breast Cancer Risk Factors

Several factors influence breast cancer's development and can increase its risk:

- Age: The risk of breast cancer rises with age.
- Familiy History [9]: In some cases, breast cancer can be hereditary. It is passed from one generation to the next due to BRCA1 and BRCA2 mutations.
- Lifestyle: Human lifestyle plays a significant role in increasing the risk of disease. Factors such as diet and climate change can vary from one region to another, contributing to differences in disease prevalence.
- Breast density refers to how the breast appears on a mammogram, based on the distribution of fat, stromal, and epithelial tissues within the breast. It can be measured by determining the proportion of the mammogram that consists of dense tissue.

2.3 Mammographic Density and the BI-RADS Standard

Mammographic density measures the amount of radiographically dense fibroglandular tissue compared to the fat tissue visible on a mammogram [10]. Breast imaging-reporting and data system (BI-RADS) is considered a standard in the medical community for classifying mammograms based on their density.

BI-RADS is a quality control system used by radiologists to evaluate mammograms, MRI scans, and ultrasound images of the breast. Mammography is a relatively fast and cost-effective imaging method capable of detecting small tumors or other abnormalities in the breast [11]. The BI-RADS system classifies breast density into four groups: almost entirely fat (A), scattered fibroglandular densities (B), heterogeneously dense (C), and extremely dense (D).

3 State of the Art

3.1 Datasets

Developing efficient systems for computer-aided diagnostics presents a serious challenge due to the limited availability of large public mammography datasets. In this context, our objective is to provide an overview of the most frequently used datasets for breast density detection, with a particular focus on those evaluated using the BI-RADS assessment. Several public datasets have been developed in the state-of-the-art, including MIAS [12], Mini-MIAS [13], DDSM [14], IRMA [15], CBIS-DDSM [16], MINI-DDSM [16], and Vindr-Mammo [17]. Moreover, various private datasets have been proposed to develop customized systems for a specific population [18–22]. Table 1 summarizes the most commonly used datasets in the state-of-the-art.

Table 1. The commonly used datasets for mammography image analysis.

Name/Owner	Mini-DDSM [16]	INBreast [23]	Vindr-Mammo [17]	AOUP [18]
Origin	USA	Portugal	Vietnam	Italy
Year	2020	2012	2022	2013
Views	MLO and CC	MLO and CC	MLO and CC	MLO and CC
Number Of Images	9684	410	20 000	7848
Acquisition Mode	SFM	FFDM	FFDM	FFDM
BI-RADS	YES	YES	YES	YES
Breast Density	YES	YES	YES	YES

When comparing the presented datasets, several characteristics should be considered to select the appropriate one for developing a robust prediction model, such as release year, distribution, quality, and representativeness. The number of samples is essential to avoid overfitting problems in deep neural networks.

Moreover, it is recommended to use recently approved medical images. On the other hand, an imbalanced distribution of samples can lead to biased models and result in poor performance. Table 2 highlights the distribution of datasets from previous studies. In some datasets, such as DDSM [14] and INBreast [15], the distribution has been artificially manipulated to create a balance between malignant and benign classes. In other investigations [19,20], the authors used a natural distribution without artificial balancing. The authors in [24] studied the effect of the distribution of BI-RADS density classes on CNN training using a large number of images. The results showed that the BI-RADS Atlas distribution (A: 10%, B: 40%, C: 40%, D: 10%) was more accurate on tests than both the uniform distribution (A: 25%, B: 25%, C: 25%, D: 25%) and the native distribution. Another important characteristic for selecting a dataset is the quality of its samples, because the complex structure of cells and tissues in medical images requires high-resolution images to extract important features. Finally, representativeness refers to the degree to which the dataset accurately reflects the target population.

Table 2. The distribution of BI-RADS density classes in the most commonly used datasets.

Datasets	A	B	C	D	Type
Mini-DDSM	1076	3020	2320	1388	patients
Vindr-Mammo	25	477	3823	675	exams
AOUP [18]	200	473	804	185	exams
Li et al. [22]	319	423	541	702	images
Wang et al. [20]	19939	85665	83852	11723	exams
INbreast	136	147	99	28	images
DDSM	279	1050	805	390	exams

3.2 Preprocessing Techniques

Preprocessing is an important step for deep learning algorithms. Its main objective is to prepare data and enhance the model's performance. In digital mammography, images are exported from scanners in a DICOM format. Preparing and preprocessing these images is necessary to make them suitable for training.

The high resolution of the original exported mammography images can lead to significant computational complexity for deep learning models. Resizing is among the exploited solutions to address this issue. In [18], the authors concluded that 650 $\times$ 650 pixels generated the most accurate model. In other studies [19,20,25], high-resolution input sizes were managed using aggressive convolution and pooling layers. Converting grayscale mammography images into RGB format presents an important preprocessing step for breast density assessment from mammography images. In [26], the magma color maps technique was applied to the generated RGB images to enhance the perceptual quality of fibroglandular and fat tissues [16,27]. To enhance the quality of mammography images, different enhancements have been used, such as contrast limited adaptive histogram equalization (CLAHE) and histogram equalization algorithms [28] and binarization of images based on Otsu's algorithms [24].

In addition to the previously cited preprocessing techniques, segmentation was exploited in different research due to its various advantages. The minimum bounding box technique was generally used to segment the breast [21,22,25,26,28]. In another investigation [24], the marching-square algorithm was used to identify the breast skin line and outline, and then the minimum bounding box technique was applied to crop the breast. To accurately identify the boundary boxes, Nguyen et al. [29] used an automated detector based on the YOLOv5 architecture. Another use case of segmentation techniques in mammography images was removing pectoral muscle from the MLO view, which can be seen as unnecessary background information. Different techniques have been exploited in the state of the art: the depth-first search (DFS) algorithm [28], a deep learning algorithm to calculate breast density based on a weight-adaptive multitask method [25], and replacing the pectoral muscle area with the mean gray value after separating it from the breast with a linear function [24].

The need for large open-source medical image datasets presents a serious challenge in building high-performing deep learning models. Moreover, DL models are prone to overfitting problems on limited volumes of data. To solve this problem, data augmentation techniques have been largely used, such as random cropping [19,20], random zoom, random rotations [24,30], random horizontal flipping, and vertical flipping [22,26].

3.3 Breast Density Prediction Using Deep Learning Methods

When analyzing mammography images using deep learning (DL) techniques, there are two distinct approaches: single-view and multi-view.

Single-View Methods. In the classic single-view method, Lizzi et al. [18] proposed an extremely deep residual convolutional neural network (CNN) architecture with 41 convolutional layers arranged in residual blocks. In a related work, the authors in [24] were inspired by this architecture but introduced data augmentation and pectoral muscle removal. They also studied the effect of dataset distribution and found that pectoral muscle removal improved the results.

Multi-view Methods. A more efficient approach is known as the multi-view approach. It takes advantage of the fact that each patient undergoes four mammography images, including two views of the left breast (LCC, LMLO) and two views of the right breast (RCC, RMLO). In this context, the authors in [19] proposed an end-to-end approach that handles multiple views using a large-scale data set. All views were processed by a deep convolutional layer, then passed through a fully connected layer and a softmax layer for classification. Regularization techniques, such as data augmentation and dropout methods, were applied to prevent overfitting. In [20], a pretrained VGGNet was used with four views as input. In another investigation, Kaiser et al. [31] exploited the VGGNet architecture with global average pooling (GAP) by introducing multichannel inputs. A novel approach is used by Liu et al. [21] to predict the malignancy of BI-RADS 4 microcalcifications. Two DL models were designed: one using only mammography images and the other utilizing both mammography images and clinical variables (age, family history, menopausal status, breast composition, etc.). Microcalcification images cropped from CC and MLO views were used to extract features based on the MobileNetV2 model. The generated features were merged and transformed, then fed into a softmax output layer to estimate the malignancy score. Pawar et al. [28] used multichannel DenseNet121 to extract features from the four views separately.

A new concept of combining extracted features was introduced in [25]. In this research, they developed a multi-view deep evidential neural network by choosing Inception ResNet-v2 as the best backbone and the view-specific late-evidential fusion strategy (VS-LEF) due to its high accuracy against FLEF, VS-FLEF, and LEF fusion strategies. To enhance feature extraction, Li et al.[22] incorporated dilated convolution and channel-wise attention with the ResNet50 model and achieved improved accuracy compared to previous models on both a private clinical dataset and the INBreast public dataset using a multi-view approach. The dilated convolution helps to maintain the spatial resolution of the feature map while allowing the network to capture a larger context, and the channel-wise attention mechanism enables the network to concentrate on the most relevant channels, reducing the influence of noise. In another multi-view approach called the TwoViewDensityNet model [26], fine-tuning was applied to the pretrained ResNet50 model. The selected features were passed through global average pooling (GAP) to reduce their dimensions. Then, they were concatenated and fed to the classification layer of the model for prediction. In a recent contribution, Nguyen et al. [29] proposed a BI-RADS density classification architecture that consists of two steps. In the first phase, CNNs (ResNet-34 and EfficientNet-B2) were used for feature extraction from the input images. In the second stage, a LightGBM classifier was implemented to predict BI-RADS and density scores. The LCC and LMLO feature vectors and the R-CC and R-MLO feature vectors were then combined and averaged. Finally, the average vectors were inputted into the LightGBM to provide density and BI-RADS scores with confidence.

Table 3 summarizes the proposed interesting approaches for breast density assessment from mammograms. According to the comparative study in [22], multi-view techniques are more efficient than single-view methods. In general, multi-view approaches received more interest compared to single-view techniques. The multi-view methods are composed of three main phases: feature extraction, fusion, and classification. Various CNN architectures, such as EfficientNet, Inception, ResNet, VGGNet, and Inception-ResNet, have been exploited for feature extraction. Besides, LightGBM, DEN, and SoftMax have been exploited for classification.

Table 3. The proposed methods in the literature for breast density detection.

	Used dataset	Approach	Feature Extractor	Classifier	Metric
[18]	AOUP	Single View	Residual CNN		Accuracy=0.894
[20]	private	Multi-View	VGGNet		MacAUC=0.842
[29]	Vindr-mammo	Multi-view	EfficientNet-B2	LightGBM	MacroF1=0.6165
	DDSM				MacroF1=0.6702
[28]	DDSM	Multi-channel	Multi-channel DenseNet		Accuracy = 0.90
[25]	Vindr-mammo	Multi-view	Inception-ResnetV2	DEN	F1-score =0.67
	Mini-DDSM				F1-score = 0.68
[24]	AOUP	Single View	Residual CNN		Accuracy=0.785
	BI-RADS atlas				Accuracy=0.831
	Uniform				Accuracy=0.778
[26]	DDSM	Dual View	ResNet 50	SoftMax	Accuracy=0.9583
	INbreast				Accuracy = 0.96
[30]	private	MLO-view	DCN with 11 layers,3 FC,output SoftMax layer		Accuracy=0.897
		CC-View			Accuracy=0.866
[22]	IN-Breast	Single View	ResNet 50 + DC + CA		Accuracy=0.70
	private				Accuracy=0.887
		Multi-view			Accuracy=0.921
[31]	private	Multi-channel	VGGNet		Accuracy=0.88

To overcome the overfitting problem on limited volumes of data, transfer learning has been largely exploited in density classification applications from mammograms. In this context, pretrained models can be exploited as feature extractors by fine-tuning them on the new classification task. Wu et al. [20] initialized the parameters of their model using the weights of the previously trained model in [19]. Feature extraction has been widely used with pretrained CNN models, where various pretrained architectures have been used: MobileNet [21], VGGNet [31], DenseNet201 [28], and ResNet50 [26].

3.4 Visualization Techniques

In deep learning applications, visualization techniques are necessary to identify the important regions in the image that impact the classification task. For breast density classification, Liu et al. [21] applied Grad-CAM to produce the saliency maps of a mammogram, and they found that the distribution of attentional regions is comparable with microcalcification areas, and the relative weights of clinical variables are unique to each patient. In the same context, Gudhe et al. [25] used the saliency map to determine the model's prediction confidence level and the specific areas that it focuses on when making a decision. In another contribution [24], the heatmap evaluation was used to compare the denser regions in the original image with the highlighted regions in heatmaps.

4 Discussion

From the above review, we concluded that the detection of breast density presents a complex challenge. Previous studies conducted in this field have produced diverse and complementary perspectives, each providing its own unique insight. Finding an open-source dataset that meets all the necessary conditions, such as being large and recent, as well as having a balanced distribution of classes, is a significant challenge. The choice of the dataset is a crucial aspect, and it must fulfill various requirements. Preprocessing steps, especially in the medical field, like breast density detection, represent a necessary step before training. It has an effect not just on the result but also on the classifier's explainability.

For the choice of architecture, in most studies, the multi-view approach was helpful. Due to the lack of large public datasets, CNNs with transfer learning can be advantageous in detecting breast density. It is often able to produce satisfactory results with smaller datasets, in contrast to training a network from scratch. Despite this, each researcher was able to choose an approach that best fit their needs and conditions. However, to discover contextual information and dependencies among the four views, more advanced deep learning models, such as vision transformers, should be exploited. Additionally, more efforts should be made in fusion techniques, where typically, the generated features from feature extraction are concatenated without focusing on common patterns among the different views (LCC, LMLO, RCC, and RMLO) or using dimensionality reduction techniques. For evaluation, the use of multiple evaluation metrics is essential to gain a comprehensive understanding of the model's performance. Generally, the proposed research focused on the macro-average F1-score and macro-average area under the curve, which are more appropriate for an unbalanced dataset. Moreover, having a visual representation by using visualization techniques, such as Grad-CAM, can significantly aid human readers and radiologists in breast density prediction. This technique can highlight the regions of interest used by the model for class prediction. For classes C and D, the generated regions should be high compared to those for classes A and B, owing to the high presence of fibroglandular tissue in classes C and D.

In conclusion, for future research on breast density prediction, researchers should prioritize multi-view techniques and pay more attention to recent data augmentation methods, such as generative adversarial networks (GANs), to address the issue of data imbalance. Furthermore, exploring novel deep learning architectures like vision transformers can enhance results.

5 Conclusion

In this paper, we investigated the latest deep learning methods and architectures for effective breast density detection using mammogram images. Despite the fact that various architectures produced satisfactory results, further research using larger and more recent data sets is necessary to improve their performance and achieve more efficient outcomes. The presented results highlight the efficiency of the multi-view approach. This approach exploits transfer learning to extract the important features and analyze the four views simultaneously. However, it is essential to note that the lack of large public datasets and the presence of class imbalances limit the research in this field. As a perspective to this work, we plan to propose a new multi-view technique by taking into consideration the class imbalance problem.

Disclosure of AI Tools Usage. The preparation of this manuscript involved the use of ChatGPT for enhancing language and improving readability. Subsequently, the authors reviewed and edited the content as necessary, and take full responsibility for the paper's content.

References

1. Sung, H., et al.: Global cancer statistics 2020: GLOBOCAN estimates of incidence and mortality worldwide for 36 cancers in 185 countries. CA: Cancer J. Clin. **71**(3), 209–249 (2021)
2. Cherif, M.H., Kara, L., Atoui, S., Boudefar, F., et al.: Données épidémiologiques du cancer dans l'est et le sud-est algérien, 2014–2017. Algerian J. Health Sci. **2**(1), 14–31 (2020)
3. Hammouda, D., Boutekdjiret, L.: Surveillance épidémiologique des cancers: Résultats préliminaires nationaux. Algerian J. Health Sci. **2**, 31–40 (2020)
4. Mustapha, B., Nabil, A.: La recherche en cancérologie dans le contexte national. Algerian J. Health Sci. **2**, 42–52 (2020)
5. Grangaud, J., et al.: L'évaluation: moteur du processus de planification sanitaire" plan national cancer 2015–2019. Algerian J. Health Sci. **2**, 52 (2020)
6. Vachon, C.M., Kuni, C.C., Anderson, K., Anderson, V.E., Sellers, T.A.: Association of mammographically defined percent breast density with epidemiologic risk factors for breast cancer (united states). Cancer Causes Control **11**, 653–662 (2000)
7. Hall, F.M.: Mammographically determined breast density and cancer risk. Radiology **248**(3), 1083–1084 (2008)
8. Lee, S.H., Moon, W.K.: Glandular tissue component on breast ultrasound in dense breasts: a new imaging biomarker for breast cancer risk. Korean J. Radiol. **23**(6), 574 (2022)
9. Loman, N., Johannsson, O., Kristoffersson, U., Olsson, H., Borg, A.: Family history of breast and ovarian cancers and BRCA1 and BRCA2 mutations in a population-based series of early-onset breast cancer. J. Natl Cancer Inst. **93**(16), 1215–1223 (2001)
10. Oliver, A., et al.: Breast density analysis using an automatic density segmentation algorithm. J. Digit. Imaging **28**, 604–612 (2015)
11. Bilal, R., Khan, B.M., Young, R.: Breast cancer detection and diagnosis. In: Emerging Developments and Practices in Oncology, pp. 1–27. IGI Global (2018)
12. Suckling, J.: The mammographic images analysis society digital mammogram database. In: Exerpta Medica. International Congress Series, vol. 1069, pp. 375–378 (1994)
13. Muštra, M., Štajduhar, A.: Segmentation masks for the mini-mammographic image analysis society (mini-MIAS) database. IEEE Consum. Electron. Mag. **9**(5), 28–33 (2020)

14. Heath, M., et al.: Current status of the digital database for screening mammography. In: Digital Mammography: Nijmegen **1998**, 457–460 (1998)
15. Moreira, I.C., Amaral, I., Domingues, I., Cardoso, A., Cardoso, M.J., Cardoso, J.S.: Inbreast: toward a full-field digital mammographic database. Acad. Radiol. **19**(2), 236–248 (2012)
16. Lekamlage, C.D., Afzal, F., Westerberg, E., Cheddad, A.: Mini-DDSM: mammography-based automatic age estimation. In: 2020 3rd International Conference on Digital Medicine and Image Processing, pp. 1–6 (2020)
17. Nguyen, H.T., et al.: VinDr-Mammo: a large-scale benchmark dataset for computer-aided diagnosis in full-field digital mammography. MedRxiv pp. 2022–03 (2022)
18. Lizzi, F., et al.: Residual convolutional neural networks for breast density classification. In: BIOINFORMATICS, pp. 258–263 (2019)
19. Geras, K.J., et al.: High-resolution breast cancer screening with multi-view deep convolutional neural networks (2017). arXiv preprint arXiv:1703.07047
20. Wu, N., et al.: Breast density classification with deep convolutional neural networks. In: 2018 IEEE International Conference on Acoustics, Speech and Signal Processing (ICASSP), pp. 6682–6686. IEEE (2018)
21. Liu, H., et al.: A deep learning model integrating mammography and clinical factors facilitates the malignancy prediction of BI-RADS 4 microcalcifications in breast cancer screening. Eur. Radiol. **31**, 5902–5912 (2021)
22. Li, C., et al.: Multi-view mammographic density classification by dilated and attention-guided residual learning. IEEE/ACM Trans. Comput. Biol. Bioinf. **18**(3), 1003–1013 (2020)
23. Lee, R.S., Gimenez, F., Hoogi, A., Miyake, K.K., Gorovoy, M., Rubin, D.L.: A curated mammography data set for use in computer-aided detection and diagnosis research. Scientific data **4**(1), 1–9 (2017)
24. Lizzi, F., Scapicchio, C., Laruina, F., Retico, A., Fantacci, M.E.: Convolutional neural networks for breast density classification: performance and explanation insights. Appl. Sci. **12**(1), 148 (2022)
25. Gudhe, N.R., Sudah, M., Mannermaa, A., Kosma, V.M., Behravan, H.: Multi-view deep evidential fusion neural network for assessment of screening mammograms. In: Eleventh International Conference on Learning Representations (2024)
26. Busaleh, M., Hussain, M., Aboalsamh, H.A., Al Sultan, S.A.: TwoViewDensityNet: two-view mammographic breast density classification based on deep convolutional neural network. Mathematics **10**(23), 4610 (2022)
27. Sickles, E.A., D'Orsi, C.J.: How should screening breast US be audited? the BI-RADS perspective. Radiology **272**(2), 316–320 (2014)
28. Pawar, S.D., et al.: Multichannel densenet architecture for classification of mammographic breast density for breast cancer detection. Front. Public Health **10** (2022)
29. Nguyen, H.T., Tran, S.B., Nguyen, D.B., Pham, H.H., Nguyen, H.Q.: A novel multi-view deep learning approach for BI-RADS and density assessment of mammograms. In: 2022 44th Annual International Conference of the IEEE Engineering in Medicine & Biology Society (EMBC), pp. 2144–2148. IEEE (2022)
30. Ciritsis, A., et al.: Determination of mammographic breast density using a deep convolutional neural network. Br. J. Radiol. **92**(1093), 20180691 (2019)
31. Kaiser, N., et al.: Mammographic breast density classification using a deep neural network: assessment based on inter-observer variability. In: Medical Imaging 2019: Image Perception, Observer Performance, and Technology Assessment, vol. 10952, pp. 156–161. SPIE (2019)

Effective Writer Identification from Text Fragments Using Lightweight Convolutional Autoencoder

Amina Briber(✉) and Youcef Chibani

Laboratoire d'Ingénierie des Systèmes Intelligents et Communicants (LISIC), Faculty of Electrical Engineering, University of Science and Technology Houari Boumediene (USTHB), 32, El Alia, Bab-Ezzouar, 16111 Algiers, Algeria
{abriber,ychibani}@usthb.dz

Abstract. Writer identification using handwritten text fragments is a challenging task due to the variability of handwriting styles and the limited amount of data available for training. This paper proposes a new open writer identification system based on a lightweight convolutional autoencoder (CAE) framework. The proposed system leverages the CAE's ability to extract discriminative features from handwritten text fragments. Notably, the system is trained on a small subset of writers, making it an open system. A new metric, based on the number of correctly classified fragments, is introduced during the design step to enhance the robustness of extracted features, which are then fed to the distance-based classifier (DBC) for writer identification. Importantly, integrating new writers into the system does not require retraining the lightweight CAE. Experimental results demonstrate that the proposed system achieves competitive identification rates of 95.86% and 89.78% on both the IFN/ENIT and IAM datasets, respectively.

Keywords: Open system · Ligthweight CAE · Text fragments · Distance-Based Classifier · IFN/ENIT · IAM

1 Introduction

Handwriting analysis has gained significant popularity in recent years due to its diverse applications [12,18]. One of the most crucial tasks within handwriting analysis is writer identification, which allows finding the writer based on handwritten documents [3]. It can be achieved using various elements, including entire documents [9], paragraphs [8], lines [1], words [2], characters [20], and, in recent years, fragments of handwritten text [6,11–13,15]. In this context, the paper aims specifically the writer identification from the limited availability of handwritten text fragments, which is useful particularly for applications such as forensic or deeds of sale [6]. Therefore, when the entire document is not available, text fragments can be used for designing an effective writer identification system [6,7], yielding promising reliability [11–13,15].

B. Djamaa et al. (Eds.): CSA 2024, LNNS 1145, pp. 80–90, 2025.
https://doi.org/10.1007/978-3-031-71848-9_8

When designing an effective identification system, feature generation is the crucial step aiming to extract the most relevant information to form a feature vector, which serves as a representation of the writer's writing style. In recent years, several studies have explored different heavy deep learning architectures [14,16,19,21]. In this paper, an alternative approach is proposed based on a lightweight Convolutional Autoencoder (CAE) framework. The proposed system uses a subset of writers for training, making it an open system, and used the CAE's ability to extract discriminative features from handwritten text fragments. CAEs are considered unsupervised learning models dedicated for various tasks such as image reconstruction. Commonly, relevance of CAE model and reconstruction quality are assessed based on reconstruction loss and accuracy calculated between the original input images and reconstructed output images without any target information. However, achieving accurately the reconstruction of images is not a guarantee to achieve a relevant feature representation. To leverage the feature extraction capabilities of CAEs, some works used CAEs as semi-supervised learning models by incorporating a classification stage during training in their systems to improve the classification results [5]. To ensure finding the most relevant feature representation in the proposed system, a new metric is defined based on the identification rate of fragments correctly classified (IRF) calculated during the design step. Finally, the extracted relevant features are then fed to the Distance-Based Classifier (DBC) for final identification.

The remainder of this paper is organized as follows. Section 2 reviews some related works. Section 3 describes the proposed writer identification system using a lightweight CAE. Section 4 describes the experimental results and comparative analysis against established methods. Finally, Sect. 5 concludes the paper and offers suggestions for future research.

2 State-of-the-Art Using Text Fragments

Few systems have been proposed for identifying writers based on handwritten text fragments. These systems can be subdivided into engineered features or deep learning features. Regarding engineered features, Hannad et al. [12,13] employed various engineered descriptors to extract features with the DBC using the Hamming distance. Hadjadji and Chibani [11] combined three engineered descriptors through a complex two-stage classification process based on Dynamic Fragment Weighting Combination and Choquet fuzzy integral techniques. For deep learning features, He et al. [15] introduced FragNet, an end-to-end system based on word or text block images containing approximately one word. FragNet aims to extract powerful features from these words not from text fragments. Briber and Chibani [6] proposed the use of a CAE for feature generation from text fragments. To create the CAE model, a subset of writers was used for training, containing only 11 writers. The trained model is then used to generate features of the entire dataset, which are fed to the DBC for writer identification. While the previous systems are considered closed systems, Briber and Chibani [6] system is considered as an open system since new writers can be added without

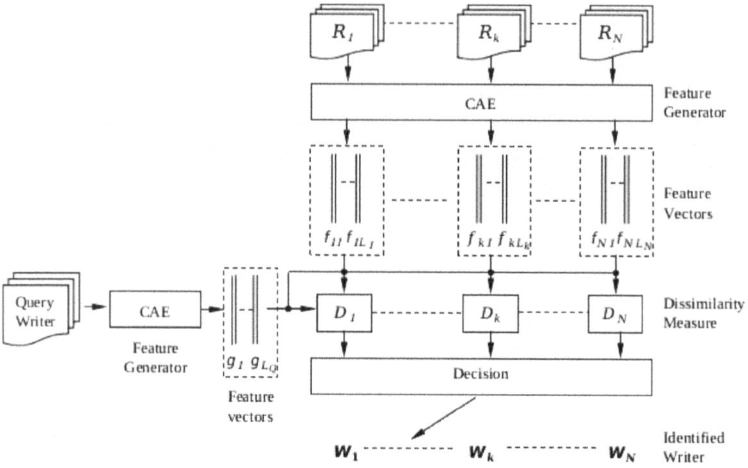

Fig. 1. System overview.

retraining the CAE. In order to improve the previous work, this paper proposes a new lightweight system based on a CAE framework associated to the DBC.

3 Proposed System

3.1 System Overview

The proposed system as depicted in Fig.1 comprises a feature generation module implemented using CAE associated to the DBC paired with a decision rule for writer identification. Hence, a writer represented by a set of their text fragments is assigned to a class offering the minimal distance [6].

Let the query document $Q = \{q_i\}_{i=1,...,L_Q}$ such as q_i is the query text fragment and L_Q is the number of fragments, each one is fed to CAE for generating features namely g_i. Similarly, let each reference writer namely $R_k = \{r_{kj}\}_{k=1,....,N,j=1,...,L_k}$ such as r_{kj} is the reference fragment fed to the CAE for generating its corresponding feature vector f_{kj} with L_k denotes the number of reference fragments for writer k, and N is the total number of reference writers. To identify the writer from its query document Q, the DBC is employed to calculate the dissimilarity measure D_k for each writer k by comparing the feature vector of the query fragment to the feature vectors of the corresponding reference fragments. Thus, the offered smallest dissimilarity measure D_{min} is considered the most likely writer of the query document.

3.2 Design of the Proposed System

To design an open system, a subset of M writers is selected from the reference writer dataset for training and determining the optimal configuration of the

Table 1. Detailed overview of the parameters used in CAE.

Encoder Layers		Decoder Layers	
Layers	#Filters(Kernels)	Layers	#Filters(Kernels)
Conv1	10(5 × 5)	UnFlatten	160(1 × 1)
BN, ReLU, MaxPool	-,-,(2 × 2)	Dconv1	160(1 × 1)
Conv2	20(3 × 3)	BN, ReLU, UpSamp	-,-,(2 × 2)
BN, ReLU, MaxPool	-,-,(2 × 2)	Dconv2	160(3 × 3)
Conv3	40(3 × 3)	BN, ReLU, UpSamp	-,-,(2 × 2)
BN, ReLU, MaxPool	-,-,(2 × 2)	Dconv3	80(3 × 3)
Conv4	80(3 × 3)	BN, ReLU, UpSamp	-,-,(2 × 2)
BN, ReLU, MaxPool	-,-,(2 × 2)	Dconv4	40(3 × 3)
Conv5	160(3 × 3)	BN, ReLU, UpSamp	-,-,(2 × 2)
BN, ReLU, MaxPool	-,-,(2 × 2)	Dconv5	20(3 × 3)
Conv6	160(1 × 1)	BN, ReLU, UpSamp	-,-,(2 × 2)
Flatten	2560	Dconv6	10(1 × 1)

lightweight CAE. The resulting feature vector is then extracted and fed to the DBC for writer identification. Selecting the optimal CAE architecture is based on several metrics including reconstruction accuracy and loss per fragment, maximum of both identification rate of fragments per writer and identification rate provided by the classifier.

Feature Generation. This step is performed using a lightweight CAE to extract distinctive features from handwritten text fragments. The CAE is a neural network architecture that uses two-stages for data representation and reconstruction, comprising an encoder and a decoder [10]. During the encoding step, a convolution operation is applied to the input image to generate a latent representation containing relevant features in a compressed form. Conversely, the decoding stage employs a deconvolution operation on the latent data allowing a reconstruction of the output image from the input image in a near-perfect way [6]. Table 1 depicts the detailed overview of the parameters used in CAE.

Classification. A dissimilarity measure is employed to compare a query document Q with a reference document R_k defined as follows [12]:

$$D_k\left(Q, R_k\right) = \frac{1}{L_Q} \sum_{i=1}^{L_Q} \min_{j=1,\ldots,L_k} \left\{distance\left(q_i, r_{kj}\right)\right\} \tag{1}$$

To assign a query writer to a class, the minimum distance is calculated between its corresponding feature vector of size P and the feature vector of the reference writers through the Euclidean distance defined as follows:

$$distance\left(q_i, r_{kj}\right) = \sqrt{\sum_{p=1}^{P}\left(g_{pi} - f_{pkj}\right)^2} \tag{2}$$

Finally, assigning a writer to a class is based on identifying the minimum distance among the offered responses from all classes as follows [12]:

$$Writer_Q = \underset{k=1,\ldots,M}{argmin}\left\{D_k\left(Q, R_k\right)\right\} \tag{3}$$

It is worth noting that the subset M is used during the design step, while N is used during the evaluation step.

Optimization Criteria. Despite their effectiveness in reconstruction, CAE may not always capture meaningful features. To address this limitation and ensure the relevance of extracted features, a new metric is introduced during the design process to assess the CAE's ability for feature extraction. Since any explicit writer-specific information provided during reconstruction process, the maximum Identification Rate of Fragments (IRF) serves as a crucial metric for selecting the optimal CAE architecture. During training step, the feature vector is retrieved on each epoch. Then, the dissimilarity between validation and reference features is calculated and the fragment is assigned to the writer with the minimum distance among all fragments using the following equation:

$$Label(q_i) = \underset{k=1,\ldots,M}{argmin}\left\{\underset{j=1,\ldots,L_k}{min}\left(distance\left(q_i, r_{kj}\right)\right)\right\} \tag{4}$$

The IRF is calculated as follows:

$$IRF = \frac{\#F}{\#TF} \tag{5}$$

$\#F$ and $\#TF$ define the number of fragment well classified and the total number of fragments, respectively. The model offered the highest IRF is selected and used for feature extraction during evaluation without retraining. While, the reconstruction quality is evaluated using the mean squared error (MSE).

4 Experimental Results

4.1 Dataset Description

The proposed system is evaluated using text fragments from the two popular datasets IFN/ENIT and IAM. The IFN/ENIT dataset is composed of Arabic handwritten documents with over 26,400 names of Tunisian cities and towns written in Arabic script [4]. Whereas, the IAM dataset is a collection of English handwritten documents containing 13,353 isolated and labeled text lines of variable content contributed by 657 different writers [17]. For a fair comparison, the paper uses the same fractions of fragment for training and testing, with a size of 100 × 100 pixels, as reported in [12]. Figure 2 shows some handwritten text fragment samples collected from the same writer from both the IFN/ENIT and IAM datasets, respectively.

(a) (b)

Fig. 2. Samples of the same writer (a) Arabic fragments (b) Latin fragments.

4.2 Evaluation Criterion

In order to make a fair comparison from the state of-the-art, the system's performance is measured using the standard Identification Rate of the Writers (IRW) namely IR in other existing systems [11,12]. It is defined as the ratio of correctly identified writers to the total number of writers.

4.3 Experimental Protocol

To optimize the proposed system's architecture and hyperparameters, a subset of writers is used during the design step, which serves as a set of reference text fragments for training the CAE. The selected text fragments from each writer are split into training (60%) and validation (40%) subsets. While, the test set of text fragments belonging to these writers is solely reserved for evaluation. The best model design is selected using various metrics such as the reconstruction accuracy, reconstruction loss, and the maximum IRF per writer. These criteria are jointly considered in conjunction with the DBC performances to find the optimal trained model. Once the model is optimized, the CAE is then dedicated to generate features for all writers of the dataset, which are then fed into the DBC for writer identification. It is important to note that no data augmentation is employed.

4.4 Experimental Design

During the design step, reconstruction loss, reconstruction accuracy, IRF, and IRW are calculated for each epoch of the model to facilitate the selection of an optimal CAE architecture, as depicted in Fig. 3. The best model exhibits a combination of high maximum IRF and IRW values with satisfactory reconstruction accuracy and minimal reconstruction loss. The choice of the best architecture is performed on the Arabic IFN/ENIT subset using only 12 writers. Figure 3 illustrates the evolution of these metrics using training and validation subsets versus Epochs within the range of [1, 100].

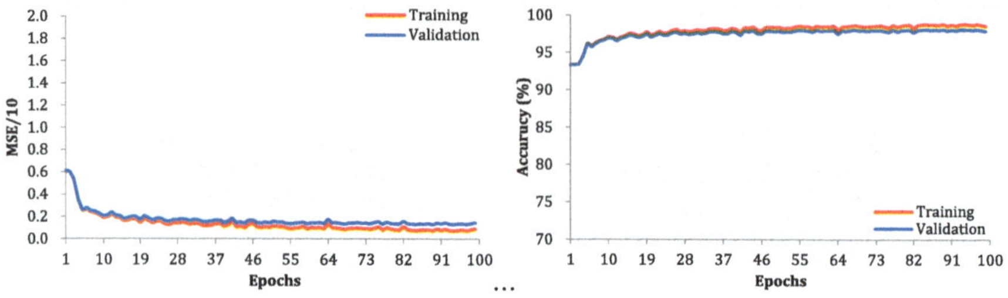

(a) MSE and Accuracy (%) of the fragment reconstruction versus Epochs.

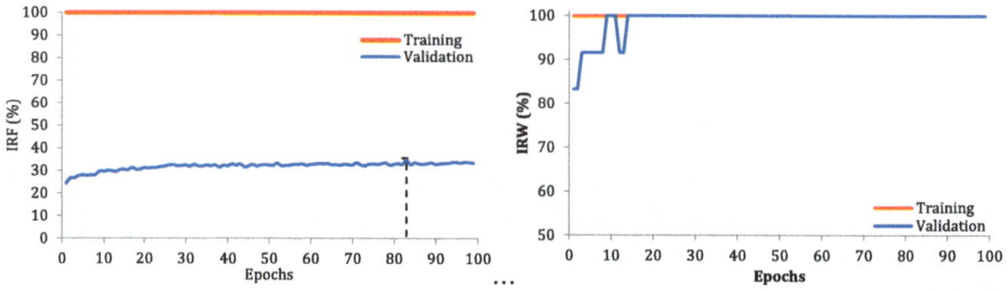

(b) IRF (%) and IRW (%) versus Epochs.

Fig. 3. Evolution of metrics versus Epochs using Training and Validation Subsets.

As can be seen, the proposed model accurately reconstructs input fragments, demonstrating successful training without overfitting. This can be explained by the fact that CAE are mainly used for image reconstruction using limited data, enabling a rapid convergence and accurate reconstruction of the image. Furthermore, the used fragments are binary images with small size, which allows facilitating reconstruction. This faithful reconstruction does not provide any information regarding the feature vector quality. Instead, the IRW and IRF metrics provide critical insights into the quality of the features by measuring classification accuracy. These metrics assess the effectiveness of feature vectors in distinguishing between writers. Consequently, the model that achieves the highest IRF is selected for the experimental evaluation.

4.5 Experimental Evaluation

The proposed system's performance is evaluated on the entire dataset using different experiments: intra-domain, cross-domain, and transfer learning. The first experiment assesses the model within the same language, the second involves evaluating the model between two different languages, and the third entails transferring the model from one domain to another.

Intra-domain Evaluation

IFN/ENIT Dataset: The model is trained using a subset of only 12 writers from the IFN/ENIT dataset, to seamlessly generate features for the entire dataset of 411 writers without requiring retraining. The evaluation yields an impressive IRW of 94.89%, indicating the effectiveness of the model in writer identification.

IAM Dataset: To assess the model's robustness for generalization to other datasets, it undergoes directly evaluation on the IAM dataset without any hyper-parameter fine-tuning. Initially, a subset of 23 writers is selected for training the model, which is then used for generating features for all 656 writers in the IAM dataset without additional training. The evaluation across the entire IAM dataset offers an IRW of 89.17%. Notably, the proposed architecture is designed using the IFN/ENIT dataset and is subsequently used directly for training the IAM dataset without finding its own and specific parameters. The non adaptation of parameters may explain the higher performance observed in the IFN/ENIT dataset compared to the IAM dataset.

Cross-Domain Evaluation

To evaluate the generalization of the trained models, a cross-domain evaluation is conducted using IFN/ENIT and IAM models to generate features for other datasets without retraining. The evaluation is performed using trained model on IFN/ENIT and tested on IAM dataset, and vice-versa. The obtained IRWs are 93.91% and 87.35%, respectively, which confirm the generalization of the proposed system when performed in a cross-domain.

Transfer Learning

To investigate the impact of Transfer Learning (TL) from a language to another language, four distinct TL experiments are designed and implemented:

- One TL (OTL) from the IFN/ENIT model to the IAM dataset and vice-versa.
- Double TL (DTL) involving a forward transfer from the IFN/ENIT model to IAM and a backward transfer back to IFN/ENIT and vice-versa.

As can be reported in Table 2, both OTL or DTL consistently enhance the IRW for both datasets compared to the No-TL. OTL from IFN/ENIT to IAM and vice versa shows the potential of transferring knowledge between related datasets. While DTL further improves the IRWs, suggesting that the transfer of knowledge across both datasets refine the feature extraction process.

Table 2. IRW using Transfer Learning.

Training	IFN/ENIT (No-TL)	IAM (OTL)	IFN/IAM (DTL)	Training	IAM (No-TL)	IFN/ENIT (OTL)	IAM/IFN (DTL)
IFN/ENIT	94.89%	94.64%	95.86%	IAM	87.35%	89.17%	89.78%

4.6 Comparative Analysis

For a fair comparison between the proposed system and the state-of-the-art, the same evaluation criterion is used. Moreover, the distance measure used is Euclidean distance since it is well adapted to the proposed system. Table 3 shows the existing systems that use handwritten text fragments for both IFN/ENIT and IAM datasets.

Table 3. IRW (%) obtained for open and closed systems against the state-of-the-art methods using text fragments for IFN/ENIT and IAM datasets.

Type of system	#Training/Testing Writers	References	Features	Classifier	IRW(%)
		IFN/ENIT dataset			
Closed	411/411	Hannad et al. [12]	LPQ	Hamming distance	94.89
	411/411	Hannad et al. [13]	HOG	Hamming distance	86.62
	411/411	Hadjadji and Chibani. [11]	LPQ	Hamming distance	94.40
	411/411	Hadjadji and Chibani. [11]	Combined	OC-K-Means	97.56
Open	11/411	Briber and Chibani. [6]	CAE	Euclidean distance	92.70
	12/411	Proposed	CAE (No-TL)	Euclidean distance	94.89
			CAE (OTL)	Euclidean distance	94.64
			CAE (DTL)	Euclidean distance	95.86
		IAM dataset			
Closed	656/656	Hannad et al. [12]	LPQ	Hamming distance	89.11
	656/656	Hadjadji and Chibani. [11]	LPQ	Hamming distance	88.26
	656/656	Hadjadji and Chibani. [11]	Combined	OC-K-Means	94.51
Open	23/656	Proposed	CAE (No-TL)	Euclidean distance	87.35
			CAE (OTL)	Euclidean distance	89.17
			CAE (DTL)	Euclidean distance	89.78

For the IFN/ENIT dataset, Hannad et al. [12] and Hadjadji and Chibani [11] achieved an IRW of 94.89% and 94.40%, respectively, using the LPQ descriptor with the Hamming distance classifier. The high IRW is obtained by Hadjadji and Chibani [11] yielded an IRW of 97.56% via a combined system using the OC-K-Means. Briber and Chibani [6] proposed a CAE for feature extraction and the Euclidian distance for classification, achieving an IRW of 92.70%. As can be observed, the selection of an appropriate distance for classification is influenced by the specific feature extraction method. Hence, authors try to select the best classification distance well adapted to the feature extraction method.

In comparison, the proposed system achieves an IRW of 94.89%, 94.64% and 95.89% on the IFN/ENIT dataset using No-TL, OTL and DTL, respectively. Regarding the IAM dataset, Hadjadji and Chibani [11], and Hannad et al. [12] achieved IRW of 89.11% and 88.26%, respectively. Hadjadji and Chibani [11] achieved an IRW of 94.5% with the combined system. In contrast, the proposed system achieves an IRW of 87.35%, 89.17% and 89.78% with a model only evaluated on IAM dataset using No-TL, OTL and DTL, respectively without exploring suitable parameters to assess its ability for generalization. The proposed method ranks second compared to the combined system proposed by

[11]. However, the proposed system offers several advantages over existing methods. The proposed system employs a lightweight CAE architecture, making it computationally efficient and less complex than the combined system proposed by [11]. Furthermore, the proposed system is an open system trained only on a small subset of writers, not exceeding 4% for both IFN/ENIT and IAM datasets, showing its ability to well performing with limited data.

5 Conclusion

This paper proposed a new open lightweight CAE system for effective writer identification through the use of text fragments trained on a small subset of writers. A new metric based on the IRF, is introduced to evaluate the robustness of the extracted features during the design step. The experimental results demonstrate that the proposed system achieves competitive IRW on both the IFN/ENIT and IAM datasets with limited training data and without data augmentation. Furthermore, the proposed system exhibits generalization to other datasets through cross-domain and transfer learning. Future endeavors will focus on developing an adapted system to various datasets and languages. Additionally, it is planned to develop a model adapted to IAM dataset and perform tests on larger dataset to explore the generalization capacity of the proposed model.

Disclosure of AI Tools Usage. The preparation of this manuscript involved the use of chatGPT and Bard to correct the spelling and grammar. Subsequently, the authors reviewed and edited the content as necessary, and take full responsibility for the paper's content.

References

1. Abdi, M.N., Khemakhem, M.: A model-based approach to offline text-independent Arabic writer identification and verification. Pattern Recogn. **48**(5), 1890–1903 (2015)
2. Abdi, M.N., Khemakhem, M., Ben-Abdallah, H.: A novel approach for off-line Arabic writer identification based on stroke feature combination. In: 2009 24th International Symposium on Computer and Information Sciences, pp. 597–600. IEEE (2009)
3. bin Abdl, K.M., Hashim, S.Z.M.: Handwriting identification: a direction review. In: 2009 IEEE International Conference on Signal and Image Processing Applications, pp. 459-463 (2009). https://doi.org/10.1109/ICSIPA.2009.5478698
4. Altwaijry, N., Al-Turaiki, I.: Arabic handwriting recognition system using convolutional neural network. Neural Comput. Appl. **33**(7), 2249–2261 (2021)
5. Bank, D., Koenigstein, N., Giryes, R.: Autoencoders. machine learning for data science handbook: data mining and knowledge discovery handbook, pp. 353–374 (2023)
6. Briber, A., Chibani, Y.: On the use of the convolutional autoencoder for Arabic writer identification using handwritten text fragments. In: International Conference on Artificial Intelligence and its Applications, pp. 301–311. Springer (2022)

7. Briber, A., Chibani, Y.: Open writer identification from handwritten text fragments using lite convolutional neural network. Int. J. Doc. Anal. Recogn. (IJDAR), pp. 1–23 (2023)
8. Bulacu, M., Schomaker, L., Brink, A.: Text-independent writer identification and verification on offline Arabic handwriting. In: Ninth International Conference on Document Analysis and Recognition (ICDAR 2007), vol. 2, pp. 769–773. IEEE (2007)
9. Chawki, D., Labiba, S.M.: A texture based approach for Arabic writer identification and verification. In: 2010 International Conference on Machine and Web Intelligence, pp. 115–120. IEEE (2010)
10. Cheng, Z., Sun, H., Takeuchi, M., Katto, J.: Deep convolutional autoencoder-based lossy image compression. In: 2018 Picture Coding Symposium (PCS), pp. 253–257. IEEE (2018)
11. Hadjadji, B., Chibani, Y.: Two combination stages of clustered one-class classifiers for writer identification from text fragments. Pattern Recogn. **82**, 147–162 (2018)
12. Hannad, Y., Siddiqi, I., El Kettani, M.E.Y.: Writer identification using texture descriptors of handwritten fragments. Expert Syst. Appl. **47**, 14–22 (2016)
13. Hannad, Y., Siddiqi, I., El Merabet, Y., El Youssfi El Kettani, M.: Arabic writer identification system using the histogram of oriented gradients (hog) of handwritten fragments. In: Proceedings of the Mediterranean Conference on Pattern Recognition and Artificial Intelligence, pp. 98–102 (2016)
14. He, K., Zhang, X., Ren, S., Sun, J.: Deep residual learning for image recognition. In: Proceedings of the IEEE Conference on Computer Vision and Pattern Recognition, pp. 770–778 (2016)
15. He, S., Schomaker, L.: Fragnet: writer identification using deep fragment networks. IEEE Trans. Inf. Forensics Secur. **15**, 3013–3022 (2020)
16. Huang, G., Liu, Z., Van Der Maaten, L., Weinberger, K.Q.: Densely connected convolutional networks. In: Proceedings of the IEEE Conference on Computer Vision and Pattern Recognition, pp. 4700–4708 (2017)
17. Marti, U.V., Bunke, H.: The IAM-database: an English sentence database for offline handwriting recognition. Int. J. Doc. Anal. Recogn. **5**, 39–46 (2002)
18. Morris, R.N.: Forensic handwriting identification: fundamental concepts and principles (2020)
19. Simonyan, K., Zisserman, A.: Very deep convolutional networks for large-scale image recognition (2014). arXiv preprint arXiv:1409.1556
20. Sreeraj, M., Idicula, S.M.: A survey on writer identification schemes. Int. J. Comput. Appl. **26**(2), 23–33 (2011)
21. Szegedy, C., Vanhoucke, V., Ioffe, S., Shlens, J., Wojna, Z.: Rethinking the inception architecture for computer vision. In: Proceedings of the IEEE Conference on Computer Vision and Pattern Recognition, pp. 2818–2826 (2016)

Computer Vision and Graphics

Efficient End-to-End Deep Learning Model for Algerian Drug Label Recognition

Mouad Bensouilah(✉), Mokhtar Taffar, and Elias Boulham

LaRIA Laboratory, Faculty of Exact Sciences and Computer Science, University of Jijel, 18000 Jijel, Algeria
{mouad_bensouilah,mokhtar_taffar}@univ-jijel.dz, elias.boulham@gmail.com

Abstract. Optical Character Recognition (OCR) is still a very active area of computer vision research. This work proposes a robust text detection/recognition framework for Algerian drug labels (ADL). It extracts the text from the drug label using the benchmark YOLOv8 object detector. Text recognition is done through our CRNN-gMLP model combining a convolutional recurrent neural network (CRNN) and a gate multi-layer perceptron (gMLP) network. The gMLP detects the spatial interaction between the target characters, which helps CRNN learn a more precise alignment at each decoding step. Our end-to-end YOLO-CRNN-gMLP framework locates label text and then recognizes words by converting text pixels into sequences of characters. To evaluate our approach, we created a new dataset specifically for ADL, which contains texts in various sizes and fonts for training and testing purposes. The experiment results showed that CRNN-gMLP achieved an acceptable text recognition rate even on the least readable labels.

Keywords: OCR · Drug Label Recognition · YOLOv8 · CRNN · gMLP

1 Introduction

Despite significant advances in optical character recognition (OCR) technology, many applications underlying the recognition of printed or handwritten text remain a challenge, including document scanning [15] and handwriting recognition (HTR) [3]. OCR tasks involve two functions: text detection and text recognition. The text detectors attempt to locate regions of text in the image. In contrast, text recognition algorithms aim to comprehend the patch contents and transcribe them into digital items of the natural language.

The difficulties that OCR models face are mainly the accuracy of locating text regions in images and discerning characters or words with high precision, especially when image scenes are complex and cluttered, like the case of Algerian drug label images illustrated in Fig. 1. These difficulties are inherent in various environments, such as illumination, low resolution, and inadequate scale, or often

B. Djamaa et al. (Eds.): CSA 2024, LNNS 1145, pp. 93–103, 2025.
https://doi.org/10.1007/978-3-031-71848-9_9

taken from a sensitive viewpoint [21]. Thus, they introduce additional challenges in word region localization tasks leading to poor word recognition performance.

Due to the complex nature of ADLs where the printed text can take on various and complex fonts or sizes, as in Fig. 1, the performance of any OCR model dedicated to drug label processing can be severely degraded. In addition, these labels present other obstacles including character overlapping, context dependency, and cluttered label images. Therefore, there is a need to implement a deep learning-based OCR model that can overcome these text variation issues and recognize the words on a drug label with acceptable accuracy.

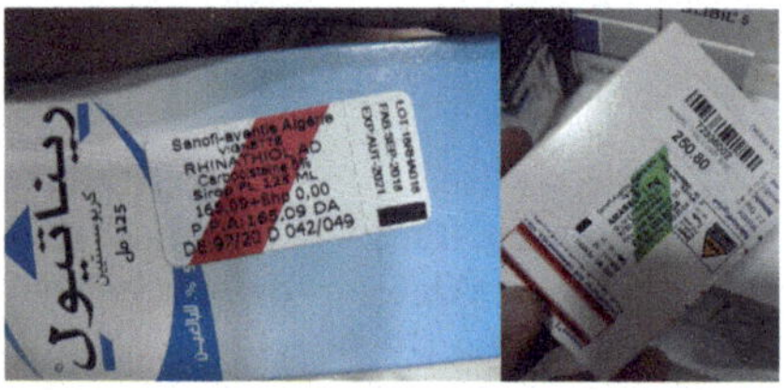

Fig. 1. Algerian drug label samples

In this paper, we present an OCR framework without explicit segmentation by designing a recent hybrid approach integrating a text detection algorithm based on the YOLOv8 object detector [9] and a deep learning-based text recognition model where the architecture is a fusion of a CRNN and a gMLP network inspired by the CRNN-gMLP model presented in [3]. The framework aims to tackle the difficulties concerned with reading drug labels. It is trained on our ADL-Dataset [2] for high-precision recognition of label texts. The originality lies in the design of a coherent CRNN-gMLP network to recognize texts. The network enables better extraction of features and models contextual information regarding the difficult nature of the drug labels, ensuring word recognition with minimal error, something wished in a trusted OCR system.

2 Related Works

Over the past decade, some high-performance computer vision and deep learning algorithms have established themselves as benchmarks, highlighting that problems are better addressed by an end-to-end OCR system rather than through a series of independent solutions for text detection, location, and recognition [24].

As mentioned above, word detection is an integral aspect of image text recognition, and various proposed solutions address this process. In [16], authors employed the SIFT features to handle text distortion and detect text. Huang *et al.* [8] presented a hybrid approach for scene text detection, integrating MSER (Maximally Stable Extremal Regions) [20] operator with Convolutional Neural Networks (CNN).

Most modern text detection algorithms are based on deep learning models. Liu *et al.* [14] suggested a real-time character detector integrating an encoding-decoding CNN-based model. Subsequently, researchers have found inspiration from object detection algorithms. For instance, Liao *et al.* [10] have adapted the Single Shot Multibox Detector (SSD) framework [12] to handle the various orientations and aspect ratios of text across TextBoxes. Furthermore, several studies have leveraged the state-of-the-art You Only Look Once (YOLO) approach for text detection. Notably, Bensouilah *et al.* [4] have adapted YOLOv3 [18] for Algerian license plate detection.

Similar to detection, research on word recognition has continued to produce increasingly efficient and reliable approaches, particularly with the advent of deep models and their ability to learn large quantities of words and their relationships and sequencing in the text. CTC-based models are widely used due to their efficiency and lower computing costs. Shi *et al.* [19] proposed the CRNN, a very efficient model for text recognition. It combines CNN which extracts highly representative features and RNN which encodes features to get a string of characters. This string is then decoded through the CTC [6] layer. In the same context, Gao *et al.* [5] developed a scene text reading model with fully convolutional sequence modeling. A full CNN extracts and encrypts image features with a CTC for decoding. This network has achieved a lower cost than those state-of-the-art that rely on recursive networks in encryption. Regarding attention-based models, they extract features better through the attention mechanism, like the character-based model proposed by Liu *et al.* [13] where hierarchical attention corrects characters. There are also end-to-end models for text detection/recognition. Zhan *et al.* [25] introduced an end-to-end trainable network for scene text recognition using iterative rectification.

3 Proposed Model

The model we propose is a YOLO-CRNN-gMLP, a word-level OCR approach. As shown in Fig. 2, the model structure consists of two modules. Indeed, the proposed architecture consists of two structures: YOLO for text detection and CRNN-gMLP for recognition.

3.1 YOLOv8 Text Detector

The text zone detection is provided by the YOLOv8 [9] state-of-the-art object detector. YOLO [17] is a CNN algorithm that is fast, effective, and easy to learn and integrate into real systems. YOLOv8 is the last optimized version of the YOLO detector family that incorporates significant improvements in accuracy while maintaining speed and performance. Its adaptability to various vision tasks, such as segmentation, tracking, and recognition, makes it a powerful general-purpose model for diverse computer vision applications.

Trained on our ADL-Dataset, our detection module enables efficient localization of text regions to obtain the entire bounding box of words to feed into

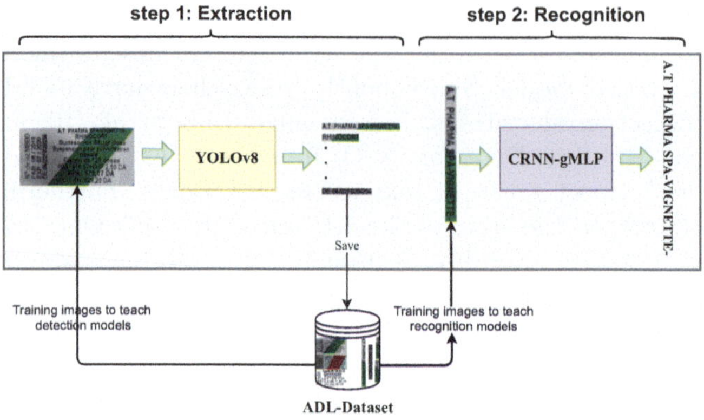

Fig. 2. The general design of our system of Algerian drug label recognition

the classification module. Regions of text containing one or more words are landmarked according to the internal representation of the text constructed by YOLOv8 during learning. Over our experiments, we will try to train, test, and compare two model versions: YOLOv8 and YOLO-NAS [1].

3.2 CRNN-gMLP Text Recognizer

We suggest an assembly of CRNN with gMLP networks to further focus the attention of our text recognition model on dealing with characters of labels. gMLP [11] is a basic MLP network with gating that we placed after the CNN to improve character prediction in the RNN (recurrent neural network) decoder by learning the inter-channels relationships and spatial interaction. This would reduce information redundancy between channels and focus on the text of the image. Therefore, we need to improve the network's ability to learn the correct alignment between image pixels and target characters and extract the most relevant information for each part of the output sequence.

Our CRNN-gMLP recognition module compiles the words of the label image and restores the text of words as strings. The CRNN-gMLP structure, shown in Fig. 3, is a coupled architecture of CNN and BLSTM (bidirectional long short-term memory) that allows capturing local and contextual information. Indeed, the CRNN-gMLP enhances recognition accuracy where gMLP enables better extraction of features and models contextual information regarding the difficult nature of the Algerian drug labels, ensuring a low word recognition error.

3.3 gMLP Networks

The gMLP network was proposed by the Google Brain team [11] as an alternative to Transformers [22] which incorporates an attention mechanism to find the relationships between data items. A gMLP is an MLP-based model without

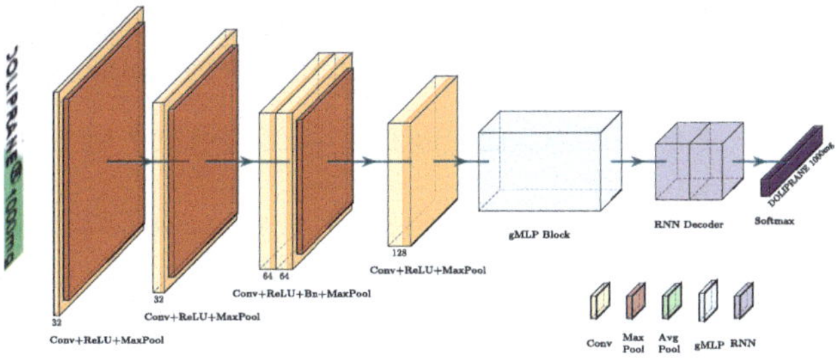

Fig. 3. The architecture of our CRNN-gMLP model

self-attention. It consists of channel and spatial projections with static parameterization. It gave close to state-of-the-art results on natural language processing (NLP) tasks but with far fewer parameters to train.

A gMLP, or basic MLP layers with gating, is a stack of L identical blocks. Let $X \in R^{n \times d}$ be the symbolic representations of sequence length n and dimension d. Each block is defined as:

$$Z = \sigma(XU), \qquad \tilde{Z} = s(Z), \qquad Y = \tilde{Z}V. \tag{1}$$

where σ is an activation function such as GeLU [7]. U and V define linear projections and $s(.)$ is the layer capturing the spatial interactions.

4 Text Detection and Recognition

The text detection process identifies and locates image regions containing textual content. This means that the text detector is an algorithm that can effectively distinguish text blobs from non-text ones and handle all text appearance variabilities in the drug label image. YOLOv8 [9] algorithm divides the image into regions to predict bounding boxes and evaluate the probabilities for each. A probability score is assigned to a box to predict text class. For each drug label, the YOLOv8 detects a set of regions that should contain text. The detected regions consisting of the words require no post-processing. Thus, the text regions are extracted as a set of rectangles, each one surrounding a word. However, YOLO-NAS [1] significantly upgrades quantization support and tradeoff accuracy/latency. In our text detection experiments of drug labels, we compare the performances of YOLO-NAS and YOLOv8 models.

To recognize words on drug labels, we use the CRNN model, which is integrated with a gMLP network. Although CRNN is a powerful network that combines the strengths of CNN for feature extraction and RNN for sequence analysis, and whose last layers constitute a fully connected (FCN) network and a decoder, the fusion with the g-MLP network provides additional power to the network by describing relationships between words in terms of spatial information. gMLP applies a linear gating procedure to translate an input of shape $[n \times d]$ (n words with the embedding dimension of d) into an attention matrix with shape $[n \times n]$.

Then, the matrix information is re-injected into the input data before being used for predictions. In general, our text recognition model includes four main parts:

a) A CNN encoder of type VGG-12 constructed to extract visual features from drug-label text regions.
b) A gMLP network is used for complementarity between spatial interaction and the global context when determining the states of the characters due to the imprecise and overlapping forms of printed drug label text.
c) A RNN for sequence labeling is built with two recurrent layers of bidirectional long-term and short-term memory (BLSTM) for transcription. They give it a strong ability to learn the global contextual information in a text sequence, and encode extracted features that facilitate transcription [23].
d) A CTC [6] layer to transcribe drug label texts into word sequences.

5 Experimental Results

5.1 Dataset and Training

To evaluate the proposed approach, we collected 773 drug label images from drug boxes of a pharmacy, with the majority captured by the camera phone. We named ADL-Dataset, our dataset for Algeria Drug Label, and published it in [2]. To ensure the diversity of labels images were captured from different angles, scales, and lighting. Then, data augmentation tools were used to expand and triple the number of images, thereby increasing the variability of label images. As shown in Fig. 4, some samples of label text images of the ADL-Dataset.

To train and test our OCR model, we extract approximately 23,000 text images from label images to augment our ADL-Dataset [2] of 2,320 images, about 10 label texts extracted from each medication label image. However, the synthetic label text images contain one or some words with a different case and font size, and where noise is injected to avoid overfitting.

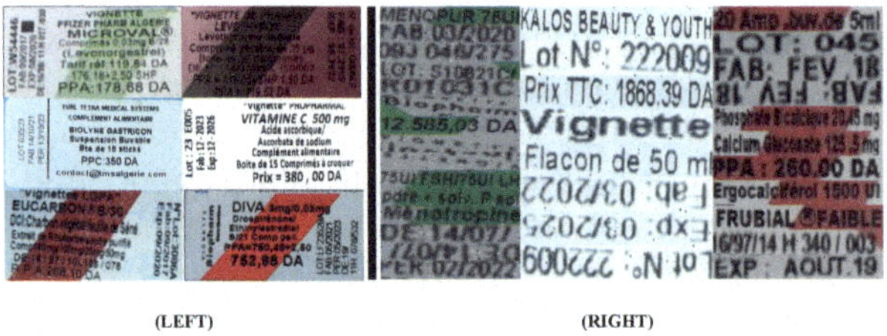

Fig. 4. Samples of ADL-Dataset images: drug label images for training YOLOv8 (left side) and label text images for training CRNN-gMLP (right side).

The ADL-Dataset [2] was annotated and divided into two groups to train the models to detect the labels and recognize words within them. For each drug vignette, a bounding box annotation was added to each set of words in the same row/column, to obtain about 8,265 label text images for detection training purposes. Then, each label image was paired with its corresponding text for recognition training. Figure 5 illustrates the steps for annotating the dataset.

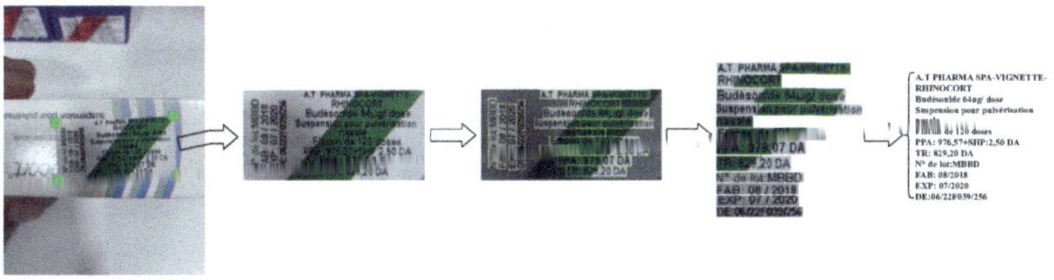

Fig. 5. Annotation process of ADL-Dataset

To train the YOLO detection model using the ADL-Dataset, which comprises 2,320 label text images, we specified an input size of 640 pixels. This size is crucial to ensure the effectiveness of the drug label detection process and enhance the model's performance. Similarly, to ensure the model's ability to recognize texts, we used the training set of 23000 label text images with the size of 400×50. The diversity of the training data aims to improve the performance of the YOLO model in drug label detection and the CRNN-gMLP for text recognition.

5.2 Text Detection

In this section, we aim to estimate the accuracy of the YOLOv8 text detector on our ADL-Dataset. Table 1 shows the results obtained by the two detection models, YOLOv8 and YOLO-NAS, tested on our ADL-Dataset, after the completion of 100 training epochs with an early stopping mechanism. Compared to YOLOv8, YOLO-NAS's detection speed is faster and always meets real-time requirements. However, in terms of mean Average Precision (mAP) performance, YOLOv8 is better, which means that the drug labels detected by YOLOv8 are closer to the ground truth.

Figure 6 illustrates the progression of mean Average Precision with mAP^{50} and mAP^{50-95}, denoting bounding box confidence at $IoU = 0.5$ and $IoU = [50, 95]$ respectively. This presentation summarizes the nuanced performance dynamics observed across epochs during training.

Table 1. Comparison of the performance of YOLOv8 and YOLO-NAS

Detection model	Inference (ms/image)	Recall	Precision	mAP^{50}	mAP^{50-95}
YOLOv8	2.5	0.985	**0.999**	**0.995**	**0.863**
YOLO-NAS	**2.36**	**0.99**	0.99	0.99	0.752

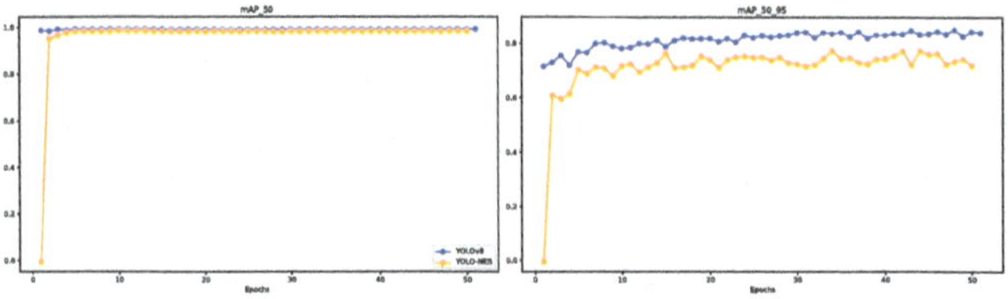

Fig. 6. Progression of mean Average Precision with mAP^{50} and mAP^{50-95} during YOLOv8 and YOLO-NAS training

5.3 Text Recognition

In this section, we present the experimental results obtained by evaluating our word recognition model on part of our ADL-Dataset.

The criterion for evaluating the accuracy of drug label text recognition is the correct recognition of all characters in the input image. In the case of recognizing letters, there may be an addition, deletion, or substitution of the correct letter in the treated text compared to the real text, which limits the process of calculating accuracy according to the sequence of letters. For this purpose, we used both the Word Error Rate (WER) and the Character Error Rate (CER) to evaluate a dedicated recognition model, in which each takes the smallest number of words/letters in the three cases above and divides it by the total number of words/letters, respectively.

Table 2 shows the performance results reached by our CRNN recognition model with and without the gMLP attention network. By Comparing the results, the CRNN-gMLP model with attention mechanism is more efficient for drug label text recognition with a CER of 0.22 on the testset images of the ADL-Dataset. This result is encouraging despite the obstacles presented by the drug labels due to their low visibility and interlacing printed characters.

Table 2. Comparison of the performance of CRNN and CRNN-gMLP.

Recognition model	Image Size	CER (%)	WAR (%)
CRNN	200×50	34.0	37.0
CRNN-gMLP	400×50	**21.6**	**23.7**

The proposed end-to-end text recognition framework integrates an efficient YOLOv8 text detector which achieves a word detection/extraction performance of approximately 99% on drug labels and a CRNN-gMLP network which achieves a word recognition ratio of 76%. We consider these performances acceptable, given the low quality and diversity of printing on the Algerian drug labels; in which the overlap of characters and the variety of fonts complicate both the extraction and recognition processes. Figure 7 shows the results of our end-to-end YOLOv8-CRNN-gMLP model for steps of detection and recognition.

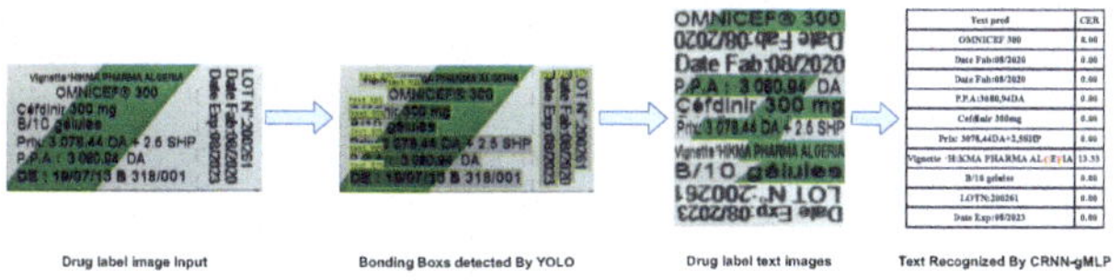

Fig. 7. Drug label text detected and recognized by our YOLOv8-CRNN-gMLP

Although the CRNN-gMLP recognizer performs well, which is appropriate for such OCR applications, it can be improved with better training on a larger and better drug-label dataset. However, our framework can increase its real-world executions with more controlled conditions and by using high-performance hardware technologies such as GPU processors and sophisticated cameras.

6 Conclusion

In this work, we presented two contributions: an end-to-end OCR framework for detecting/recognizing texts and a new image dataset of the Algerian drug labels (ADL-Dataset). The proposed model is a montage of the YOLOv8 text detector and a text recognition CRNN-gMLP network trained on our ADL-Dataset. It is a significant contribution in addition to an effective scientific approach to text recognition. This framework is also an OCR tool and a technological device dedicated to the pharmaceutical sector. Future works are committed to designing a backbone architecture composed of a block of layers whose role is to define a one-stage text detection/recognition process.

Disclosure of AI Tools Usage. The authors disclose that the content of this paper is not generated using any AI tools. We claim that the developed model is completely original, giving the work legal content.

References

1. Aharon, S., et al.: Eran-Deci: Deci-AI/super-gradients: 3.0.8 (2023). https://doi.org/10.5281/zenodo.7789328
2. Bensouilah, M.: Algeria drug label ADL-dataset (2024). https://drive.google.com/drive/folders/1YIYe-pEPUmX3mARJZq5EngVV357gthGI. Accessed 01 Mar 2024
3. Bensouilah, M., Taffar, M., Zennir, M.N.: gMLP guided deep networks model for character-based handwritten text transcription. Multimedia Tools Appl. **83**, 13557–13575 (2023). https://doi.org/10.1007/s11042-023-15293-1
4. Bensouilah, M., Zennir, M.N., Taffar, M.: An ALPR system-based deep networks for the detection and recognition. In: ICPRAM, pp. 204–211 (2021). https://doi.org/10.5220/0010229202040211
5. Gao, Y., Chen, Y., Wang, J., Tang, M., Lu, H.: Reading scene text with fully convolutional sequence modeling. Neurocomputing **339**, 161–170 (2019). https://doi.org/10.1016/j.neucom.2019.01.094

6. Graves, A., Fernández, S., Gomez, F., Schmidhuber, J.: Connectionist temporal classification: labeling unsegmented sequence data with recurrent neural networks. In: 23rd International Conference on ML, pp. 369–376. Association for Computing Machinery (2006)
7. Hendrycks, D., Gimpel, K.: Gaussian error linear units (gelus). arXiv preprint arXiv:1606.08415 (2016)
8. Huang, W., Qiao, Yu., Tang, X.: Robust scene text detection with convolution neural network induced MSER trees. In: Fleet, D., Pajdla, T., Schiele, B., Tuytelaars, T. (eds.) ECCV 2014. LNCS, vol. 8692, pp. 497–511. Springer, Cham (2014). https://doi.org/10.1007/978-3-319-10593-2_33
9. Jocher, G., Chaurasia, A., Qiu, J.: Ultralytics yolov8 (2023). https://github.com/ultralytics/ultralytics. Accessed 05 Mar 2024
10. Liao, M., Shi, B., Bai, X., Wang, X., Liu, W.: Textboxes: a fast text detector with a single deep neural network. In: Proceedings of the AAAI Conference on AI, vol. 31 (2017). https://doi.org/10.1609/aaai.v31i1.11196
11. Liu, H., Dai, Z., So, D., Le, Q.V.: Pay attention to MLPs. In: Advanced in NIPS, vol. 34, pp. 9204–9215 (2021)
12. Liu, W., et al.: SSD: single shot multibox detector. In: Leibe, B., Matas, J., Sebe, N., Welling, M. (eds.) ECCV 2016. LNCS, vol. 9905, pp. 21–37. Springer, Cham (2016). https://doi.org/10.1007/978-3-319-46448-0_2
13. Liu, W., Chen, C., Wong, K.Y.: Char-net: a character-aware neural network for distorted scene text recognition. Proceedings of the AAAI Conference on AI, vol. 32, no. 1 (2018). https://doi.org/10.1609/aaai.v32i1.12246
14. Liu, Z., Li, Y., Ren, F., Goh, W.L., Yu, H.: Squeezedtext: a real-time scene text recognition by binary convolutional encoder-decoder network. In: Proceedings of the AAAI Conference on AI, vol. 32, no. 1 (2018). https://doi.org/10.1609/aaai.v32i1.12252
15. Paliwal, S.S., et al.: Tablenet: deep learning model for end-to-end table detection and tabular data extraction from scanned document images. In: ICDAR, pp. 128–133 (2019). https://doi.org/10.1109/ICDAR.2019.00029
16. Phan, T.Q., Shivakumara, P., Tian, S., Tan, C.L.: Recognizing text with perspective distortion in natural scenes. In: ICCV (2013). https://doi.org/10.1109/ICCV.2013.76
17. Redmon, J., Divvala, S., Girshick, R., Farhadi, A.: You only look once: unified, real-time object detection. In: Proceedings of the IEEE Conference on CVPR (2016)
18. Redmon, J., Farhadi, A.: Yolov3: an incremental improvement. arXiv preprint arXiv:1804.02767 (2018)
19. Shi, B., Bai, X., Yao, C.: An end-to-end trainable neural network for image-based sequence recognition and its application to scene text recognition. IEEE Trans. PAMI **39**(11), 2298–2304 (2017). https://doi.org/10.1109/TPAMI.2016.2646371
20. Tabassum, A., Dhondse, S.A.: Text detection using MSER and stroke width transform. In: Fifth International Conference on CSNT, pp. 568–571 (2015)
21. Uchida, S.: Text Localization and Recognition in Images and Video, pp. 843–883. Springer London (2014). https://doi.org/10.1007/978-0-85729-859-1_28
22. Vaswani, A., et al.: Attention is all you need. In: Advanced in NIPS, vol. 30. Curran Associates, Inc. (2017)
23. Williams, R.J., Zipser, D.: Gradient-based learning algorithms for recurrent networks and their computational complexity. In: Backpropagation, pp. 433–486. Psychology Press, London (2013)

24. Ye, Q., Doermann, D.: Text detection and recognition in imagery: a survey. IEEE Trans. PAMI **37**(7), 1480–1500 (2015). https://doi.org/10.1109/TPAMI.2014.2366765
25. Zhan, F., Lu, S.: Esir: end-to-end scene text recognition via iterative image rectification. In: Proceedngs of the IEEE/CVF Conference on CVPR, pp. 2059–2068 (2019)

Exploring the Extremes of Two-Layer Convolutional Autoencoders Architecture in Image Inpainting

Youcef Azri(✉), Abdenour Amamra, Yacine Amara, and Fethi Ourghi

Ecole Militaire Polytechnique, Bordj El-Bahri BP 17, Algiers, Algeria
azriyoucefuu@gmail.com

Abstract. Inpainting, the process of filling in missing or damaged regions of an image, has witnessed significant progress with the advent of deep learning, particularly neural networks. This investigation delves into the outermost capabilities of two-layer Convolutional Autoencoders (2L-CAEs) architectures applied to image inpainting. The study scrutinizes the boundaries of performance and feasibility within this specific neural network configuration. The overarching objective is to provide an intricate understanding of the potential and limitations inherent in pushing a 2L-CAE architecture to its extremes, particularly when tasked with the intricate process of mending missing or impaired regions within an image.

Keywords: Inpainting · Neural Networks (NNs) · Network Structure Design · Inpainting Performance · Convolutional Neural Networks (CNNs) · Convolutional Autoencoders (CAEs) · Computational Considerations

1 Introduction

Deep learning's explosive growth in diverse architectures [1] fuels its remarkable success across domains. This momentum is further fueled by breakthroughs in hardware advancements. Equipped with exceptional representation learning capabilities, neural networks dominate image inpainting, seamlessly reconstructing missing information and generating realistic textures.

Inpainting, the art of replenishing image segments, plays a crucial role in applications like restoration, object removal, and virtual reality. Yet, traditional methods struggle with real-world complexities [2]. This has spurred interest in neural network-powered inpainting [3], achieving impressive results.

However, current deep learning methods face a trade-off: extraordinary results versus substantial computational resources. These methods typically rely on multi-layered encoder-decoder architectures. The damaged image is first compressed into a latent space code through a convolutional neural network (CNN) encoder. This code is then reconstructed back into the image using a decoder,

B. Djamaa et al. (Eds.): CSA 2024, LNNS 1145, pp. 104–113, 2025.
https://doi.org/10.1007/978-3-031-71848-9_10

often incorporating additional techniques such as the attention mechanism and transformers. Their reliance on multi-layered architectures demands significant power and memory, becoming a hurdle for real-time scenarios, resource-constrained environments like the Internet of Things (IoT), and scenarios with stringent storage limitations.

This necessitates seeking innovative, lightweight solutions. This paper presents the two-layer convolutional autoencoder (2L-CAE), challenging the paradigm with its minimal two-layer design. We investigate the impact of varying filter counts in the encoder on inpainting results using the ImageNet-1k dataset and metrics like SSIM and PSNR. Additionally, we analyze the model's performance across different missing patch sizes and mask ratios, revealing its range of capabilities.

The remainder of this paper is structured as follows. Section 2 provides a comprehensive review of existing image inpainting techniques. In Sect. 3, we delve into the details of our proposed 2L-CAE model, explaining its architecture, components, and design rationale. Section 4 describes the experimental setup that paves the way for showcasing the 2L-CAE inpainting performance. Finally, Sect. 5 offers a concise summary of our findings and concludes the paper.

2 Background and Related Work

Deploying powerful DNNs on resource-constrained devices like mobile phones or embedded systems remains a significant challenge [4]. This section explores key approaches for building efficient DNNs.

2.1 Network Design Strategies

A collection of techniques that optimize model architecture and parameters for efficiency. These strategies fall under two main categories: kernel factorization and model compression.

Kernel Factorization: Restructures standard convolutions for reduced computational demands while preserving accuracy.

- **Depthwise Separable Convolutions:** This popular approach splits standard convolutions into two steps - depthwise and pointwise - significantly reducing parameters and computations. This has become a cornerstone of efficient architectures like MobileNet-V2 [5].
- **Group Convolutions:** This technique introduces sparsity by dividing channels into groups and performing convolutions within each group [6], alleviating the computational burden of standard convolutions.

Model Compression: Unlocks efficiency by sculpting existing giants.

- **Pruning and Fine-Tuning:** Identifying and removing redundant connections while fine-tuning the remaining network can significantly shrink model

size and improve inference speed. Han et al. [7] has demonstrated impressive results, achieving a 9x reduction in AlexNet's [8] parameters (from 61 million to 6.7 million), all while maintaining accuracy.
- **Quantization:** Dramatically reduces memory footprint and computational demands by converting weights and activations to lower precision formats (e.g., integers or low-bit floats). Jiaxiang et al. [9] demonstrated a 4x–6x speedup in computation and 15x-20x compression on the ILSVRC-12 benchmark using fixed-point quantization, with only a minor 1% accuracy loss.
- **Knowledge Distillation:** Transferring knowledge from a larger, pre-trained model to a smaller one enables comparable performance with fewer resources [10].

2.2 Task-Specific Architectures

While general-purpose designs have proven powerful, maximizing efficiency for specific applications requires a tailored approach.

Tailoring for Low-Level Vision: This realm showcases the transformative potential of task-specific architectures, as evidenced by the Single Image Dehazing [11].

Leveraging Specialized Operations: The advancement of task-specific architectures is further fueled by innovative operations, as evidenced by the Mixed-Convolution Layers [12].

3 Our Approach

Conventional methods face a limitation - complex DNNs excel in quality but lack efficiency, while task-specific solutions sacrifice adaptability. Our novel two-layer convolutional autoencoder (2L-CAE) disrupts this paradigm. At its core, the 2L-CAE leverages autoencoders with just two convolutional layers. This minimalist design compresses images into a compact latent space, capturing key features and discarding redundancies. The decoder efficiently reconstructs the original image, even in missing regions. Figure 1 illustrates the architecture.

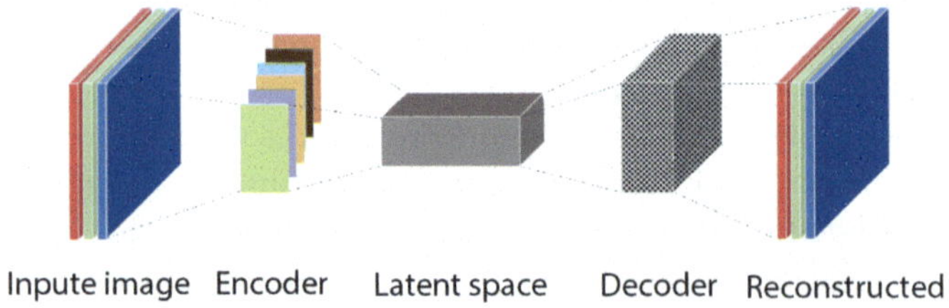

Fig. 1. 2L-CAE architecture

Network Architecture. A crucial factor in the 2L-CAE is the number of filters in the encoding layer. These determine the latent space's detail capture and efficiency. We rigorously investigated filter configurations (3 to 2700) through extensive experimentation, testing various patch sizes and missing area proportions (10% to 70%). This comprehensive approach helped us identify the optimal filter count that balances high-quality reconstruction with computational efficiency.

Patch Sizes and Missing Area Proportions. To assess the 2L-CAE flexibility, we devised intricate experiments encompassing varying patch sizes and missing area proportions. Images were divided into non-overlapping patches for controlled evaluation. We then employed random sampling to mask regions, ensuring a uniform distribution of missing areas. To evaluate the model's ability to handle diverse image complexities, we tested missing area sizes from 10% to 70% and patch sizes from single pixels to 21×21 pixels. This comprehensive approach allowed us to rigorously assess the 2L-CAE's adaptability and identify its optimal operating parameters.

Loss Function. Unlike traditional pixel-wise loss functions, such as Mean Squared Error (MSE) or Peak Signal-to-Noise Ratio (PSNR), we opted for the Structural Similarity Index Measure (SSIM) loss function. While PSNR focuses solely on minimizing the pixel-wise difference between the reconstructed and original images, it has been criticized for not aligning well with human perception of image quality.

Specifically, PSNR prioritizes minimizing absolute pixel value differences, which can lead to visually unappealing inpainting results despite achieving high PSNR values. In contrast, SSIM incorporates measures of luminance, contrast, and structure, better mimicking how humans perceive image similarity.

Furthermore, we selectively apply SSIM to unmasked regions, guiding the 2L-CAE to accurately regenerate missing areas based on surrounding information without being overly influenced by existing pixels. This approach helps ensure visually coherent and perceptually-relevant inpainting results.

4 Performance Evaluation

Our extensive experiments, combining various filter counts, patch sizes, and missing area proportions, yielded a rich dataset for identifying the best 2L-CAE design for different inpainting challenges. The following sections will delve into the quantitative and qualitative results, revealing how these choices influence inpainting quality and computational efficiency.

4.1 Dataset: ImageNet-1K

The 2L-CAE thrives on the extensive visual lexicon offered by the ImageNet-1K [13] dataset, boasting over 1.2 million images across 1,000 categories. This variety fosters robust and generalizable representations within the model. Images are standardized to 210×210 pixels for consistent processing, utilizing the dataset's pre-existing divisions for training (1.28 million), validation (50,000), and unseen testing (100,000).

4.2 Experimental Setup

Hardware Platform. A meticulously chosen hardware setup orchestrated the 2L-CAE learning journey. An Intel Xeon W-2235 CPU coupled with 64 GB of DDR4 RAM provided the processing power, while an NVIDIA Ampere Tesla A40 GPU accelerated training with lightning speed. Fast SSD storage minimized bottlenecks and a stable Linux environment based on CentOS-8 ensured smooth operation.

Metrics: Quantifying Success. We employed a comprehensive evaluation framework focused on image quality. We used Peak Signal-to-Noise Ratio (PSNR) for pixel-level accuracy and Structural Similarity Index Measure (SSIM) for structural fidelity. This dual approach ensures both fine detail preservation and overall visual coherence.

4.3 Results and Discussion

To illuminate the intricate relationships between filter count, patch size, missing area proportion, and inpainting quality, we utilize scatter plots (Fig. 2). Each mask ratio (10%-70%) has two plots: one for SSIM and one for PSNR. Each dot represents a unique combination of mask ratio, filter count, and patch size, with color indicating patch size and **vertical** position indicating filter count. This visualization allows us to identify trends and optimal configurations for different inpainting scenarios.

Deciphering SSIM Trends. Our analysis of SSIM values paints a fascinating picture of the 2L-CAE capabilities and limitations. As expected, smaller missing regions (lower mask ratios) generally lead to higher SSIM scores, demonstrating the model's effectiveness in utilizing surrounding information for reconstruction.

Interestingly, SSIM often rises with increasing filter count, indicating that additional filters enhance the model's ability to extract and encode meaningful features. Notably, for mask ratios below 50%, a "sweet spot" filter count exists that achieves SSIM exceeding 59, showcasing the model's potential for high-quality inpainting. However, this threshold drops for larger mask ratios, suggesting a performance ceiling beyond 50% missing area, where even with increased filter complexity, the model struggles to recover lost information effectively.

Contrary to expectations, SSIM exhibits a positive correlation with patch size for moderate missing areas (10% to 50%). This seemingly counter-intuitive trend suggests that larger patches reduce redundancy in available information, forcing the model to delve beyond simple pixel-level extrapolation and learn deeper image characteristics. However, for mask ratios exceeding 50%, the sheer

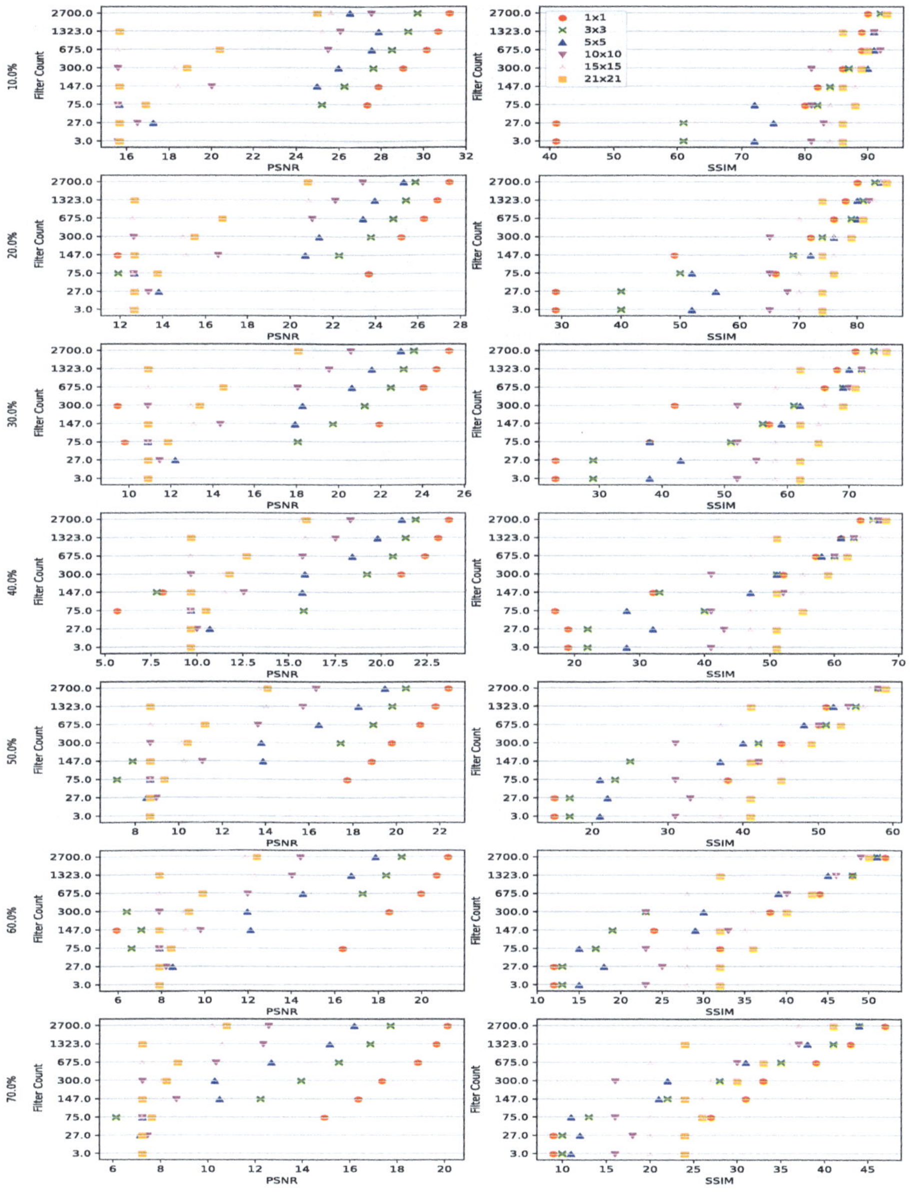

Fig. 2. A comparative study on ImageNet with varying patch sizes, mask ratios, and filter counts (color-coded by patch size)

amount of missing information overwhelms the model, leading to a decline in SSIM even with larger patches.

These SSIM trends offer valuable insights into the 2L-CAE's strengths and weaknesses, highlighting its ability to leverage surrounding information, extract

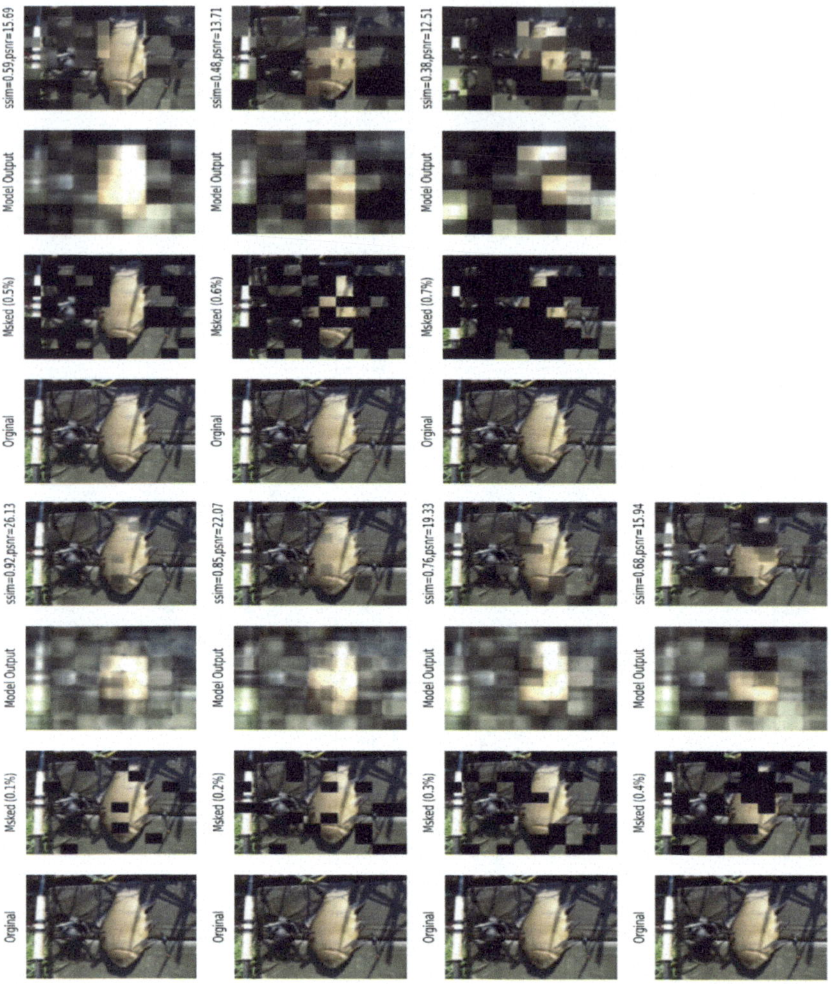

Fig. 3. Inpainting: 2700 filters, 21 × 21 patch. Shows reconstruction at various missing ratios (10%-70%). (1) Original, (2) Masked, (3) Prediction, (4) Restored (Prediction + Unmasked regions)

meaningful features, and learn beyond low-level statistics for successful inpainting. However, they also reveal limitations at high missing area ratios, suggesting the need for further exploration and optimization strategies for more challenging scenarios.

Deconstructing PSNR. Analyzing PSNR paints a picture similar to SSIM, with smaller missing areas (lower mask ratios) leading to higher pixel accuracy (PSNR) thanks to abundant surrounding information. Additionally, increasing filter count often boosts PSNR, implying improved feature extraction for faithful pixel replication.

Table 1. Peak SSIM/PSNR for filter counts, mask ratios, and patch sizes (rows represent filter count, columns represent mask ratio, values indicate highest achieved SSIM and PSNR for each pairing with optimal patch size)

	10%		20%		30%		40%		50%		60%		70%	
	SSIM	PSNR	SSIM	PSNR	SSIM	PSNR	SSIM	PSNR	SSIM	PSNR	SSIM	PSNR	SSIM	PSNR
3	86	15.68	74	12.68	62.0	10.91	51.0	9.66	41.0	8.69	32.0	7.9	24.0	7.23
27	86.0	17.27	74.0	13.82	62.0	12.21	51.0	10.68	41.0	8.97	32.0	8.51	24.0	7.45
75	88.0	27.34	76.0	23.67	65.0	18.05	55.0	15.8	45.0	17.73	36.0	16.35	27.0	14.94
147	88.0	27.87	76.0	22.28	65.0	21.92	55.0	15.72	45.0	18.85	35.0	12.11	31.0	16.35
300	90.0	29.04	79.0	25.22	69.0	21.22	59.0	21.08	49.0	19.78	40.0	18.5	33.0	17.36
675	92.0	30.15	81.0	26.28	71.0	24.04	62.0	22.37	53.0	21.09	44.0	19.97	39.0	18.87
1323	92.0	30.69	83.0	26.91	74.0	24.66	64.0	23.08	56.0	21.79	48.0	20.69	43.0	19.66
2700	93.0	31.22	85.0	27.47	76.0	25.26	68	23.66	59.0	22.38	52.0	21.21	47.0	20.13

However, unlike SSIM, PSNR paints a stark picture of the increasing challenges posed by larger missing regions. While at 10% and 20% mask ratios, a suitable filter count can push PSNR above 20 dB (considered acceptable) for all patch sizes, this window of opportunity shrinks rapidly. At 30% mask ratio, only smaller patches (1×1 to 10×10) benefit from high PSNR configurations, and this pool further dwindles as mask ratios escalate. By 50%, only the 1×1 patch size can barely reach a PSNR above 20 dB, highlighting the model's struggle with extensive pixel-level reconstruction at higher mask ratios.

Beyond Pixels: A Divergence in Perspectives - PSNR vs. SSIM. While both PSNR and SSIM assess inpainting quality, they do so through distinct lenses. PSNR, a pixel-focused metric, rewards precise pixel matching, but struggles with large missing areas where pixel discrepancies can overwhelm the overall reconstruction. SSIM, on the other hand, takes a broader view, evaluating how well the reconstructed image captures the original's structure and texture, making it more tolerant of larger gaps.

This divergence highlights the importance of considering both metrics when evaluating the 2L-CAE. PSNR reveals the model's pixel-level precision, while SSIM shines a light on its ability to maintain visual coherence and integrity. Together, they paint a nuanced picture of the 2L-CAE performance, allowing us to identify areas for potential improvement.

Unveiling the Best SSIM and PSNR Values. Table 1 lays bare the highest SSIM and PSNR values attainable, meticulously selecting and showcasing the most successful patch size for each filter count and mask ratio pairing. Its rows ascend from baseline to maximum filter counts, while columns mirror mask ratios from 10% to 70%. Each intersection reveals the peak performance attainable for a given filter count and mask ratio under the most advantageous patch size, painting a vivid portrait of potential and limitations. To complement this numerical data, Fig. 3 showcases the inpainting prowess of the 2700 filter count/21×21 patch size combo (highest performance!), with varying mask ratios (10%–70%).

Diminishing Returns Guide Filter Count Choice. Our search for the optimal number of filters ended at 2700, a choice carefully balanced between performance and efficiency. While increasing filters initially boosted SSIM scores, the gains dwindled and plateaued beyond 2700. This phenomenon held true across mask ratios and patch sizes, implying that further increasing filters wouldn't significantly improve image quality, but would significantly inflate training time and resource demands.

5 Conclusion and Future Research Directions

Our exploration into the intricacies of 2L-CAE's inpainting prowess reveals its impressive capability to recover missing information, particularly for smaller regions. The interplay between filter count, patch size, and mask ratio paints a dynamic canvas of potential and limitations. We observed how increasing filter count unlocks richer feature extraction, and how larger patch sizes often enhance performance for moderate missing areas. However, the model faces a formidable challenge beyond 50% mask ratios, where pixel-level accuracy struggles to contend with the vastness of missing information.

These revelations pave the way for exciting future research directions. One key aspect to explore is whether the observed performance can be generalized to networks with more layers. While the 2L-CAE has demonstrated promising results, investigating deeper architectures could potentially unlock further capabilities, particularly for handling larger missing areas. Furthermore, exploring alternative network architectures, loss functions, or pre-processing strategies specifically geared towards extensive missing regions remains crucial. This quest for improved inpainting robustness could unlock a wider range of application domains where significant reconstruction tasks are necessary.

Furthermore, the rich feature maps learned by 2L-CAE's encoder offer a tantalizing avenue for exploration beyond inpainting. Leveraging these outputs as inputs for another model trained for object recognition or image classification could be a fascinating next step. This potential synergy between reconstruction and high-level understanding could open doors to innovative applications in the realms of medical imaging, autonomous vehicles, and even creative image manipulation.

In conclusion, while 2L-CAE shines a light on the remarkable potential of deep learning for image inpainting, it also illuminates the exciting challenges that lie ahead. By relentlessly pursuing advancements in model architectures, learning strategies, and cross-modal applications, we can unlock the full potential of image completion and revolutionize our ability to reconstruct visual information in countless domains.

References

1. Krizhevsky, A., Sutskever, I., Hinton, G.E.: Imagenet classification with deep convolutional neural networks. In: Pereira, F., Burges, C., Bottou, L., Weinberger, K. (eds.) Advances in Neural Information Processing Systems, vol. 25. Curran Associates, Inc. (2012). https://proceedings.neurips.cc/paper_files/paper/2012/file/c399862d3b9d6b76c8436e924a68c45b-Paper.pdf
2. Bertalmio, M., Sapiro, G., Caselles, V., Ballester, C.: Image inpainting. In: Proceedings of the 27th Annual Conference on Computer Graphics and Interactive Techniques. SIGGRAPH '00, pp. 417–424. ACM Press/Addison-Wesley Publishing Co., USA (2000)
3. Xu, Z., et al.: A review of image inpainting methods based on deep learning. Appl. Sci. **13**(20) (2023)
4. He, K., Sun, J.: Convolutional neural networks at constrained time cost. In: Proceedings of the IEEE Conference on Computer Vision and Pattern Recognition (CVPR) (2015)
5. Howard, A., Zhmoginov, A., Chen, L.C., Sandler, M., Zhu, M.: Inverted residuals and linear bottlenecks: mobile networks for classification, detection and segmentation. In: CVPR (2018)
6. Zhang, X., Zhou, X., Lin, M., Sun, J.: Shufflenet: an extremely efficient convolutional neural network for mobile devices. In: Proceedings of the IEEE Conference on Computer Vision and Pattern Recognition (CVPR) (2018)
7. Han, S., Pool, J., Tran, J., Dally, W.: Learning both weights and connections for efficient neural network. In: Cortes, C., Lawrence, N., Lee, D., Sugiyama, M., Garnett, R. (eds.) Advances in Neural Information Processing Systems, vol. 28. Curran Associates, Inc. (2015)
8. Krizhevsky, A., Sutskever, I., Hinton, G.E.: ImageNet classification with deep convolutional neural networks. In: Advances in Neural Information Processing Systems, vol. 25. Curran Associates, Inc. (2012)
9. Wu, J., Leng, C., Wang, Y., Hu, Q., Cheng, J.: Quantized convolutional neural networks for mobile devices. In: Proceedings of the IEEE Conference on Computer Vision and Pattern Recognition (CVPR) (2016)
10. Hinton, G., Vinyals, O., Dean, J.: Distilling the knowledge in a neural network (2015)
11. Zhang, J., Tao, D.: Famed-net: a fast and accurate multi-scale end-to-end dehazing network. IEEE Trans. Image Process. **29**, 72–84 (2020)
12. Zhang, Z., Wang, X., Jung, C.: DCSR: dilated convolutions for single image super-resolution. IEEE Trans. Image Process. **28**(4), 1625–1635 (2019)
13. Russakovsky, O., et al.: ImageNet large scale visual recognition challenge. Int. J. Comput. Vision **115**(3), 211–252 (2015). https://doi.org/10.1007/s11263-015-0816-y

Enhancing Diabetic Retinopathy Identification Through Novel Image Representation and Hybrid CNN-Transformer Segmentation

Mahdi Hadef(✉), Lotfi Gana, Said Yacine Boulahia, and Abdenour Amamra

Artificial Intelligence and Virtual Reality Lab, Ecole Militaire Polytechnique, Algiers, Algeria
d_hadef.mahdi@emp.mdn.dz, hadef.mahdi@gmail.com

Abstract. Diabetic retinopathy, an insidious complication arising from diabetes, emerges as a predominant catalyst for vision impairment. Early identification of this condition remains a pressing need to prevent the risk of blindness. Many existing approaches for diabetic retinopathy identification focus on the processing of retinal images via standard preprocessing techniques before feeding them to classification models, without exploring alternative representations. Additionally, recent segmentation techniques, which may particularly contribute to diabetic retinopathy detection or progression, face challenges related to high computational and time consumption. To address diabetic retinopathy diagnosis from performance improvement and computational time reduction perspectives, we have focused on two primary contributions. The first involves diabetic retinopathy identification utilizing a novel fundus retinal image transformation approach. This novel representation aims to extract meaningful retinal information while discarding insignificant borders, effectively reducing image size. The second contribution entails retinal part segmentation using a hybrid approach that leverages the strengths of both convolutional neural networks and Vision Transformers in terms of rapidity and accuracy, respectively. The results obtained by the diabetic retinopathy identification using the novel representation approach on both datasets, namely EyePACS [5] and APTOS [6], as well as those obtained through the proposed segmentation approach on the IDRiD Dataset [9], are promising.

Keywords: Computer vision · Diabetic Retinopathy classification · Fundus image segmentation · CNN-Transformer · pixel-by-pixel traversal approach · EyePACS · APTOS · IDRiD

1 Introduction

Computer vision, an interdisciplinary field encompassing image processing, pattern recognition, and segmentation, holds a crucial role across diverse domains,

B. Djamaa et al. (Eds.): CSA 2024, LNNS 1145, pp. 114–125, 2025.
https://doi.org/10.1007/978-3-031-71848-9_11

notably in medical imaging and diagnosis. Numerous imaging methods are employed for this purpose, including both 2D and 3D techniques [3]. Among the multitude of applications, the analysis of retinal images holds particular significance in the diagnosis and monitoring of ocular conditions, notably diabetic retinopathy [15].

The early diagnosis of diabetic retinopathy through deep learning methods is a testament to the profound impact of technology on healthcare [1]. It not only enhances diagnostic accuracy and assesses the risk of diabetic retinopathy (DR) progression but also plays a critical role in preventing vision loss, thus significantly improving the quality of life for individuals with diabetes.

Despite advances in imaging technology, retinal image analysis remains a critical component in diabetic retinopathy diagnostics, facing numerous challenges. Chief among these challenges is the extraction of pertinent information from retinal images, with a prevalent difficulty arising from the presence of irrelevant extraneous elements bordering the retinal region in imaging techniques. These parts can interfere with essential details and hinder the accuracy of diagnostic interpretation. Current literature methods often mishandle unnecessary image content, leading to potential misinterpretation or overlooking critical diagnostic information. This underscores a clear gap in retinal image analysis techniques, a gap that demands innovative solutions to enhance diagnostic accuracy and efficiency.

From the other side, diabetic retinopathy can induce changes in the retina, including the formation of lesions such as microaneurysms and exudates, as well as alterations in critical regions like the optical disc and blood vessels. Segmentation of these areas allows clinicians and automated systems to detect subtle changes or abnormalities more effectively, serving as indicators of diabetic retinopathy progression. Effective segmentation of these parts is crucial for the accurate diagnosis of diabetic retinopathy. Most current segmentation approaches utilize deep learning methods, and more recently, Vision Transformers, which are derived from attention mechanisms. However, despite their effectiveness, these emerging segmentation techniques often face challenges, including computational inefficiency and significant time consumption.

To address diabetic retinopathy diagnosis from performance and computational time perspectives, we have proposed two main contributions. The first introduces an innovative approach designed to concentrate on relevant areas within retinal images by excluding non-essential regions. Unlike traditional methods, this approach provides a more targeted analysis, overcoming one of the fundamental hurdles in retinal image processing. It also transforms the way retinal images are processed, reducing the image size by selectively retaining pixels that represent retinal information while discarding extraneous black borders. This technique allows for a concentrated focus on the retina, refining the analysis process for enhanced accuracy and computational efficiency.

The second contribution introduces a novel hybrid segmentation approach, aiming mainly to significantly reduce computational time by leveraging the speed of convolutional neural networks and the enhanced precision of Vision Transformers.

The subsequent sections of the paper are organized as follows: Sect. 2 provides an overview of recent literature on diabetic retinopathy classification and segmentation methods. In the first part of Sect. 3, we introduce our proposed diabetic retinopathy classification approach. In the second part of the same section, we present our novel hybrid segmentation approach. Section 4 details the experimental results and discusses the obtained outcomes. Finally, Sect. 5 provides conclusions and outlines future works.

2 Literature Review

In recent years, deep learning based techniques have emerged as a promising approach for automated diabetic retinopathy detection in retinal fundus images. In this literature review, our aim is to provide an overview of existing research in this field, emphasizing the image treatment approaches adopted and the challenges encountered.

Samanta et al. [12] proposed an approach to identify four diabetic retinopathy stages using pretrained convolutional neural network architecture on limited data. The images, predominantly reddish with a black background, underwent various image processing steps aimed at enhancing contrast while maintaining brightness and reducing noise. This was followed by employing a pre-trained model to classify each image into the corresponding stage. While the image processing approach used in this study seems beneficial for highlighting certain features, they might inadvertently suppress or alter other subtle yet critical details in the images, which is a significant concern in medical imaging.

The need for more meaningful image information to enhance image exploitation led to the study by Oh et al. [8], which introduces a diabetic retinopathy detection system utilizing ultra-wide-field (UWF) fundus photography rather than fundus photography. This study employed an automatic segmentation approach to selectively focus on areas likely to be relevant for diabetic retinopathy detection. While these techniques may appear advantageous in concentrating on key retinal areas, they may overlook crucial peripheral regions not encompassed by their approach.

In a recent study, Tang et al. [14] proposed a convolutional-based approach while applying transfer learning method for detecting neovascularization, a crucial indicator of proliferative diabetic retinopathy, in retinal fundus images. The authors utilized a combination of pre-trained convolutional neural network models on retinal fundus image patches instead of the entire images. These images underwent green channel extraction and (CLAHE) pre-processing. The proposed image transformation yielded good results; however, analyzing patches, instead of the entire retinal fundus image might overlook global features or contextual information essential for accurate diagnosis.

The presented works demonstrate a collaborative effort to improve diagnostic accuracy. However, it encounters a range of challenges, chief among them is that they predominantly concentrates on conventional preprocessing and classification models, often overlooking alternative representations of retinal images. In the following section, we will provide an overview of existing researches in diabetic retinopathy segmentation field, emphasizing the approaches adopted and the challenges encountered.

Before the widespread adoption of neural network methods, researchers explored different classical approaches in the segmentation of diabetic retinopathy. Soumil [4] proposed a method for detecting diabetic retinopathy by focusing on the segmentation of exudates in retinal fundus images. The main idea is to exploit the uniform characteristics of healthy regions in the retina, rather than focusing on the diseased areas, for segmentation. They used a filter to enhance the edges in the image, making high-intensity areas, such as the optical disc, more discernible. The detection utilizes the distinct properties of the optical disc, such as its high intensity and circular geometric structure. While they achieved relatively good segmentation results, they suffer from increasing segmentation algorithmic complexity and computational power requirements.

The continuous improvement in this field has led to the exploitation of convolutional neural networks for the segmentation task. Sun et al. [13] introduced a novel approach to optic disc segmentation in retinal fundus images using deep object detection networks. This study redefined the segmentation problem as an object detection task, employing Faster R-CNN as the primary detection mechanism. Their methodology involves initially detecting the optic disc as a bounding box, which is then transformed into a vertical, non-rotated ellipse to approximate the optic disc boundary. However, the reliance on bounding box detection may oversimplify the optic disc's shape, potentially missing nuanced irregularities crucial in disease diagnosis leading to a lower accuracy.

With the emergence of Transformers as a viable alternative to convolutional neural networks in various segmentation tasks, Wang et al. [16] introduced an approach that integrates hyperbolic embeddings with a Vision Transformer (ViT) architecture. This model incorporates a spatial prior module to improve local continuity in the images and leverages hyperbolic embeddings for pixel-level classification, addressing the inherent complexity of diabetic retinopathy lesions. However, the integration of hyperbolic space and ViT-Adapters results in a complex model, potentially presenting challenges in terms of computational efficiency and practical deployment.

The presented segmentation approaches underscore the considerable potential for advancing diabetic retinopathy classification and segmentation. Current approaches often center on conventional preprocessing techniques for retinal image processing before input into classification models, overlooking exploration of alternative representations. Furthermore, recent segmentation techniques crucial for diabetic retinopathy detection encounter challenges associated with elevated computational demands and time consumption. Our work aims to address these challenges by proposing novel approaches that tackle these traditional limitations.

As we transition into the subsequent approach section, our focus will delve into a more detailed exploration of the choices made in our image transformation approach along with the hybrid segmentation approach.

3 Proposed Approaches

In this section, after providing an overview of the related works and the challenges faced, we will present our proposed approaches to tackle these challenges, especially those related to the lack of exploration of novel representations of the retinal eye that discard irrelevant parts of raw retinal images in the field of diabetic retinopathy classification, and those related to the high computational and inference time consumption in the field of image segmentation, particularly in the context of recent Transformer-based methodologies.

In order to address the first challenge, related to the novel representations of the retinal images for diabetic retinopathy classification, we propose an approach which focuses on transforming the retinal image into another form while maintaining only the meaningful parts of the images. The resultant image issued from this transformation is relatively smaller size, ensuring that the produced images lack irrelevant pixels. The process of transformation involves removing border regions while retaining only the informative pixels, essentially capturing the retinal disc in a new image representation. Figure 1 illustrates our novel representation approach.

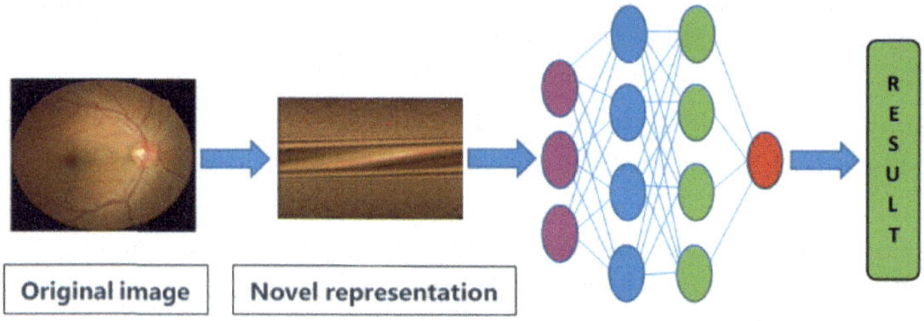

Fig. 1. Novel representation approach.

As illustrated in Fig. 1, our approach involves applying image transformation method to the original input image, using a pixel-by-pixel traversal method. This method is aimed at refining the visual attributes of the image, thereby enhancing its key features. In order to define the dimensions of the resultant transformed image, we proceed to a statistical study of data images sizes, which serves at adopting an adequate reduced size of the image without a significant loss of information, retaining only the maximum of pixels representing the retina. The transformation process is specifically designed to preserve the most relevant information for diagnosis while eliminating noise and irrelevant regions from the data. This enables the model to concentrate on the critical features without succumbing to overfitting, potentially enhancing its performance. Additionally, by transforming the images into a lower dimension, as presented in Fig. 1, the computational load on the models is effectively reduced. This can result in a faster and more memory-efficient training process.

Following the transformation phase, the image undergoes further refinement through appropriate processing techniques. Additionally, augmentation operations are implemented to address data imbalance and scarcity issues. Subsequently, these newly processed images are fed into classification models designed to identify various stages of diabetic retinopathy. The models categorize each image into a specific stage, thus reflecting the severity and distinctive characteristics of the condition as depicted in the image.

In addressing the second challenge of computational efficiency and inference time in using Vision Transformers for segmentation tasks, we introduce a pioneering image segmentation approach. This approach operates directly on original retinal images without relying on the previously introduced novel representation. This latter is specifically designed to overcome the limitations commonly encountered in existing Transformer-based techniques.

Recent advancements in the field have marked the effectiveness of Transformer-based approaches in segmentation tasks, particularly their proficiency in segmenting images with high accuracy. Despite their success in achieving precise segmentation, there remains a critical gap in optimizing these methods for faster computational processing and reduced inference time. Our proposed approach aims to bridge this gap, enhancing not just the efficiency but also maintaining the accuracy of image segmentation processes.

Our strategy involves synergizing the rapid processing capabilities of convolutional neural networks with the precision of Vision Transformers, aiming to achieve rapid and efficient image segmentation by leveraging the strengths of both models. Figure 2 illustrates our hybrid segmentation approach.

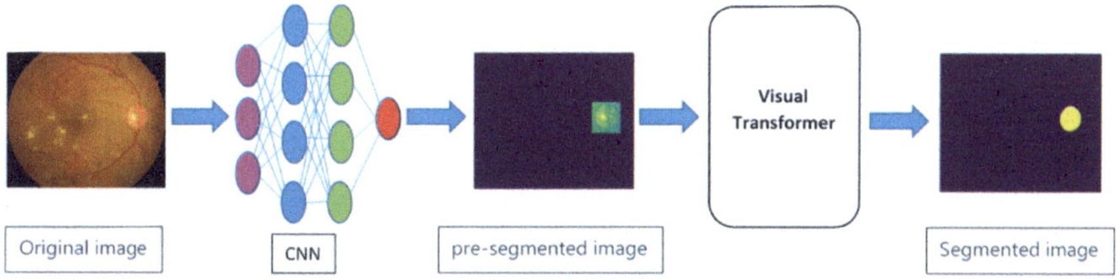

Fig. 2. CNN-Vision Transformer approach.

This approach involves the utilization of two distinct models. Initially, we employ a fast convolutional neural network to deliver a preliminary segmentation of the image. The convolutional neural network rapidly processes the image, dividing it into various parts, highlighting the main sections, and rendering the remainder in black. This step establishes a basic understanding of the targeted areas, laying the groundwork for further segmentation refinement. After that, we proceed to enlarge the areas of interest to ensure better coverage of the targeted regions. This marks the transition phase between the application of the convolutional neural network model and the Vision Transformer. Once this operation is achieved, the outcome is a pre-segmented image that is ready to be fed to the next phase for segmentation refinement, as shown in the flow process in Fig. 2.

Subsequently, we apply a Vision Transformer, renowned for its precision, even though it comes with a longer processing time when compared to convolutional neural networks. The Transformer is specifically tasked with processing the non-black portions of the image, which are the areas where detailed and accurate segmentation is essential. This targeted application ensures that the Transformer's capabilities are optimally utilized where they are most needed, rather than processing the entire image. This operation outcome is the final refined segmented image.

As we progress into the next section, dedicated to discussing our experimental results, we will step into detailing the implementation of our approaches, along with discussing and analyzing their outcomes.

4 Experimental Results and Discussion

In this section, we will present and discuss the results obtained using our approaches. For this purpose, we will start by introducing our selected datasets and the supporting models utilized. Afterward, we will outline the evaluation protocols to which we adhered. Following this, we will provide comprehensive details concerning the image processing methodology employed in our classification approach. Finally, we will elaborate on the methodology employed in conducting our classification and segmentation experiments and discuss the obtained results.

4.1 Datasets and Evaluation Protocols

Within the scope of our research, we have utilized two public, challenging and widely used datasets in this research area, namely: EyePACS [5] and APTOS [6]. The EyePACS [5] dataset is a comprehensive collection, assembled for the study of computer vision and machine learning in diabetic retinopathy field.

Both datasets encompass annotations for five distinct stages of diabetic retinopathy, specifically: no DR, mild DR, moderate DR, severe DR, and proliferative DR. This stratification facilitates the classification of each image into one of the aforementioned five gradations.

Regarding the segmentation task, we utilized IDRiD (Indian Diabetic Retinopathy Image Dataset) [9], which comprises 516 retinal images. Its primary segmentation focus encompasses lesions associated with diabetic retinopathy (microaneurysms, hemorrhages, soft and hard exudates), optic disc, and fovea, along with severity level assessments for both diabetic retinopathy and Diabetic Macular Edema (MDE).

For the evaluation phase of the diabetic retinopathy classification, we have employed three supporting models, namely VGG-16, InceptionV3, and DenseNet-121. While for the segmentation task, we have employed Segmenter and YOLACT segmentation models:

Regarding evaluation protocols and metrics, our primary objective is to ensure a thorough and impartial assessment of our models. To achieve this goal, we have adopted a distribution split of 80% for training and 20% for testing. Also, we employed the F-score metric due to its appropriateness in handling imbalanced classes [2]. Regarding segmentation, we employed two performance metrics to evaluate our model, namely Jaccard Index (Intersection over Union (IoU)) and Dice Coefficient (DC).

4.2 Classification Results

In this section, we present the diabetic retinopathy classification results obtained by employing pretrained models on the two previously mentioned datasets, incorporating the novel image representation generated through the pixel-by-pixel image traversal approach. Which is designed for a better exploitation of the image resolution in the context of diabetic retinopathy.

For the classification task, we employed three supporting models, namely VGG-16, InceptionV3, and DenseNet-121. These models were applied to both EyePACS [5] and APTOS [6] datasets. To evaluate the effectiveness of our approach, we applied them to both preprocessed and non-preprocessed images, choosing the F-score as the evaluation -metric.

Regarding the EyePACS [5] dataset, both VGG-16 and DenseNet-121 yielded the best F-score, reaching 55% with preprocessing. For each of the pretrained models, performance improvement is observed after applying image preprocessing. For the APTOS [6] dataset, InceptionV3 demonstrated improved performance with preprocessing, achieving a notable F-score of 66%, which surpassed those of VGG-16 and DenseNet-121. Additionally, the performance of each pretrained model was enhanced after undergoing image preprocessing.

These results underscore the performance of the supporting models in diabetic retinopathy classification on the newly constructed datasets using the novel image representation approach. They also highlight the influence of preprocessing on overall performance, benefiting specific architectures, as reflected in the improved F-scores.

The evaluation of our diabetic retinopathy classification results, along with those of the state-of-the-art works, is detailed in Table 1.

Table 1 presents a comparison of various approaches applied to the classification of diabetic retinopathy across two datasets: EyePACS [5] and APTOS [6]. Each entry in the table represents a reference work. The F-score metric is used to gauge the accuracy of the different classification systems. The approaches assessed range from established architectures like SEDensnet and MXception to ensemble methods and convolutional neural network-based strategies.

For the EyePACS [5] dataset, Qummar et al. [11] and Pratt et al. [10] achieved an F-scores of 53% using an ensemble approach and 41% using a convolutional neural network-based approach, respectively. The highest F-score achieved on this dataset is 55% using the novel "Pixel-by-pixel traversal" approach introduced in our work. The latter also performs well on the APTOS [6] dataset, obtaining an F-score of 66%, which is very competitive with other high-performing approaches

Table 1. Comparison with state-of-the-art works.

Ref	Approach	DR Datasets		F-score
		EyePACS[5]	APTOS[6]	
[11]	Ensemble Approach	✓	x	**0.53**
[10]	CNN-based	✓	x	0.41
[7]	SEDensnet classification	x	✓	0.64
[7]	MXception Multitask	x	✓	**0.66**
Our		✓	x	**0.55**
		x	✓	**0.66**

Note: Evaluation of the different works is mostly under the same conditions. Our study aims mainly to tackle diabetic retinopathy classification from different perspective views more than outperforming the best state-of-the-art performance.

such as SEDensnet [7] and MXception Multitask [7]. Achieving an F-scores of 64% and 66%, respectively, on the same APTOS [6] dataset.

The overall results emphasize the variability in performance across different models and datasets, suggesting that the choice of an appropriate model and image processing techniques are crucial for optimizing classification performance. Additionally, our image transformation approach demonstrated highly competitive results compared to other existing studies. Furthermore, our results were validated by applying our approach to two different challenging datasets, namely EyePACS [5] and APTOS [6], in contrast to other works that specialize in a single dataset. This distinction may suggest a tailored fit of the models to the specific characteristics of the data.

It is worth noting that the evaluation protocols employed in our research are aligned with those of comparable studies, ensuring that our findings are directly relevant to the existing body of work in the field. Additionally, our objective is to explore diabetic retinopathy classification from a novel perspective, focusing on methodological innovation rather than solely on surpassing existing state-of-the-art performance metrics.

4.3 Segmentation Results

In this section, we present and discuss the results of our proposed hybrid approach for image segmentation based on both convolutional neural network and Transformers, using three evaluation metrics, namely: Jaccard Index (IoU), Dice Coefficient (DC), and the total inference time. The experiments are conducted on the IDRiD Dataset [9].

Our hybrid image segmentation approach was achieved by combining two neural network architectures, namely YOLACT and Segmenter. YOLACT is a fast and less accurate convolutional neural network employed in the first phase. Its principal role is to highlight the regions of interest, which, in our case, is the optical disc, followed by the enlargement process of the area of interest before being fed to the Transformer Segmenter. Which is a Transformer-based neural network that is heavier but more precise. An example of the segmentation results of the proposed approach is illustrated in Fig. 3.

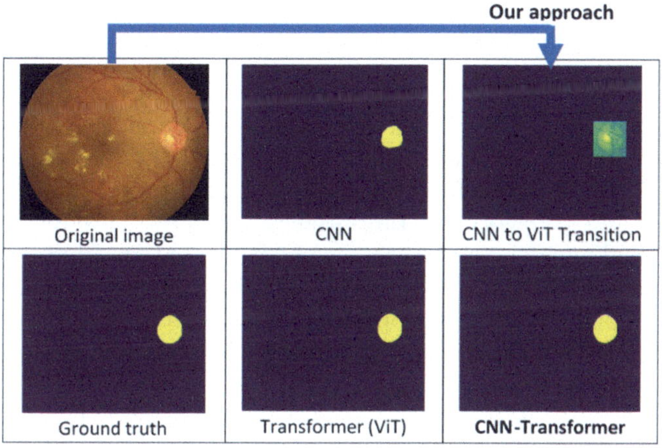

Fig. 3. Example of the output image after applying CNN-Transformer, compared to CNN-only and Transformer-only results.

Initially, the original retinal images undergo an initial segmentation using the YOLACT convolutional neural network. This process yields a pre-segmented image with a border containing some background. However, the precision of this segmentation is not optimal and necessitates further refinement for improved quality. To enhance segmentation accuracy, we extend the pre-segmented areas proportionally to their size, ensuring greater coverage of the objects of interest. Subsequently, we forward this pre-segmented image to the Vision Transformer Segmenter for additional refinement. Implementing this refinement method significantly enhances the segmentation quality, delivering more precise and reliable results. Notably, when the object of interest is well-covered in the pre-segmented regions, our approach achieves results equivalent to those obtained by directly applying the Vision Transformer. Importantly, this is achieved with considerably lower computational and time consumption.

The results obtained following our hybrid segmentation approach are illustrated in Table 2, along with a comprehensive comparison of segmentation results on the IDRiD Dataset [9]. This comparison evaluates segmentation performance based on various metrics and inference time across different hybrid configurations. It serves as a valuable insight into the trade-offs among diverse combinations of architectures, highlighting the balance between accuracy and speed.

Table 2 illustrates the segmentation results obtained for different hybrid configurations. For Transformer-only segmentation, it yielded outstanding results

Table 2. Presentation of the segmentation results in terms of IOU, DICE scores and inference time for different architectures

Metrics	CNN	Transformer	CNN-Transformer	Transformer-CNN	CNN- CNN	Transformer- Transformer
IOU	0.92	0.98	**0.97**	0.91	0.92	0.98
DICE	0.92	0.97	**0.96**	0.91	0.92	0.97
Inference Time (s)	0.06	4.12	**0.62**	4.24	0.10	4.94

with IOU and DICE scores of 98% and 97%, respectively, albeit with a longer inference time of 4.12 s. While the convolutional neural network architecture demonstrated the fastest inference time, it delivered less accurate segmentation with scores of 92% and 92% for IOU and DICE, respectively.

CNN-CNN and Transformer-Transformer architectures produced results similar to their single-model counterparts, suggesting that stacking identical models didn't significantly enhance performance. Impressively, our hybrid CNN-Transformer approach struck an optimal balance between segmentation accuracy and computational efficiency. With a short inference time of only **0.62** s, it achieved competitive scores of **97%** and **96%** for IOU and DICE, respectively, slightly below the Transformer-only architecture but approximately more than six times faster. This compromise ensures very high precision within a relatively shorter time.

5 Conclusion

In this paper, we have addressed the diabetic retinopathy diagnosis improvement, considering both performance and computational time aspects. Our primary focus was to tackle two main challenges, namely the lack of exploration of new image representations to be fed to the model and the computational and inference time of image segmentation. We faced the first challenge by utilizing a novel image transformation approach. Initially, we applied our pixel-by-pixel image traversal transformation approach to retinal images on two datasets, namely EyePACS [5] and APTOS [6]. Subsequently, we trained our models on these resulting datasets. Afterward, we evaluated our models using the F-score metric and compared our results with similar works from the literature that share aligned evaluation protocols. For the second challenge, we applied a fast convolutional neural network model to obtain pre-segmented images highlighting the regions of interest. These images were refined by a Vision Transformer to achieve high-quality segmentation while considerably reducing computational and time consumption. Overall, our results align commendably with contemporary state-of-the-art findings, underscoring the potential of our proposed approaches in advancing diabetic retinopathy classification and segmentation. Future work includes investigating the further impact of diverse image transformation approaches and the hybrid segmentation approach, not only on the identification of various retinal pathologies but also on the diagnosis of a wider range of diseases beyond ocular conditions.

References

1. Abràmoff, M.D., et al.: Improved automated detection of diabetic retinopathy on a publicly available dataset through integration of deep learning. Invest. Ophthalmol. Vis. Sci. **57**(13), 5200–5206 (2016)
2. Bej, S., Davtyan, N., Wolfien, M., Nassar, M., Wolkenhauer, O.: Loras: an oversampling approach for imbalanced datasets. Mach. Learn. **110**(2), 279–301 (2021)
3. Boudegga, H., Elloumi, Y., Akil, M., Hedi Bedoui, M., Kachouri, R., Abdallah, A.B.: Fast and efficient retinal blood vessel segmentation method based on deep learning network. Comput. Med. Imaging Graph. **90**, 101902 (2021)
4. Chugh, S., Kaur, J., Mittal, D.: Exudates segmentation in retinal fundus images for the detection of diabetic retinopathy. Int. J. Eng. Res. Technol. **3**(10), 673–7 (2014)
5. Gulshan, V., et al.: Development and validation of a deep learning algorithm for detection of diabetic retinopathy in retinal fundus photographs. JAMA **316**(22), 2402–2410 (2016)
6. Khalifa, N.E.M., Loey, M., Taha, M.H.N., Mohamed, H.N.E.T.: Deep transfer learning models for medical diabetic retinopathy detection. Acta Informatica Medica **27**(5), 327 (2019)
7. Majumder, S., Kehtarnavaz, N.: Multitasking deep learning model for detection of five stages of diabetic retinopathy. IEEE Access **9**, 123220–123230 (2021)
8. Oh, K., Kang, H.M., Leem, D., Lee, H., Seo, K.Y., Yoon, S.: Early detection of diabetic retinopathy based on deep learning and ultra-wide-field fundus images. Sci. Rep. **11**(1), 1897 (2021)
9. Porwal, P., et al.: Indian diabetic retinopathy image dataset (idrid) (2018)
10. Pratt, H., Coenen, F., Broadbent, D.M., Harding, S.P., Zheng, Y.: Convolutional neural networks for diabetic retinopathy. Procedia Comput. Sci. **90**, 200–205 (2016). 20th Conference on Medical Image Understanding and Analysis
11. Qummar, S., et al.: A deep learning ensemble approach for diabetic retinopathy detection. IEEE Access **7**, 150530–150539 (2019)
12. Samanta, A., Saha, A., Satapathy, S.C., Fernandes, S.L., Zhang, Y.D.: Automated detection of diabetic retinopathy using convolutional neural networks on a small dataset. Pattern Recogn. Lett. **135**, 293–298 (2020)
13. Sun, X., Xu, Y., Zhao, W., You, T., Liu, J.: Optic disc segmentation from retinal fundus images via deep object detection networks. In: 2018 40th Annual International Conference of the IEEE Engineering in Medicine and Biology Society (EMBC), pp. 5954–5957. IEEE (2018)
14. Tang, M.C.S., Teoh, S.S., Ibrahim, H., Embong, Z.: A deep learning approach for the detection of neovascularization in fundus images using transfer learning. IEEE Access **10**, 20247–20258 (2022)
15. Uppamma, P., Bhattacharya, S., et al.: Deep learning and medical image processing techniques for diabetic retinopathy: a survey of applications, challenges, and future trends. J. Healthc. Eng. (2023)
16. Wang, Z., Lu, H., Yan, H., Kan, H., Jin, L.: Vision transformer adapter-based hyperbolic embeddings for multi-lesion segmentation in diabetic retinopathy. Sci. Rep. **13**(1), 11178 (2023)

Enhancing Breast Mass Cancer Detection Through Hybrid ViT-Based Image Segmentation Model

Faycal Touazi(✉), Djamel Gaceb, Nesrine Boudissa, and Siham Assas

LIMOS Laboratory, Computer Science Department, University of Boumerdès, Boumerdès, Algeria
{f.touazi,d.gaceb,n.boudissa,s.assas}@univ-boumerdes.dz

Abstract. Breast cancer, primarily affecting women, is characterized by uncontrolled cell growth in breast tissue, leading to the formation of tumors. Although its exact causes remain elusive, factors such as age, genetics, and hormonal influences increase susceptibility. Early diagnosis plays a pivotal role in reducing mortality rates, emphasizing the importance of intelligent screening, diagnosis, and treatment methods. In recent years, artificial intelligence, particularly deep learning, has gained prominence in various industries, including medicine, due to its remarkable success in solving complex problems.

This study focuses on leveraging the power of deep learning to accurately segment abnormal breast regions, a critical step in breast cancer diagnosis. We explore and compare the performance of two distinct deep learning architectures: A-UNet and Trans-UNet. These architectures have demonstrated impressive potential in improving breast cancer detection. A-UNet achieved a Dice coefficient of 90.21% and 82.11% of IoU. Notably, Trans-UNet outperformed the A-UNet with a Dice coefficient of 94.13% and an IoU of 88.92%. These results showcase the promising capabilities of deep learning models in enhancing breast cancer detection, underscoring their potential to contribute significantly to early diagnosis and improved patient outcomes in the battle against this challenging disease.

Keywords: Breast Cancer · Deep Learning · ViT · UNet · TransUNet

1 Introduction

Advances in the field of medical imaging, particularly digital, have profoundly reshaped recent medical practices. These advances have introduced a new dimension related to the methods of processing and analyzing medical images, as well as the information derived from these methods.

However, despite these technological advances in medicine, the study of breast cancer remains a complex challenge and a subject of ongoing research. Mammographic imaging, a widely used modality discussed here, has become

B. Djamaa et al. (Eds.): CSA 2024, LNNS 1145, pp. 126–135, 2025.
https://doi.org/10.1007/978-3-031-71848-9_12

an indispensable tool in breast cancer examinations. Nevertheless, the interpretation of these images is becoming increasingly difficult for radiologists. Thus, the accuracy of the interpretation is crucial to ensure reliable diagnoses in the medical field.

Breast cancer stands out for its prevalence among women, with an ever-increasing mortality rate, making it an important global problem. In Algeria, this disease arouses notable apprehension among women, accounting for about half of all cases of female cancer [2].

Artificial intelligence (AI) has emerged as a transformative force in medical research, particularly in the realm of deep learning and vision transformers. These cutting-edge technologies offer unprecedented opportunities for the segmentation of breast cancer masses from medical images. The integration of deep learning and vision transformers holds immense potential in enhancing the early detection and diagnosis of breast cancer, offering clinicians invaluable insights into tumor morphology and facilitating personalized treatment strategies. Notably, while architectures like UNet have shown promising results in semantic segmentation tasks, combining this architecture with vision transformers has demonstrated the potential to achieve even better outcomes in accurately segmenting breast cancer masses.

We organized the paper in this order: Sect. 2 is dedicated to explore the relevant literature, providing a comprehensive overview of prior studies to establish the context and importance of our research. Section 3 presents our methodology, detailing the approaches and techniques utilized in our study. The main contribution is presented in Sect. 4, In this section, we present our findings and thoroughly discuss insights and interpretations of the results. Following this. Section 5 is dedicated to summarizing primary discoveries, their implications, and suggesting potential investigation avenues.

2 Related Works

In this section, we provide a summary of recent research endeavors concerning breast cancer tumor segmentation, leveraging deep learning methodologies. Ayana et al. [1] proposed a vision-transformer-based transfer learning approach for early breast cancer detection using a DDSM dataset. Their model, integrating vit-base, vit-large, swin-small, swin-base, pvt-medium, and pvt-large architectures, demonstrated remarkable performance in terms of accuracy, AUC, F1-Score, precision, recall, MCC, and Kappa values on the DDSM dataset.

Sun et al. [10] introduced the AUNet for breast mass segmentation using CBIS-DDSM and INBreast datasets. Their model achieved high Dice similarity coefficients of 81.8% and 79.1% on these datasets.

Kamran et al. [5] presented SWIN-SFTNet, a U-Net transformer-based architecture for breast micro-mass segmentation. Their model employed SWIN transformers and achieved competitive results on multiple datasets.

Liu et al. [7] developed TrEnD, a transformer-based encoder-decoder model for mass segmentation in mammograms. They introduced adaptive patch embedding (APE) and attention-gated modules, improving Dice and IOU scores on CBIS-DDSM and INBreast datasets.

Su et al. [9] suggested a YOLO segmentation model based on transformers for detecting and segmenting breast masses. They used CBIS-DDSM and INBreast datasets. Their model outperformed previous ones on both datasets by preserving local resolution through efficient data preprocessing methods.

Touazi et al. [11] presented a two-stage deep learning approach for breast cancer detection and mass segmentation. Their Vit-based model, SegNest, demonstrated a remarkable 90.15% Dice coefficient for semantic segmentation.

The fusion of U-Net with Vision Transformer (ViT) holds great promise for breast cancer detection and tumor segmentation, as seen in recent studies. This hybrid approach consistently outperforms others, showcasing its potential to improve both object detection and segmentation tasks. Integrating an attention mechanism into the U-Net architecture, a hallmark of transformers, is essential for maximizing its capabilities. This integration could unlock the full potential of deep learning for enhancing breast cancer detection and tumor segmentation.

3 Methodology

Following an extensive survey of existing literature, it is evident that the prevailing approach to breast cancer mass segmentation relies heavily on established architectures such as U-Net, YOLO, Res-UNet, and A-UNet. These architectures have demonstrated their efficacy in tackling the complex task of breast cancer mass segmentation. However, our research endeavors to introduce a fresh perspective by going through the TransUNet architecture [4], traditionally employed in organ segmentation tasks. TransUNet harnesses the combined strength of a convolutional neural network (CNN) and a vision transformer as its encoder, while its decoder employs a cascaded upsampler with skip connections.

In our proposed study, we aim to harness the potential of two distinct deep learning architectures: U-Net and TransUNet, for the segmentation of breast cancer images. Furthermore, we recognize the important role played by data preprocessing in enhancing model performance. As a result, our investigation will encompass a comprehensive analysis and comparison of these aforementioned architectures.

Through this research, we endeavor to contribute significant insights to the field of breast cancer mass segmentation, highlighting the effectiveness of TransUNet when compared to conventional architectures and underscoring the importance of data preprocessing in optimizing segmentation outcomes.

3.1 Attention U-Net

Inspired by the success of UNet and its variation to improve overall performance, the Attention UNET [8] represents a Convolutional Neural Network (CNN) often employed in image segmentation tasks (see Fig. 1). The Attention U-Net first presented in is a novel variation of the U-Net architecture that incorporates attention gates (AGs) to improve the performance of medical image segmentation tasks. The U-Net is a popular convolutional neural network (CNN) architecture that has been widely used for image segmentation in the medical field. However, the U-Net relies on multi-stage cascaded CNNs when the target organs exhibit large inter-patient variation in terms of shape and size.

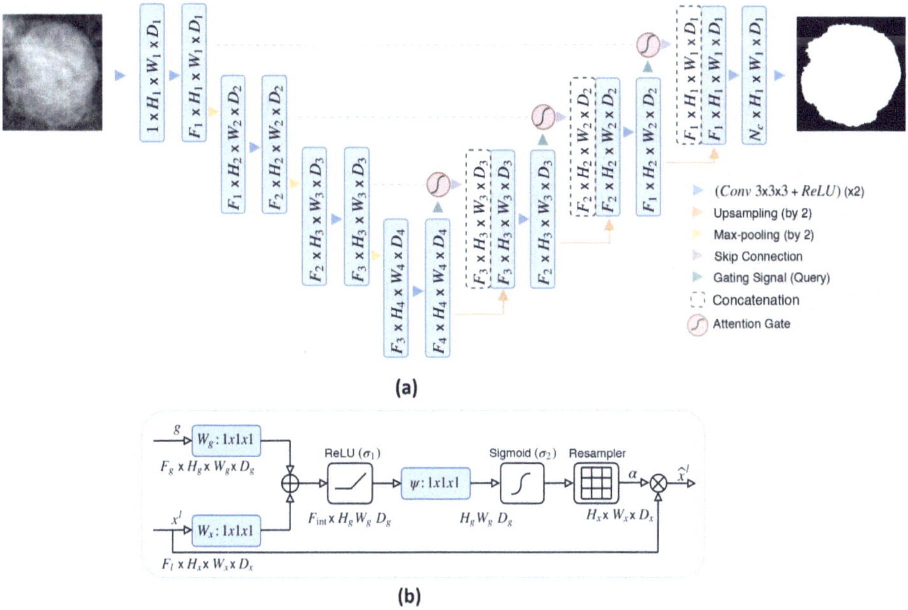

Fig. 1. (a): A block diagram of the proposed Attention U-Net segmentation model. (b): Attention Gate. Adapted from [8]. The figure is licensed under a Creative Commons Attribution 4.0 International License (CC BY 4.0).

The attention U-Net addresses this limitation by introducing attention gates, which automatically learn to focus on target structures of varying shapes and sizes. These attention gates allow the model to suppress irrelevant regions in an input image while highlighting salient features that are useful for the specific segmentation task. By doing so, the Attention U-Net eliminates the need for explicit external tissue/organ localization modules, which are commonly used in cascaded CNNs.

The attention gates in the attention U-Net can be easily integrated into the standard U-Net architecture with minimal computational overhead. They increase the model's sensitivity and prediction accuracy while preserving computational efficiency. Experimental results have shown that the Attention U-Net consistently improves the prediction performance of the U-Net across different datasets and training sizes.

3.2 TransUNet

TransUNet represents an architecture that combines the foundational principles of U-Net with Vision Transformers (ViT) (see Sect. 3.2), as originally introduced in the work by Chen et al. [4]. This architecture can be dissected into two essential components: the encoder and the decoder (see Fig. 2). For the encoder segment, the authors employ a ResNet-50 as a backbone to generate feature maps for the input data. These feature maps are tokenized and embedded to be the input sequence of ViT. The transformer will encode the input sequence in order to extract the contextual informations using 12 transformer layers. This fusion of ResNet-50 and ViT referred to both are pre-trained on ImageNet, leveraging the strengths of both architectures to encode the input information effectively. Moving on to the decoder component, TransUNet introduces an innovative approach known as the Cascaded Upsampler (CUP). The CUP consists of multiple upsampling blocks designed to restore the output to its full resolution, transforming the dimensions from (H/P)×(W/P) to H×W, where H and W represent height and width, and P denotes the upsampling factor. Each block within the CUP incorporates a sequence of operations: a 2x upsampling operation, a 3×3 convolutional layer, and a ReLU activation layer. These operations are applied successively, enabling the decoder to efficiently reconstruct the high-resolution output.

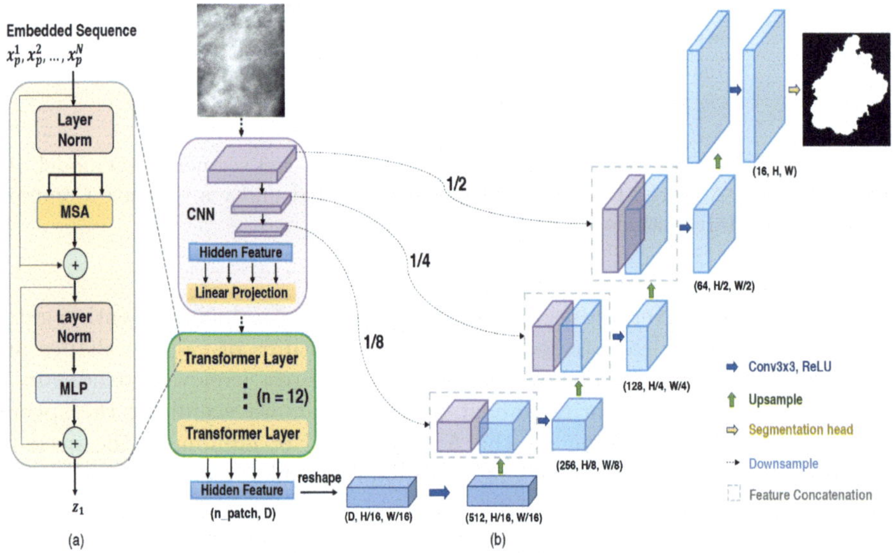

Fig. 2. Architecture of TransUnet. Adapted from [4]. The figure is licensed under a Creative Commons Attribution 4.0 International License (CC BY 4.0).

4 Experimental Setup and Findings

4.1 Dataset Description

For our study, we utilized the CBIS-DDSM dataset, a subset of the Digital Database for Screening Mammography (DDSM) [6]. This dataset serves as a valuable resource for breast cancer research, offering a diverse range of digital mammography images depicting both normal and abnormal cases. Notably complex, the dataset features richly annotated images, presenting challenges such as lesion detection and classification due to the presence of subtle lesions, variations in image quality, and diverse lesion types. Our dataset consisted of 1253 images for training and 363 images for testing.

4.2 Data Pre-processing

To prepare the dataset for analysis, we performed several pre-processing steps:

- **Image Cropping:** We focused on specific regions of interest (ROI) within the images by employing image cropping. Utilizing mask images provided within the dataset, we extracted and cropped white regions from the original images.
- **Resizing:** All cropped images were resized to dimensions of 224×224 pixels to suit the model input.
- **Contrast Limited Adaptive Histogram Equalization (CLAHE):** This technique enhances image contrast by redistributing intensity values, improving visibility of details. It ensures a more uniform distribution of pixel values, enhancing distinction between dark and bright regions.
- **Normalization:** Pixel values of images were normalized to a range of 0 to 1. This process entailed subtracting the minimum value from each pixel and dividing by the pixel value range (maximum minus minimum). Normalization improves performance, reduces numerical instabilities, and enables consistent comparisons.

4.3 Evaluation Metrics and Loss Functions

Intersection over Union (IoU). IoU assesses segmentation quality by calculating the ratio of the intersection area between predicted and reference masks to their union area.

$$IoU = |A \cap B|/|A \cup B| \tag{1}$$

Dice Score (DSC). Measures similarity between two sets, commonly used in medical image segmentation. It is calculated as twice the intersection area divided by the sum of both sets' areas.

$$DSC = (2 * |A \cap B|)/(|A| + |B|) \tag{2}$$

where $|A|$ and $|B|$ denote the cardinalities of sets A and B, respectively.

Dice Loss. The inverse of DSC, is utilized for training segmentation models, aiming to minimize this loss to enhance segmentation quality.

$$\mathcal{L}_{Dice} = 1 - DSC \tag{3}$$

Binary Cross-Entropy (BCE). BCE, a standard loss function for binary classification, quantifies the disparity between model predictions and true labels.

$$\mathcal{L}_{BCE} = -(y * log(p) + (1 - y) * log(1 - p)) \tag{4}$$

where y is the true binary label (0 or 1) and p is the probability predicted by the model for this label.

Focal Binary Cross-Entropy (Focal Loss). Focal Loss, a variant of BCE, mitigates challenges in training deep networks on imbalanced datasets by down weighting contributions of easy-to-classify examples.

$$\mathcal{L}_{BCE}^{Focal} = -\alpha_{t_i}(1 - p_{t_i})^{\gamma} \log(p_{t_i}) \tag{5}$$

where:

$$p_t = \begin{cases} p & \text{if } y = 1 \\ 1 - p & \text{otherwise} \end{cases}, \quad \alpha_t = \begin{cases} \alpha & \text{if } y = 1 \\ 1 - \alpha & \text{otherwise} \end{cases}$$

Combined Loss Function (Combo Loss). Combo Loss simultaneously minimizes Dice loss and a modified version of cross-entropy loss, balancing between segmentation accuracy and classification effectiveness.

$$\mathcal{L}_{comb} = \alpha \cdot \mathcal{L}_{BCE} + (1 - \alpha) \cdot \mathcal{L}_{Dice} \tag{6}$$

We also considered a focal version of the combined loss function, integrating focal binary cross-entropy:

$$\mathcal{L}_{comb}^{Focal} = \alpha \cdot \mathcal{L}_{BCE}^{Focal} + (1 - \alpha) \cdot \mathcal{L}_{Dice} \tag{7}$$

4.4 Results Discussion

A-U-Net: The results of AUNet model are in the table below (Table 1).

Table 1. Mass segmentation results using AUNet model

Loss function	Dice	IoU
$\mathcal{L}_{Focal}$ Eq. 5	88.10 %	78.66%
$\mathcal{L}_{comb}^{Focal}$ Eq. 7 ($\alpha = 0.5$)	90.09%	81.91%
$\mathcal{L}_{comb}$ Eq. 6 ($\alpha = 0.5$)	**90.21%**	**82.11%**

The choice of loss function significantly impacts the performance of the AUNet model in mass segmentation. The $\mathcal{L}_{comb}$ loss function with $\alpha = 0.5$ stands out as the most effective choice, achieving the highest Dice and IoU scores, which suggests accurate and well-defined mass segmentation.

Trans-UNet: The results of TransUNet model are in the table below (Table 2).

Table 2. TransUNet training results

Loss function	Dice	IoU
$\mathcal{L}_{comb}^{Focal}$ Eq. 7 ($\alpha = 0.5$)	92.48%	85.39%
$\mathcal{L}_{comb}$ Eq. 6	**94.13%**	**88.92%**

The TransUNet model trained with the $\mathcal{L}_{comb}$ loss function without the focal term and $\alpha = 0.5$ achieves the highest performance among the presented options. It provides the best Dice and IoU scores, indicating that this loss function configuration leads to the most accurate and precise segmentation results.

Table 3 displays a comparative analysis of different models alongside their corresponding performance metrics concerning breast cancer mass segmentation.

Table 3. Comparative Analysis with Previous Studies

Paper	Model	Dice	IoU
Kamran et al. [5]	Swin-SFTNet	24.13%	
Su et al. [9]	YOLO LOGO	74.52%	64.04%
Bouzar-Benlabiod et al. [3]	U-Net SE-RestNet-101	75%	64%
Wang et al. [12]	AM-MSP-cGAN	84.49%	
Liu et al. [7]	TrEnD	89.48%	81.34%
Touazi et al. [11]	SegNest	90.15%	-
Our	A-UNet	90.21%	82.11%
Our	Trans-UNet	**94.13%**	**88.92%**

Among the listed models, Swin-SFTNet has the lowest IoU and dice score at 24.13%, suggesting potential underperformance. YOLO LOGO demonstrates a balanced performance, achieving a Dice score of 74.52% and an IoU of 64.04%. U-Net SE-RestNet-101 performs decently, with a Dice score of 75% and an IoU of 64%, comparable to YOLO LOGO. AM-MSP-cGAN achieves a relatively high Dice score of 84.49%, indicating strong performance in its intended task. TrEnD shows superior results with a Dice score of 89.48% and an IoU of 81.34%. A-UNet outperforms all models with a Dice score of 90.21% and an IoU of 82.11%.

Our study results indicate that our proposed Trans-UNet model achieved exceptional performance, boasting the highest Dice score of 94.13% and an IoU of 88.92% among all listed models. Our Trans-UNet model leverages the strengths of both UNet and Vision Transformer (VIT) architectures to achieve superior performance in breast cancer mass segmentation.

UNet is renowned for its effectiveness in capturing fine-grained features and spatial information through its symmetric encoder-decoder structure, making it well-suited for semantic segmentation tasks. On the other hand, VIT excels in capturing global contextual information by utilizing self-attention mechanisms, allowing it to efficiently process and understand complex patterns in large-scale images. By integrating these two architectures, Trans-UNet harnesses the spatial awareness of UNet with the global contextual understanding of VIT, resulting in a holistic approach to image segmentation. This hybrid architecture enables our model to effectively capture both local and global features, enhancing its ability to accurately delineate breast cancer masses from medical images. Thus, Trans-UNet represents a novel fusion of complementary architectural features, offering a promising solution for improving the accuracy and efficiency of breast cancer diagnosis through automated segmentation.

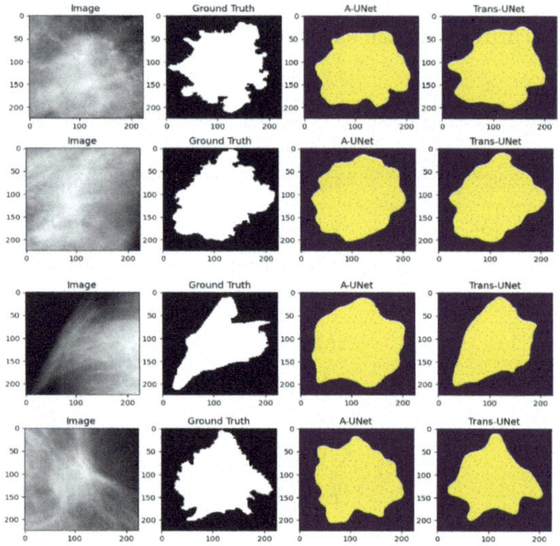

Fig. 3. Segmentation result of our models

Result Samples. Our models, A-UNet and TransUNet, demonstrate a remarkable capability to align closely with the ground truth mask. As illustrated in Fig. 3, it is evident that A-UNet tends to exhibit a slight tendency to outline the boundary of the object (in this case, breast masses) slightly beyond its actual boundaries the mask in comparison to TransUNet. This subtle distinction highlights TransUNet's superior precision in its predictions, yielding results that more accurately capture the intricacies of the underlying data. The contrast between these two models underscores TransUNet's ability to provide finer-grained and more reliable predictions.

5 Conclusion

This research introduces Trans-UNet and U-Net for mammographic mass segmentation, emphasizing the significance of early breast cancer detection. Trans-UNet outperformed A-U-net, achieving a Dice coefficient of 94.13% and an IoU of 88.92%, while A-U-net achieved 90.21% and 82.11%, respectively. Our future work involves a two-stage architecture with YOLO for mass detection and Trans-UNet for segmentation, aiming to enhance accuracy and efficiency. While our initial training focused on breast masses, we recognize the importance of calcifications and plan to improve related data. Future exploration includes the Connected U-Net architecture for more accurate mammographic mass segmentation in breast cancer diagnosis.

References

1. Ayana, G., et al.: Vision-transformer-based transfer learning for mammogram classification. Diagnostics **13**(2), 178 (2023)
2. Bounedjar, A., Melzi, M.A., Idir, H., Heba, N.: General oncology care in Algeria. In: Al-Shamsi, H.O., Abu-Gheida, I.H., Iqbal, F., Al-Awadhi, A. (eds.) Cancer in the Arab World, pp. 15–30. Springer, Singapore (2022). https://doi.org/10.1007/978-981-16-7945-2_2
3. Bouzar, L., Harrar, K., Yamoun, L., Khodja, M.Y., Akhloufi, M.A.: A novel breast cancer detection architecture based on a CNN-CBR system for mammogram classification. Comput. Biol. Med. **163**, 107–133 (2023)
4. Chen, J., et al.: Transunet: transformers make strong encoders for medical image segmentation. arXiv preprint arXiv:2102.04306 (2021)
5. Kamran, S.A., Hossain, K.F., Tavakkoli, A., Bebis, G., Baker, S.: Swin-sftnet: spatial feature expansion and aggregation using swin transformer for whole breast micro-mass segmentation. In: 2023 IEEE 20th International Symposium on Biomedical Imaging (ISBI), pp. 1–5. IEEE (2023)
6. Lee, R.S., Gimenez, F., Hoogi, A., Miyake, K.K., Gorovoy, M., Rubin, D.L.: A curated mammography data set for use in computer-aided detection and diagnosis research. Sci. data **4**(1), 1–9 (2017)
7. Liu, D., Wu, B., Li, C., Sun, Z., Zhang, N.: Trend: a transformer-based encoder-decoder model with adaptive patch embedding for mass segmentation in mammograms. Med. Phys. **50**(5), 2884–2899 (2023)
8. Oktay, O., et al.: Attention U-net: learning where to look for the pancreas. arXiv preprint arXiv:1804.03999 (2018)
9. Su, Y., Liu, Q., Xie, W., Hu, P.: Yolo-logo: a transformer-based yolo segmentation model for breast mass detection and segmentation in digital mammograms. Comput. Methods Programs Biomed. **221**, 106903 (2022)
10. Sun, H., et al.: Aunet: attention-guided dense-upsampling networks for breast mass segmentation in whole mammograms. Phys. Med. Biol. **65**(5) (2020)
11. Touazi, F., Gaceb, D., Chirane, M., Herzallah, S.: Two-stage approach for semantic image segmentation of breast cancer: deep learning and mass detection in mammographic images. In: IDDM 2023, pp. 121–135 (2023)
12. Wang, Y., Wang, S., Chen, J., Wu, C.: Whole mammographic mass segmentation using attention mechanism and multiscale pooling adversarial network. J. Med. Imaging **7**(5), 054503–054503 (2020)

New Adaptive PolSAR Speckle Filtering Based on Sigma Edge Detection

Souhila Boutarfa[1,2(✉)] and Youcef Smara[2]

[1] Laboratory of Innovative Technologies, Department of Electrical Engineering and Industrial Computing, National Higher School of Advanced Technologies (ENSTA), Dergana, Bordj El Kiffan 16087, Algiers, Algeria
s.boutarfa@hotmail.fr

[2] Laboratory of Image Processing and Radiation, Department of Telecommunications, Faculty of Electrical Engineering, University of Sciences and Technology Houari Boumediene (USTHB), Bab Ezzouar, 16111 Algiers, Algeria

Abstract. Radar imaging encounters a significant hurdle with the presence of noise, commonly referred to as speckle. This noise complicates the interpretation of images by diminishing the accuracy of the conveyed information. To address the speckle effect and enhance image quality, filtering becomes essential. In this paper, we introduce a novel polarimetric filtering approach that integrates edge detection with the improved sigma filter. This integration aims to enhance the effectiveness of the method in reducing speckle noise while retaining crucial information. This approach relies on calculating the probability density of the radar signal to determine a confidence interval for pixels exhibiting similar characteristics in a homogeneous region while safeguarding heterogeneous areas. This will conserve not only the edges but also the polarimetric information. The newly applied method enables filtering elements within the polarimetric covariance or coherency matrix. To assess the effectiveness of the filtering process, we employed the following criteria: smoothing homogeneous regions, preserving edges, and retaining polarimetric information.

Keywords: PolSAR · Speckle · filtering · Edge detection · Sigma range

1 Introduction

The formation of speckle arises from the coherent summation process during SAR data measurement. This particular noise significantly disrupts the processing of radar data and the interpretation of their outcomes, as it corrupts both the amplitude and the phase of the images. It manifests as a granular texture affecting radar images, displaying itself as random variations [1, 2]. The presence of speckle poses challenges in interpretation and information extraction, leading to degradation in segmentation performance and a reduction in target detectability. To mitigate speckle in PolSAR images and apply information extraction techniques effectively, the implementation of polarimetric filtering becomes necessary [3–5].

B. Djamaa et al. (Eds.): CSA 2024, LNNS 1145, pp. 136–145, 2025.
https://doi.org/10.1007/978-3-031-71848-9_13

The objective is to reduce speckle and enhance the readability of the images. In practical terms, there is no one-size-fits-all filter suitable for all problems. Some filters may prove effective for specific images but entirely inappropriate for others. Therefore, the filtering process involves estimating the ideal radar reflectivity based on the noisy observation. It essentially requires preprocessing images with suitable filtering methods before progressing with their analysis.

In this paper, our objective is to enhance the robustness of the filtering process and achieve a more effective balance between smoothing and preserving details. To achieve this, we introduced a contour detection procedure into the improved sigma filtering. This adaptation makes the sigma filter more flexible, resulting in enhanced preservation of linear structures. As a consequence, the outcome is a clearer image with well-defined contours.

In the following, we define the principle of the sigma filter and its improved version. Next, we present our proposed method, based on the adaptability of the improved sigma filter. Finally, we showcase the obtained results along with our conclusions on the performance of the proposed filter.

2 Sigma Filter

The sigma filter proposed by Lee [6, 7] was designed for filtering SAR (single-channel) images. Lee defined a two-sigma probability interval, so that 95.5% of the pixels in the image are distributed within this interval around the mean. During the filtering process, local statistics (specifically the mean) are estimated from the pixels included in this interval.

This filter is adapted to PolSAR images by processing the elements of the covariance or coherency matrix according to the principle of polarimetric filtering [8]. To do this, the sigma interval $[I_1, I_2]$ is calculated from the Span image (Eq. 4), as follows:

$$I_1 = y - 2\sigma_v y \tag{1}$$

and

$$I_2 = y + 2\sigma_v y \tag{2}$$

with

$$\sigma_v = \frac{1}{\sqrt{L}} \tag{3}$$

and

$$Span = |S_{HH}|^2 + 2|S_{HV}|^2 + |S_{VV}|^2 \tag{4}$$

where:

- S_{HH}, S_{HV} and S_{VV} are the polarimetric signals (H: Horizontal, V: Vertical).
- $|.|$ represents the amplitude of the complex value.
- y denotes the value of the central pixel within the moving window.

- σ_ν stands for the standard deviation of the noise.
- L is the number of looks.

Derived from the Span, 95.5% of the pixels within the study window falling within the sigma interval are specifically chosen. Subsequently, the average of these selected pixels is computed to derive the filtered pixel for each element within the covariance or coherency matrix. Pixels located outside this interval are treated as outliers and are excluded from the mean calculation. This process facilitates the separation of homogeneous areas with high contrast, resulting in the reduction of speckle.

The sigma filter has been widely used in SAR data denoising due to its simplicity and effectiveness in reducing speckle in homogeneous areas. However, shortcomings have been identified, necessitating correction or compensation. The limitations of this filter, as outlined in [7], include:

- It does not maintain a constant mean value for the image. To address this issue, a new sigma range must be calculated to preserve the mean.
- Strong scatterers are blurred and, in some cases, even removed. Preserving these scatterers is crucial for target detection and structure conservation. Strong scatterers are less affected by speckle, a property not considered by the sigma algorithm. Adaptive algorithms, such as the Minimum Mean Square Error (MMSE) filter [9, 10], demonstrate moderate effectiveness.
- Isolated dark pixels may appear in the filtered data due to very low-intensity pixels resulting in extremely narrow intervals, and in extreme cases, the interval may be zero, causing unfiltered dark areas. These noisy dark pixels could pose challenges for specific applications. Sigma filter has incorporated the 3×3 local mean to address this issue. While this technique has assisted in mitigating the impact of isolated dark pixels, some traces may persist.

3 Improved Sigma Filter

The improved sigma filter [7] aims to overcome the limitations of the previous filter. The problem of mean variation has been addressed by redefining the sigma range based on the speckle probability density function, selecting 90% of the pixels. An estimator of the MMSE is incorporated for adaptive speckle reduction.

In this case, the sigma interval is adjusted to $\left[I_1\hat{x}, I_2\hat{x}\right]$, with the newly estimated values of I_1 and I_2 for each number of looks as outlined in [7]. The value of $\hat{x}$ is determined through the MMSE within a 3×3 pixels window centered on the moving window. This adjustment serves to minimize isolated dark pixels and enhance the overall quality of the filtering process.

$$\hat{x} = \overline{y} + b(y - \overline{y}) \tag{5}$$

with

$$b = \frac{var(x)}{var(y)} \tag{6}$$

and

$$var(x) = \frac{var(y) - \overline{y}^2 {\sigma_v}^2}{1 + {\sigma_v}^2} \tag{7}$$

where b is a coefficient ranging from 0 to 1, which needs to be computed from the Span. In Eq. (5), y and $\overline{y}$ represent, respectively, the value of the central pixel of the sliding window and the mean within the new sigma interval of the noisy image.

In Eqs. (6) and (7), $\overline{y}$ and $var(y)$ denote the mean and variance, respectively, of the pixels within the new sigma interval of the Span image. Within the moving window, pixels falling within the new sigma interval become part of the filtering process, and the MMSE filter is reapplied to these selected pixels. Nevertheless, during the application of the MMSE filter in the final stage of filtering, it was observed that the speckle standard deviation must be adjusted [7]. This adjustment is necessary because the pixels chosen by the MMSE method create a heavily filtered image. As a result, the standard deviation value needs to be modified by the new value $\widetilde{\sigma}_v$. Specifically, in the MMSE filter and in Eq. (5), σ_v is substituted with $\widetilde{\sigma}_v$.

To simplify the filtering process, Lee et al. [7] provided a visualization of the values of σ_v and the adjusted standard deviation $\widetilde{\sigma}_v$ for each number of looks.

Thus, the improved sigma filter provides better smoothing than its predecessor. However, the challenge of preserving edges remains. We propose to tackle this issue with an algorithm for detecting edges, which we will delve into in the following.

4 New Adaptive Sigma Filter

For this approach, we will integrate the edge detection process with the improved sigma filter. This involves initially applying the edge detection and subsequently incorporating the characteristics of the improved sigma filter.

The edge detection relies on selecting adjacent pixels with diffusion features resembling the central pixel. A suitable filtering process is then applied using a straightforward and efficient algorithm, preserving edge sharpness by utilizing a non-square window aligned with the edge direction [8].

When implementing the proposed filter, the initial step is to select one of the eight edge-aligned windows based on the edge direction [8], for filtering the central pixel. The chosen non-square window contains pixels with radiometric properties akin to the central pixel, offering superior noise filtering. Conversely, a square window would encompass pixels from diverse scattering environments, resulting in image blurring. The process of selecting the window aligned with the edge and following the edge direction is computed step by step.

Initially, the 7×7 moving window is partitioned into nine 3×3 sub-windows, and their respective means are computed. These mean values are then stored in a 3×3 matrix as follows:

$$m = \begin{bmatrix} m_{11} & m_{12} & m_{13} \\ m_{21} & m_{22} & m_{23} \\ m_{31} & m_{32} & m_{33} \end{bmatrix} \tag{8}$$

The detection of edge direction involves the multiplication of the 3 × 3 matrix by the following four masks:

$$M_1 = \begin{bmatrix} -1\,0\,1 \\ -1\,0\,1 \\ -1\,0\,1 \end{bmatrix} \tag{9}$$

$$M_2 = \begin{bmatrix} 0 & 1 & 1 \\ -1 & 0 & 1 \\ -1 & -1 & 0 \end{bmatrix} \tag{10}$$

$$M_3 = \begin{bmatrix} 1 & 1 & 1 \\ 0 & 0 & 0 \\ -1 & -1 & -1 \end{bmatrix} \tag{11}$$

$$M_4 = \begin{bmatrix} 1 & 1 & 0 \\ 1 & 0 & -1 \\ 0 & -1 & -1 \end{bmatrix} \tag{12}$$

In this case, we will have:

$$Mul(n) = m.M_n \tag{13}$$

The sum of the elements in each resulting 3 × 3 matrix from the multiplication is calculated as follows:

$$Som(n) = \sum\nolimits_{(i=1)}^{3} \sum\nolimits_{(i=1)}^{3} Mul(n)_{ij};\text{ with } n = 1, 2, 3, 4 \tag{14}$$

Then, the maximum absolute value among these sums is determined as follows:

$$MaxSom = Max\{|Som(1)|, |Som(2)|, |Som(3)|, |Som(4)|\} \tag{15}$$

This maximum value corresponds to one of the masks used for edge direction selection. For each direction, we can choose one of the two aligned-edge windows based on the average closest to the central average of the 3 × 3 matrix (Eq. 8). These averages are calculated using nine sub-windows, represented as a superimposed 3 × 3 window. If MaxSum (Eq. 15) is equal to |Sum2|, a choice must be made between the two aligned-edge windows [8]. To decide, we compare the two average values with the central average value. The value closest to the central average allows for the selection of the appropriate window.

Upon detecting the edge window, the improved sigma filter method is applied for filtering, considering only pixels within the sigma interval $\left[I_1\hat{x}, I_2\hat{x}\right]$ in the bright areas of the identified edges, where I_1, I_2 are defined from [7], and the a priori mean $\hat{x}$ is determined by applying the MMSE criterion on a 3 × 3 window. The established interval is then used to filter only pixels belonging to it using a moving window. In this window, each central pixel is substituted with the a priori mean $\hat{x}$ calculated through Eq. (5). The parameter b (Eq. 6) is computed using the variance of x, expressed through

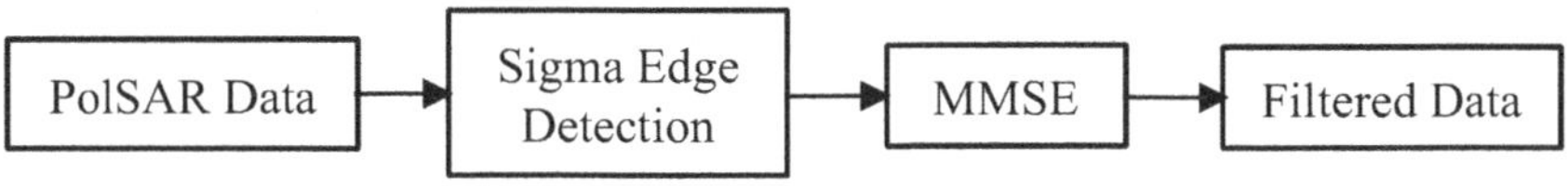

Fig. 1. Simplified block diagram of the proposed method.

Eq. (7). This calculation involves replacing the value of the speckle standard deviation σ_v with its adjusted value $\widetilde{\sigma}_v$ [7]

To enhance comprehension of the proposed adaptive sigma filtering technique, a simplified block diagram has been devised (Fig. 1).

Therefore, the proposed method allows obtaining filtered images with better preservation of linear structures.

5 Results and Discussion

The filters are tested using a complex single-look polarimetric SAR image captured by the E-SAR sensor in 2001. This image encompasses the Oberpfaffenhofen region near Munich, Germany, and has dimensions of (2300 × 925) pixels, operating in the P-band. The image is diverse, featuring distinct areas of forests, buildings, and roads, all regarded as edges and targets.

The assessment of each filter is guided by the following criteria: effectiveness in smoothing homogeneous regions, capacity to retain edges, and preservation of polarimetric information. This evaluation involves a visual examination through the display of images with a RGB (Red, Green, Blue) color composition of the three polarimetric components. To validate this visual evaluation and enhance our analysis, we later conducted a statistical assessment.

5.1 Visual Evaluation

In the comparative study, we presented the original and filtered images, with a size of 400 × 400 pixels (Fig. 2), along with zooms (Fig. 3) on a region of buildings to thoroughly illustrate the benefits of filtering. These images are displayed in RGB color composition of the polarimetric channels $R = |S_{HH}|$, $G = |S_{HV}|$, $B = |S_{VV}|$.

For a more accurate characterization of objects and a detailed comparison, we also presented optical images of the same scenes obtained from the Google Earth application along with the results.

Based on the obtained results, we notice that the adaptive sigma filter allows for effective smoothing of homogeneous areas, better preservation of linear structures, especially in buildings, and a good retention of strong scatterers. Thanks to the improvement added to the calculation procedure, it provides a better compromise between smoothing and preserving the image's characteristics, and the granular aspect of the speckle in the original images has been significantly reduced.

However, the sigma filter exhibits weak smoothing and some unfiltered pixels, and the improved sigma filter offers strong smoothing that blurs fine structures, resulting in poor information preservation.

The preservation of colors after filtering proves that the polarimetric information has not been degraded. The displayed zooms containing fine structures help detect differences between the filters and clearly show the good balance between smoothing and edge preservation achieved by the proposed method.

Visual evaluation alone is not sufficient to validate the obtained results; therefore, a statistical evaluation is necessary to better analyze the performance of the proposed filtering.

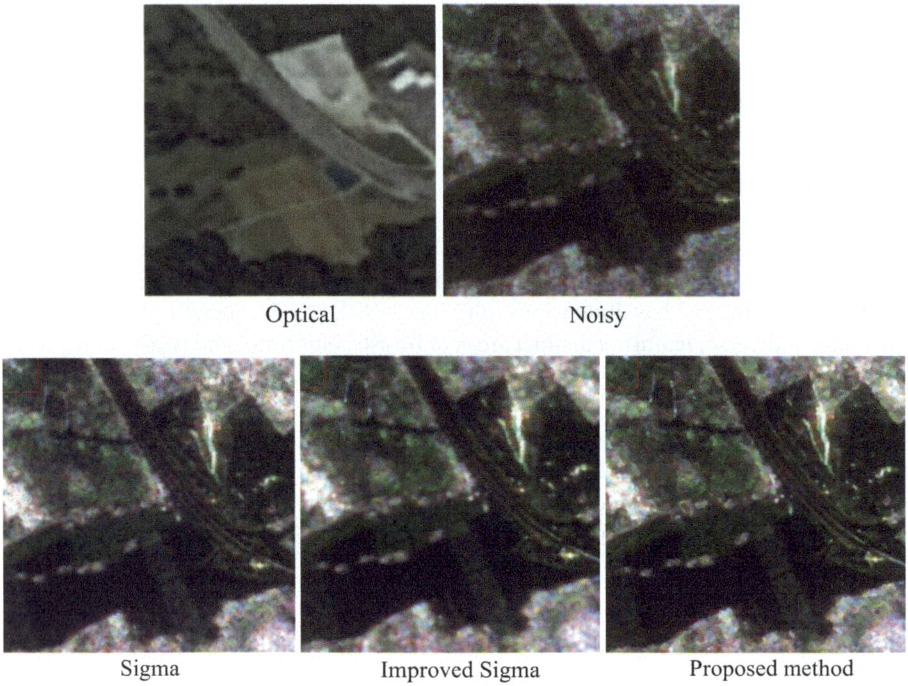

Fig. 2. Images before and after filtering.

5.2 Statistical Evaluation

Statistical computations are conducted on the Span image (Eq. 4) derived from the polarimetric channels both before and after the application of filters. To gauge the effectiveness of each filter in smoothing homogeneous regions, we perform statistical analyses on homogeneous zones measuring 20 × 20 pixels, extracted from the image both before and after the filtering process.

The calculated statistical parameters include the standard deviation σ, the mean μ, the coefficient of variation Cv, and the Equivalent Number of Looks (ENL), the global coefficient of variation Cvg, and the polarimetric entropy H [11].

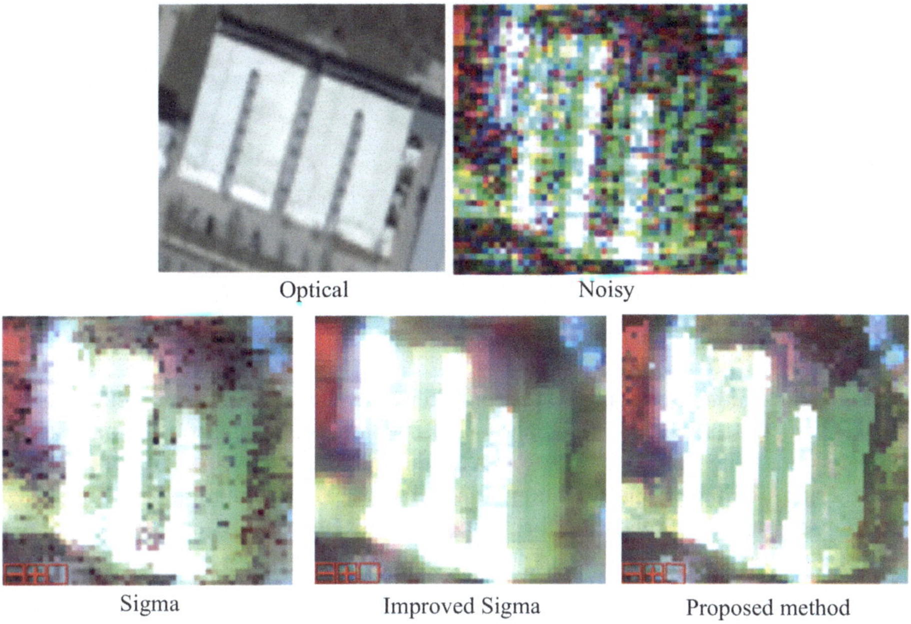

Fig. 3. Zoom images before and after filtering.

Effective filtering in homogeneous areas is achieved through, involving a reduction in the value of σ, preservation of the mean μ, a low coefficient of variation (Cv), and an increase in the value of ENL.

In terms of preserving edges, the assessment of different studied filters involves calculating the global coefficient of variation Cvg, which is the average of the coefficients of variation from various heterogeneous zones sized 10×10 pixels. The most effective filter for edge preservation is the one that produces the highest Cvg coefficient.

The computation of entropy H relies on estimating the eigenvalues and eigenvectors derived from the polarimetric coherency matrix. H delineates the mechanism of random scattering: if $H = 0$, the target exhibits coherence, indicating a single scattering mechanism; whereas, if $H = 1$, the target is incoherent, signifying scattering from multiple mechanisms (heterogeneous zone) [12].

To validate the criterion of preserving polarimetric information, we calculate the average entropy in a region of interest before and after filtering. It is crucial to maintain this average, ensuring that its value does not undergo excessive variations [11]. According to [12], in the noisy image, calculating H requires applying a boxcar filter (BF) before entropy estimation. In our case, we applied this filter with a small sliding window of size 3×3. Thus, this filter does not provide effective speckle reduction, allowing the preservation of most details.

The statistical results are illustrated in the following table:

According to Table 1, we observe that the Equivalent Number of Looks (ENL) values increase after filtering for all three methods. Consequently, the coefficient of variation decreases, ensuring the maintenance of the mean, and the standard deviation

Table 1. Statistical results.

Filters	σ	μ	*Cv*	*ENL*	*Cvg*	*H*
Noisy	1125.970	1753.550	0.642	2.425	-	0.778
Sigma	395.510	1487.120	0.265	14.137	1.109	0.835
Improved sigma	205.050	1721.100	0.119	70.451	1.135	0.827
Proposed method	313.523	1703.710	0.184	29.529	1.415	0.822

decreases. This indicates that the evaluation criteria are met. The proposed method shows an ENL value just slightly lower than that of the improved sigma filter. This implies adequate smoothing in homogeneous areas. Additionally, the preservation of the mean after filtering signifies the preservation of information. The calculation of *Cvg* revealed that the proposed filter stands out for its ability to establish a satisfactory compromise between smoothing homogeneous areas and preserving edges, hence its high *Cvg* value. This demonstrates that the addition of the proposed sigma edge detection in the new version of the filter has made the algorithm adaptive and optimal.

From Table 1, we observe that the mean values of entropy *H* for the three methods concerning the original image are preserved. The closest value to the original average is provided by our proposed method.

We also observe that the statistical results align with the conclusions drawn from the visual evaluation.

6 Conclusion

The newly proposed PolSAR filtering method outperforms the other two filters, namely sigma and improved sigma, by achieving an optimal equilibrium between smoothing homogeneous areas and preserving contours. This method demonstrates remarkable effectiveness, providing precise detection of linear structures within the image, well-defined delineation of targets, accentuation of details, and, crucially, the preservation of polarimetric information. As a result, the filtered images obtained through this method can be employed across a spectrum of applications, including classification tasks, interferogram creation, Digital Elevation Model (DEM) generation, and various other applications where high-quality, enhanced imagery is essential.

References

1. Lee, J.S., Pottier, E.: Polarimetric Radar Imaging: From Basics to Applications. CRC Press, Boca Raton, FL, USA (2009)
2. Chen, C.H.: Signal and Image Processing for Remote Sensing. CRC Press, Boca Raton, FL, USA (2012)

3. Vasile, G., Trouve, E., Lee, J.S., Buzuloiu, V.: Intensity-driven adaptive-neighborhood technique for polarimetric and interferometric SAR parameters estimation. IEEE Trans. Geosci. Remote Sens. **44**(6), 1609–1621 (2006)
4. Jiang, C., He, X.F., Yang, L.J., Jiang, J., Wang, D.Y., Yuan, Y.: Despeckling PolSAR images based on relative total variation model. Int. Arch. Photogram. Remote Sens. Spatial Inf. Sci. **42**(3), 641–644 (2018)
5. Boutarfa, S., Bouchemakh, L., Smara, Y.: Polarimetric synthetic aperture radar speckle filtering by multiscale edge detection. J. Appl. Remote. Sens. **13**(2), 024507 (2019)
6. Lee, J.S.: Digital image noise smoothing and the sigma filter. Comput. Vis. Graph. Image Process. **24**(2), 255–269 (1983)
7. Lee, J.S., Wen, J.H., Ainsworth, T.L., Chen, K.S., Chen, A.J.: Improved sigma filter for speckle filtering of SAR imagery. IEEE Trans. Geosci. Remote Sens. **47**(1), 202–213 (2008)
8. Lee, J.S., Grunes, M.R., De Grandi, G.: Polarimetric SAR speckle filtering and its implication for classification. IEEE Trans. Geosci. Remote Sens. **37**(5), 2363–2373 (1999)
9. Lee, J.S.: Digital image enhancement and noise filtering by use of local statistics. IEEE Trans. Pattern Anal. Mach. Intell. **2**, 165–168 (1980)
10. Kuan, D.T., Sawchuk, A.A., Strand, T.C., Chavel, P.: Adaptive noise smoothing filter for images with signal-dependent noise. IEEE Trans. Pattern Anal. Mach. Intell. **2**, 165–177 (1985)
11. Boutarfa, S., Smara, Y., Youcef, C., Ballah, R.: Evaluation techniques for PolSAR speckle filtering. In: International Conference on Earth Observation and Geo-Spatial Information (ICEOGI), pp. 1–6. IEEE, Algeria (2023)
12. Cloude, S.R., Pottier, E.: A review of target decomposition theorems in radar polarimetry. IEEE Trans. Geosci. Remote Sens. **34**(2), 498–518 (1996)

Deep Learning for Effective Tuberculosis Detection from Chest Radiographs: A Comparative Study

Nesrine Boudoukhani(✉), Zakaria Elberrichi, and Latefa Oulladji

EEDIS Laboratory, Djillali Liabes University of Sidi Bel Abbes, Sidi Bel Abbes, Algeria
nesrine.boudoukhani@univ-sba.dz

Abstract. This paper presents a comparative study to assess the performance of the visual transformer (ViT) alongside several convolutional-based models in distinguishing tuberculosis (TB) from healthy and various other lung conditions using chest radiographs. Seven deep learning models including Xception, InceptionNetV3, VGG16, ResNet50, DenseNet201, EfficientNetV2 and the ViT model are selected in this study for comparison on the TBX11K dataset. Empirical results revealed the superiority of established convolutional models, especially ResNet50, over the ViT model in TB recognition by reaching 99.22%, 98.27%, 99.08%, and 99.62% for accuracy, recall, precision and specificity, respectively. This study contributes to the medical field by emphasizing the crucial role of deep learning (DL) model selection in medical image analysis for automatic TB diagnosis from chest x-rays.

Keywords: Tuberculosis Detection · Chest X-Ray · Deep Learning · Visual Transformer · Convolutional Neural Networks

1 Introduction

The challenge of TB detection from chest x-rays (CXRs) has prompted many researchers in the field of artificial intelligence (AI) to contribute in automating the interpretation of CXRs using advanced computer vision techniques.

The first attempts focused mainly on differentiating TB cases from healthy ones by collecting datasets composed of TB-infected and healthy CXRs [1, 2]. Consequently, a significant portion of the existing literature on TB detection used these binary classification data to train and evaluate their DL models [3–6]. This approach resulted in building computer-aided detection (CAD) systems capable of accurately differentiating between TB and healthy cases. However, they do not guarantee maintaining this high performance in real-world scenarios involving CXRs of various lung conditions. To fill this gap, recent studies expanded the scope to include CXRs of other pulmonary conditions in addition to TB and healthy samples via multi-class classification [7–9].

Deep convolutional neural networks (CNNs) are well known in computer vision for image processing. However, a novel transformer-based deep neural network referred by

B. Djamaa et al. (Eds.): CSA 2024, LNNS 1145, pp. 146–155, 2025.
https://doi.org/10.1007/978-3-031-71848-9_14

the Visual Transformer (ViT) has recently gained recognition [10] without being yet extensively explored in CXR interpretation for TB detection.

In light of these considerations, this paper presents a comparative study between convolutional-based and transformer-based DL models in TB detection from CXRs. Our objective is to assess the performance of the ViT model [10] alongside several established CNN models including VGG16 [11], ResNet50 [12], InceptionNetV3 [13], Xception [14], DenseNet201 [15] and EfficientNetV2 [16] in identifying and distinguishing TB cases from healthy and various other lung diseases using CXRs. The main contributions of this paper are:

1 Reviewing the recent contributions using DL for automatic TB detection from CXRs.
2 Conducting a comparative study between six convolutional-based models and the ViT in TB detection and differentiation from other pulmonary diseases using CXRs.
3 Emphasizing the best DL model in classifying CXRs for effective TB detection and diagnosis.

The rest of this paper is structured as follow: the related work is presented in Sect. 2. The description of the DL models selected for comparison is outlined in Sect. 3. Section 4 presents the experimental setup employed in this study. Section 5 presents and discusses the results of experimentations and Sect. 6 summarizes the findings of this study.

2 Related Work

Current research on computer-aided diagnosis of TB from CXRs is focused on using DL techniques. Indeed, the existing literature has shown a variety of methods, from which two main approaches are followed:

2.1 Binary Classification: Normal vs. Tuberculosis

The first approach consists of identifying TB samples from healthy ones. In the early research by [1], the authors conducted an experimental study combining image preprocessing, data augmentation, lung segmentation and DL to build an effective TB detector based on DenseNet201 [15].

Ensemble learning (EL) techniques were broadly employed in literature to resolve TB detection from CXRs. The authors in [17] proposed an average-based ensemble composed of modality-specific DL models built by training ImageNet computer vision models on large CXR datasets and fine-tuning them on a set of TB/Normal CXRs. Bayesian ensemble frameworks were also proposed by employing dropout in identical DL models [4] or applying varied augmentation and preprocessing techniques [5]. The researchers in [6] combining type-1 Sugeno fuzzy integral ensembling and meta-heuristic optimization to introduce an ensemble framework composed of three base learners trained on TB/Healthy CXRs.

To prompt the effective learning of TB features from CXRs, the authors in [18] introduced a stochastic learning artificial neural network (ANN) to classify CXRs as healthy or TB-infected trained by incorporating randomness functions during training.

Other recent contributions proposed building multi-stage frameworks for automatic TB detection. The main approach followed is to build double-stage frameworks composed of a segmentation network to segment lungs within CXRs and a classification model to detect TB in these lung segmented images [19, 20]. A recent study by [3] proposed an effective multi-stage framework composed of a segmentation network to localize TB infected areas in CXRs, followed by multiple feature extractors for extracting relevant features, and finally a classification network to classify the resulted feature vectors as healthy or TB.

2.2 Multi-class Classification: TB Vs. Other Lung Diseases

The second approach aims at differentiating TB cases from other pulmonary diseases, whereby DL models are trained to recognize several lung conditions alongside TB via multi-class classification [7].

Novel DL frameworks were proposed in literature to achieve an effective recognition of TB cases using CXRs. The contribution by [21] introduced new modified and hybrid classifiers using EfficientNets [22] and Visual Transformers [10] to differentiate between TB, pneumonia (PN), and healthy cases. The researchers in [23] proposed a new normalization-free network with progressive image resizing, achieving high performance on different binary and multiclass TB detection challenges.

Notable works reported the use of multi-stage frameworks. The researchers in [24] leveraged DL for feature extraction from the region-of-interest (ROI) in CXRs. These features were then used with machine learning (ML) and EL models to distinguish TB from covid-19 and healthy CXRs. The research by [9] introduced a new double-stage framework using CNNs and grid search optimization. The first step consists of detecting the diseased cases and the second step aims at identifying the specific lung disease among Covid-19, lung opacity (LO), lung cancer (LC), TB, and PN.

Recent contributions shifted towards multiclass CXR classification by including several lung diseases in their datasets. The research presented by [25] proposed a multichannel EfficientNet [22] model followed by a stacked EL model of ML classifiers for detecting multiple lung diseases in CXRs, including TB. A comparative study by [26] highlighted the outperformance of DenseNet201 [15] among eight pretrained CNNs in differentiating between PN, covid-19, pneumothorax, TB and healthy cases in measure of AUC (Area Under Curve)[1], accuracy, recall and precision. The authors in [8] used mobile inverted residual bottleneck blocks [22, 27] and wavelet features to build a novel framework referred by the DWTMBConvNet for the identification of multiple lung diseases alongside TB (see Table 1).

3 Materials and Methods

This section explains in detail the building process of DL models selected for comparison. These models are: VGG16 [11], ResNet50 [12], InceptionNetV3 [13], Xception [14], DenseNet201 [15] EfficientNetV2 [16], and the ViT [10].

[1] Area Under Receiver Operating Characteristic (ROC) curve

Table 1. Literature review summary.

Work	Method	TB images	Image shape	Task	Classes	Dataset	Test set	Accuracy (%)
[21]	Modified and Hybrid ViT and EfficientNet	698	320× 320	Multi-class	3	TB, PN and Healthy CXRs	3680	97.72
[25]	Multi-channel EfficientNets + EL	800	240× 240, 260× 260, 300× 300	Multi-class	3	TB, Healthy and other lung disease CXRs	2520	99.00
[26]	Pretrained CNNs	700	227× 227, 224× 224, 256× 256, 299× 299	Multi-class	5	TB, PN, Pneumothorax, Covid-19 and Healthy CXRs	5-fold cross-valid (3500)	97.20
[23]	Normalization-Free Network + progressive image resizing	4334	224× 224	Multi-class	3	TB, Healthy and other lung disease CXRs	2670	96.91
				Binary	3	Combination of public CXRs	1400	95.91
[9]	CNN + Grid search optimization	1400	224× 224	Multi-class	6	TB, LO, LC, PN, Covid-19 and Healthy CXRs	9415	98.75
[8]	Wavelet features + MBConvNet	1837	224× 224	Multi-class	5	TB, bacterial PN, Viral PN, Covid-19 and Healthy CXRs	13020	95.50

The convolutional models used [11–16] are pretrained on the ImageNet dataset[2]. On the other hand, we used the ViT model [10] pretrained on the ImageNet-21k dataset[3]. Employing transfer learning from natural images aims to compensate for the lack of large CXR dataset for TB detection, enabling effective training of large DL models.

The convolutional based models [11–16] are constructed by sequentially adding three layers on top of the final convolutional block output. These layers are a concatenation of global average pooling (GAP) and global max pooling (GMP), a dropout and a dense connected layer of three neurons with softmax activation (i.e. output layer). Combining GMP and GAP layers enables capturing richer discriminative representation from images with reduced dimensionality to avoid model overfitting.

---

[2] ImageNet dataset is composed of 1 million images belonging to 1000 classes. (http://www.image-net.org/challenges/LSVRC/2012/)

[3] ImageNet-21k dataset is composed of 14 million images with 21000 classes (http://www.image-net.org/).

The ViT model [10] is built by stacking a dropout layer on top of the final ViT network's classification output (i.e. CLS output) followed by the output layer.

The incorporation of dropout before the output layer in all the DL models aims to promote learning relevant and generalizable representations by introducing randomness during training.

4 Experimental Setup

Experiments conducted in this study are carried out in Google Colaboratory providing a 12GB RAM and a Tesla T4 GPU. All the DL models are built using Tensorflow and Keras by instantiating the convolutional models [11–16] from Keras applications and the ViT model [10] from Hugging face library. The following of this section presents the details related to training and evaluation process.

4.1 Dataset

The dataset used for experiments namely, TBX11K dataset[4] is created by [7] for TB detection. This dataset comprises 800 TB CXRs, 3800 healthy CXRs and 3800 CXRs belonging to various other pulmonary diseases (see Fig.1 below). During experiments, this dataset is randomly split into 70%, 10% and 20% (see Table 2) to compose the training, validation and test sets, respectively.

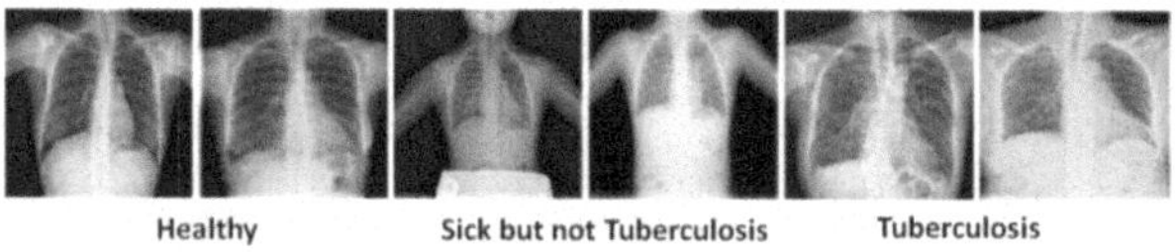

Fig.1. Chest X-Rays from TBX11K dataset [7].

Table 2. Detail of training, validation and test sets.

Dataset	Type	Training Set	Validation Set	Test Set	Total
TBX11K	Normal	2660	380	760	3800
	Sick but not TB	2660	380	760	3800
	TB	560	80	160	800
	Total	5880	840	1680	8400

[4] https://www.kaggle.com/datasets/usmanshams/tbx-11

4.2 Training Hyperparameters

All the DL models are trained for 25 epochs with a batch size set to 16 and categorical cross entropy as loss function. The Adam optimizer is employed with a scheduled learning rate (LR) that starts from 1e-4, and it is decreased by a factor of 0.1 after every 3 training epochs without loss improvement until reaching a minimum value fixed to 1e-7. All CXRs are resized to a common dimension of 224×224 pixels and scaled by dividing each pixel by 255. The dropout rate is set to 0.3 for all the DL models. These specifications are set empirically.

4.3 Performance Evaluation Metrics

The comparative study relies on seven performance metrics including accuracy, precision, recall, specificity, f1-score, false positive (FPR) and false negative (FNR) rates. These metrics are defined in the equations (1), (2), (3), (4), (5), (6), and (7), respectively. In addition, we plot the confusion matrices and the ROC curves.

$$\text{Acc} = (\text{TP} + \text{TN}) / (\text{TP} + \text{TN} + \text{FP} + \text{FN}) \tag{1}$$

$$\text{Rec} = (\text{TP}) / (\text{TP} + \text{FN}) \tag{2}$$

$$\text{Prec} = (\text{TP})/(\text{TP} + \text{FP}) \tag{3}$$

$$\text{F1 - score} = (2 \times \text{Rec} \times \text{Prec}) / (\text{Rec} + \text{Prec}) \tag{4}$$

$$\text{Spec} = (\text{TN}) / (\text{TN} + \text{FP}) \tag{5}$$

$$\text{FPR} = (\text{FP}) / (\text{FP} + \text{TN}) \tag{6}$$

$$\text{FNR} = (\text{FN}) / (\text{FN} + \text{TP}) \tag{7}$$

5 Results and Discussion

This section presents the experimental results of the comparative study between the convolutional-based and the transformer-based models in TB detection from CXRs.

Table 3 below presents the accuracy, recall, precision, f1-score and specificity scores recorded on the test set for each DL model. These results highlight the superiority of the convolutional-based models [11–16] over the ViT [10] in CXR classification for TB detection. ResNet50 [12] achieved the highest scores with no statistically significant difference when compared to other convolutional models.

Despite the high performance of the ViT model [10] reported in various computer vision tasks, but its generalizability to CXR classification for TB detection seems to be limited. Indeed, pulmonary disease manifestations are observed as local spots and increased opacity areas in CXRs. In such a context, the ViT model [10] notably effective in detecting potential relations between distant regions within images showcased to be less suitable. Conversely, the deep convolutional-based models using deep hierarchical feature extraction performed the best, as they can perfectly capture these kinds of patterns.

Table 3. Accuracy, recall, precision, specificity and F1-score values.

Model	Acc (%)	Rec (%)	Prec (%)	Spec (%)	F1-Score (%)
Xception	99.04	97.81	98.61	99.54	98.21
InceptionNetV3	98.92	97.56	98.86	99.48	98.21
VGG16	98.98	97.93	98.41	99.49	98.17
ResNet50	**99.22**	**98.27**	**99.08**	**99.62**	**98.67**
DenseNet201	99.04	97.81	98.78	99.54	98.29
EfficientNetV2S	98.75	97.10	98.37	99.40	97.73
ViT	98.09	95.63	97.68	99.08	96.64

The ROC curves presented in Fig. 2 below demonstrate the high performance of the convolutional-based models in differentiating between 'TB', 'sick' and 'healthy' classes by achieving a maximum area under the ROC curves for all classes. In contrast, the ViT model ROC curves (blue curves) highlight a lower performance in recognizing TB samples compared to other classes.

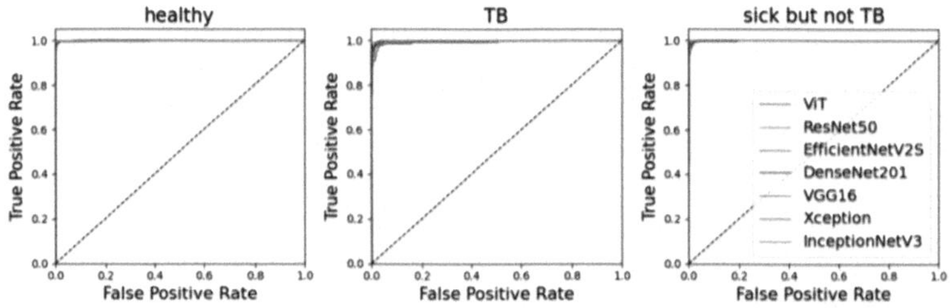

Fig. 2. ROC Curves of DL models.

Table 4 below presents the FNRs and FPRs calculated per class. The convolutional models achieved the best rates while the ViT model obtained the worst rates. Furthermore, the worst FNRs for all the DL models are recorded on the 'TB' class and the worst FPRs are obtained on the 'Sick' class. This is mainly owed to the similarity between 'TB' and 'Sick' classes, making the DL models face difficulty on recognizing TB CXRs and misclassify several of them as 'Sick'.

The comparison of confusion matrices presented in Fig. 3 showcases the low performance of the ViT model which failed even on differentiating between several diseased and healthy CXRs. On the other hand, assessing these matrices reveals that the primary error pattern of the DL models is either misclassifying TB samples as other pulmonary diseases ("sick but not TB" class) or vice versa. These errors arise from the low inter-class variability between these classes because of the similarity between lesions caused by TB and those caused by other pulmonary pathologies when observed in CXRs.

Table 4. False positive and false negative rates.

Model	False Positive Rate (%)			False Negative Rate (%)		
	Healthy	TB	Sick	Healthy	TB	Sick
Xception	**0.00**	0.26	1.30	0.52	5.62	**0.39**
InceptionNetV3	0.43	**0.13**	1.30	0.52	6.25	0.52
VGG16	0.10	0.32	1.19	**0.39**	5.00	0.78
ResNet50	0.21	**0.13**	**0.97**	**0.39**	**4.30**	**0.39**
DenseNet201	0.21	0.19	1.19	**0.39**	5.62	0.52
EfficientNetV2S	0.43	0.26	1.41	**0.39**	7.50	0.78
ViT	1.08	0.32	1.84	0.92	11.25	0.92

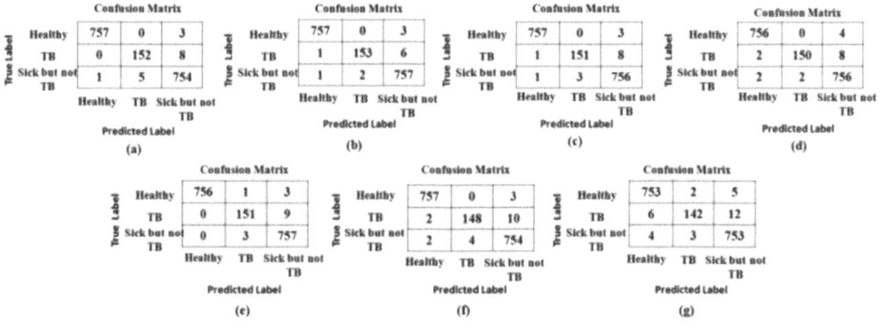

Fig. 3. Confusion matrices: (a) VGG16 [11], (b) ResNet50 [12], (c) DenseNet201 [15], (d) InceptionNetV3 [13], (e) Xception [14], (f) EfficientNetV2S [16], and (g) ViT [10].

6 Conclusion

This paper presented a comparative study between the convolutional-based and transformer-based DL models in TB detection from CXRs. This study focused on assessing the effectiveness of DL models in distinguishing TB from other lung diseases. Seven DL models were selected: VGG16, InceptionNetV3, ResNet50, DenseNet201, Xception, EfficientNetV2 and the ViT. These models were trained and evaluated on CXRs belonging to three classes: healthy, TB and other lung disease.

Empirical results revealed the superiority of the convolutional-based models over the ViT in CXR classification for TB detection in measure of all the performance evaluation metrics. Indeed, the convolutional networks enable capturing intricate patterns in images via deep hierarchical feature extraction. This qualifies them the best fit for automatic TB diagnosis using CXRs, whereby subtle details may indicate a potential abnormality. In contrast, the ability of the ViT model to learn spatial relationships in images was demonstrated to be less effective for this task.

Due to the similarity between manifestations of pulmonary diseases in CXRs, all the DL models tested in this study showcased confusion between TB and other lung diseases. This underscores the need for ongoing research to overcome this limitation by investigating the adaptability of DL models pretrained on medical images using large and balanced CXR datasets.

Disclosure of AI Tools Usage. The preparation of this manuscript involved the use of ChatGPT 3.5 to check grammar errors and improve English style. Subsequently, the authors reviewed and edited the content as necessary, and take full responsibility for the paper's content.

Disclosure of Interests. The authors have no competing interests to declare.

References

1. Rahman, T., et al.: Reliable tuberculosis detection using chest X-ray with deep learning, segmentation and visualization. IEEE Access **8**, 191586–191601 (2020)
2. Jaeger, S., et al.: Automatic tuberculosis screening using chest radiographs. IEEE Trans. Med. Imag. **33**(2), 233–245 (2013)
3. Roopa, N.K., Mamatha, G.S.: CLBO: chef leader-based optimization enabled deep learning for tuberculosis detection using x-ray images. SIViP **18**, 877–887 (2023). https://doi.org/10.1007/s11760-023-02732-7
4. Abideen, Z.U., et al.: Uncertainty assisted robust tuberculosis identification with Bayesian convolutional neural networks. IEEE Access **8**, 22812–22825 (2020)
5. Tasci, E., Uluturk, C., Ugur, A.: A voting-based ensemble deep learning method focusing on image augmentation and preprocessing variations for tuberculosis detection. Neural Comput. Appl. **33**(22), 15541–15555 (2021)
6. Dey, S., Roychoudhury, R., Malakar, S., Sarkar, R.: An optimized fuzzy ensemble of convolutional neural networks for detecting tuberculosis from Chest X-ray images. Appl. Soft Comput. **114**, 108094 (2022)
7. Liu, Y., Wu, Y.H., Ban, Y., Wang, H., Cheng, M.M.: Rethinking computer-aided tuberculosis diagnosis. In: Proceedings of the IEEE CVF Conference on Computer Vision and Pattern Recognition, 2646–2655 (2020)
8. Bhosale, R.D., Yadav, D.M.: Customized convolutional neural network for pulmonary multi-disease classification using chest X-ray images. Multimedia Tools Appl. **83**(6), 1–35 (2023)
9. Ashwini, S., Arunkumar, J.R., Prabu, R.T., Singh, N.H., Singh, N.P.: Diagnosis and multi-classification of lung diseases in CXR images using optimized deep convolutional neural network. Soft. Comput. **28**(7), 1–15 (2023)
10. Dosovitskiy, A., et al.: An image is worth 16x16 words transformers for image recognition at scale. arXiv preprint arXiv:2010.11929 (2020)

11. Simonyan, K., Zisserman, A.: Very deep convolutional networks for large-scale image recognition. arXiv preprint arXiv:1409.1556 (2014)
12. He, K., Zhang, X., Ren, S., Sun, J.: Deep residual learning for image recognition. In: 2016 IEEE Conference on Computer Vision and Pattern Recognition (CVPR), 770–778 (2016). https://doi.org/10.1109/CVPR.2016.90
13. Szegedy, C., Vanhoucke, V., Ioffe, S., Shlens, J., Wojna, Z.: Rethinking the inception architecture for computer vision. In: Proceedings of the IEEE Conference on Computer Vision and Pattern Recognition, 2818–2826 (2016)
14. Chollet, F.: Xception deep learning with Depthwise separable convolutions. In: Proceedings of the IEEE Conference on Computer Vision and Pattern Recognition (CVPR) (2017)
15. Huang, G., Liu, Z., Van Der Maaten, L., Weinberger, K.Q.: Densely connected convolutional networks. In: Proceedings of the IEEE Conference on Computer Vision and Pattern Recognition, 4700–4708 (2017)
16. Tan, M., Le, Q.: Efficientnetv2: smaller models and faster training. In: International Conference on Machine Learning, 10096–10106 (2021)
17. Rajaraman, S., Antani, S.K.: Modality-specific deep learning model ensembles toward improving TB detection in chest radiographs. IEEE Access **8**, 27318–27326 (2020)
18. Urooj, S., Suchitra, S., Krishnasamy, L., Sharma, N., Pathak, N.: Stochastic learning-based artificial neural network model for an automatic tuberculosis detection system using chest X-ray images. IEEE Access **10**, 103632–103643 (2022)
19. Nafisah, S.I., Muhammad, G.: Tuberculosis detection in chest radiograph using convolutional neural network architecture and explainable artificial intelligence. Neural Comput. Appl. **36**(11), 1–21 (2022)
20. Iqbal, A., Usman, M., Ahmed, Z.: Tuberculosis chest X-ray detection using CNN-based hybrid segmentation and classification approach. Biomed. Sig. Process. Control **84**, 104667 (2023)
21. Duong, L.T., Le, N.H., Tran, T.B., Ngo, V.M., Nguyen, P.T.: Detection of tuberculosis from chest X-ray images boosting the performance with vision transformer and transfer learning. Expert Syst. Appl. **184**, 115519 (2021)
22. Tan, M., Le, Q.: Efficientnet: rethinking model scaling for convolutional neural networks. In: International Conference on Machine Learning, 6105–6114 (2019)
23. Acharya, V., et al.: AI-assisted tuberculosis detection and classification from chest X-rays using a deep learning normalization-free network model. Comput. Intell. Neurosci. **2022**(1), 2399428 (2022)
24. Mehrrotraa, R., et al.: Ensembling of efficient deep convolutional networks and machine learning algorithms for resource effective detection of tuberculosis using thoracic (chest) radiography. IEEE Access **10**, 85442–85458 (2022)
25. Ravi, V., Acharya, V., Alazab, M.: A multichannel EfficientNet deep learning-based stacking ensemble approach for lung disease detection using chest X-ray images. Cluster Comput. **26**(2), 1181–1203 (2023)
26. Karaddi, S.H., Sharma, L.D.: Automated multi-class classification of lung diseases from CXR-images using pre-trained convolutional neural networks. Expert Syst. Appl. **211**, 118650 (2023)
27. Sandler, M., Howard, A., Zhu, M., Zhmoginov, A., Chen, L.C.: MobileNetV2: inverted residuals and linear bottlenecks. In: 2018 IEEE/CVF Conference on Computer Vision and Pattern Recognition, pp. 4510–4520 (2018). https://doi.org/10.1109/CVPR.2018.00474

On the Joint Use of CNN and OC-PCA Classifier for Cross-Domain Potato Disease Identification

Ghania Boukhlifa(✉) and Youcef Chibani

Laboratoire d'Ingénierie des Systèmes Intelligents et Communicants (LISIC), Faculty of Electrical Engineering, University of Science and Technology Houari Boumediene (USTHB), 32, El Alia, Bab-Ezzouar 16111, Algiers, Algeria
{gboukhlifa,ychibani}@usthb.dz

Abstract. Convolutional Neural Networks (CNNs) are powerful tools for image classification when performed on the same dataset. However, when trying to use the trained CNN to classify images on unseen datasets, the performance is decreased during testing. This paper proposes an approach that integrates the One-Class Principal Component Analysis (OC-PCA) classifier with CNN-extracted features to create a flexible identification system. A comparative study between CNN and OC-PCA is conducted using various datasets. The combined system shows notable performance in handling varying unseen datasets, eliminating the need for architectural tuning and data augmentation. Extensive experiments performed across intra and inter-domains demonstrate the proposed system's ability to achieve accuracies from 82% to 100%. The proposed approach presents an improvement of the CNN adaptability to new datasets without the need for a new architecture and data augmentation while addressing challenges linked to the diversity of plant diseases.

Keywords: CNN · Oc-pca classifier · Plant village · Potato leaf disease · Plantdoc · Cross-domain evaluation

1 Introduction

Diseases affecting potato crops pose a significant threat to global agriculture, impacting food security and economic stability. Accurate and timely classification of potato leaf diseases is crucial. However, challenges arise from the diverse morphology of affected leaves, regional variations in symptoms, and fluctuating data acquisition conditions. CNNs have proven to be powerful tools in image classification tasks including plant disease identification [3]. Their ability to automatically extracting features enables an accurate classification. However, the main limitation of CNNs is their lack of flexibility in handling new datasets that are not used during training [13]. When faced with novel datasets, CNNs require retraining or architectural modifications, which can be time-consuming

B. Djamaa et al. (Eds.): CSA 2024, LNNS 1145, pp. 156–166, 2025.
https://doi.org/10.1007/978-3-031-71848-9_15

and resource-intensive. To address this limitation, this work proposes the use of OC-PCA [4] in conjunction with the pretrained CNN. OC-PCA used the PCA method and the principle of one-class classifiers [6], offering advantages in terms of flexibility, making it suitable for the dynamic nature of plant leaf diseases. By applying OC-PCA, a flexible classification can be developed by solely using the data of the unseen class, independently of others. Moreover, the efficiency of PCA in dimensionality reduction plays a crucial role in selecting the relevant features that correctly represent the class. By identifying the principal components contributing to the class variability, OC-PCA builds a robust model, leading to accurate classification.

The remaining of this paper is structured as follows: Sect. 2 reviews the existing works related to potato leaf disease identification. Section 3 provides a detailed desciption of the proposed method. Section 4 presents the results and the comparative analysis. Finally, the conclusion summarizes the key findings.

2 Related Works

Various studies have explored the application of deep learning (DL), data augmentation (DA), and processing techniques for potato leaf disease identification, performed primarily on datasets like PlantVillage (PV) [17] and Potato Leaf Diseases (PLD) [12]. Notable approaches include the use of DA with VGG-19 for obtaining an accuracy of 99.2% [19]. Shirale and Sayyad achieving 97% accuracy with a CNN-based approach[15], and Nishad et al. [10] combining K-means clustering and CNN for 96% accuracy. Optimized methods like Efficient-PNet, image processing with CNN [5,9,20], and comparisons of VGG16/VGG19 [16] by other researchers have also been successful. Similar efforts on the PLD dataset include Baihaki et al.'s comparison of CNN architectures [1]. Umi Khultsum and Ghofa's use DenseNet201 [7]for 95–99.6% accuracy. Saeed et al. [13] proposed a depthwise separable CNN with 99.8% and 99.7% accuracies on PLD and PV datasets. These works show that DL architectures showcase accurate classification when performed on the same dataset. However, when applied to a new dataset with distinct classes, the performance is dropped. Hence, it can be enhanced involving retraining CNNs on the new dataset, requiring then the redesign of the architecture. This approach is obviously time consuming.

In this paper, an alternative approch is proposed based on the One class classification principle. A previous study [11] applied this principle using One-Class Support Vector Machines (OC-SVMs). Exclusively trained on healthy or diseased leaves, OC-SVMs excel at detecting texture deviations indicative of disease. When faced with a new leaf image, the model compares it to learned classes, triggering outlier detection upon significant variations. This process handles the high variability in the data. Using this principle, an alternative One-Class classifier is proposed based on the use of OC-PCA classifier having an only adjustment parameter. Hence, the pretrained CNN on a dataset is used for feature extraction and fed to the OC-PCA classifiers for classification. This approach avoids to retrain the CNN when a new dataset is used for potato leaf disease classification.

3 Proposed System

3.1 Proposed System Overview

The proposed system is a CNN having two primary components CNN-FEB (Feature Extraction Block) and CNN-FRB (Feature Reduction Block) as depicted in Fig. 1. The resulting flattened features are fed to the classification block composed either a fully connected layer namely CNN-FCB (Fully Connected Block) or the OC-PCA classifier. For better clarity, each block is described in the next sections.

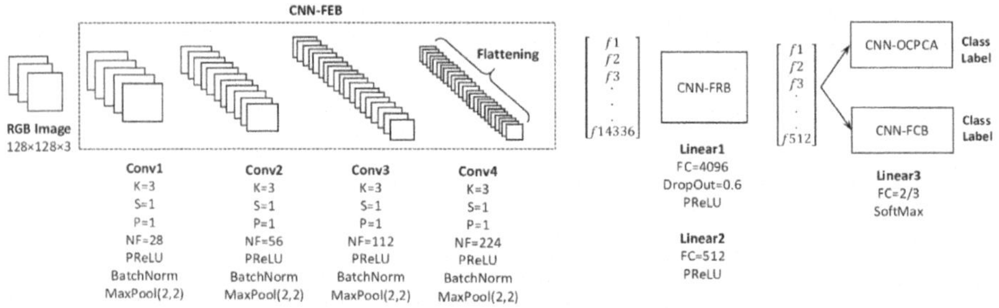

Fig. 1. Overview of the Proposed System.

3.2 CNN Architecture

The proposed CNN is divided into three main blocks, namely CNN-FEB, CNN-FRB and CNN-FCB. CNN-FEB comprises four convolutional layers (Fig. 1), each one is defined by parameters such as K (Kernel size), S (Stride), P (Padding), and NF (Number of filters) as well as PReLU activation. Each convolution is followed by a MaxPool layer for extracting relevant features in a hierarchical manner. Additionally, BatchNorm is incorporated after each convolution to enhance training stability. After flattening the feature maps, the resulting vector has 14336 elements which is reduced by the CNN-FRB to 512 elements via the $Linear1$ and $Linear2$, respectively. While, the CNN-FCB is dedicated to the classification composed of a fully connected layer $Linear3$, that allows generating the model's outputs. These outputs are then subjected to a softmax activation in order to get probabilities associated with classes.

3.3 Joint Use of the CNN Feature Extractor and OC-PCA Classifier

PCA is a statistical method for simplifying and extracting important information from datasets. It identifies significant variations by analyzing eigenvectors of the covariance matrix (CM) for selecting principal axes with the largest eigenvalues PC. PCA operates on a data matrix X, where each row represents a data point x. The CM namely C of the data matrix X is calculated as:

$$C = \frac{1}{n-1}\left(X - \bar{X}\right)^{T}\left(X - \bar{X}\right) \quad (1)$$

where n is the number of data samples and $\bar{X}$ is the mean of the data matrix. Then, a W matrix of PCs is formed by selecting the top k eigenvectors from the Q matrix obtained during the eigenvalue decomposition of C [4]. During the evaluation step, a given data point undergoes data reduction namely $x_{reduced}$ using the established W matrix, as follows:

$$x_{reduced} = W^{T}(x - \bar{X}) \quad (2)$$

This reduced representation is then transformed back into its original space via the following equation:

$$\widehat{x} = W x_{reduced} + \bar{X} \quad (3)$$

In order to appreciate the effective reduction of the data, a similarity measure based on the Pearson correlation coefficient is calculated between the original data x and its reconstructed counterpart $\widehat{x}$ as:

$$r = \frac{cov\left(x, \widehat{x}\right)}{\sigma_{x}\sigma_{\widehat{x}}} \quad (4)$$

where cov is the covariance, while σ_x and $\sigma_{\widehat{x}}$ are the standard deviations of the data and its reconstructed version, respectively. When performed on the classification, each disease class is represented by an individual OC-PCA model with its own projection matrix W_i where i $\in \{1, M\}$ and M is the number of classes. When classifying a data point x, the system assesses its similarity to each disease class model by calculating the similarity criterion r_i. The class label for the input data is then determined by selecting the output corresponding to the maximum similarity based on the Pearson ratio (Fig. 2) as:

$$Class\left(x\right) = argmax(r_i)_{i\in\{1,M\}} \quad (5)$$

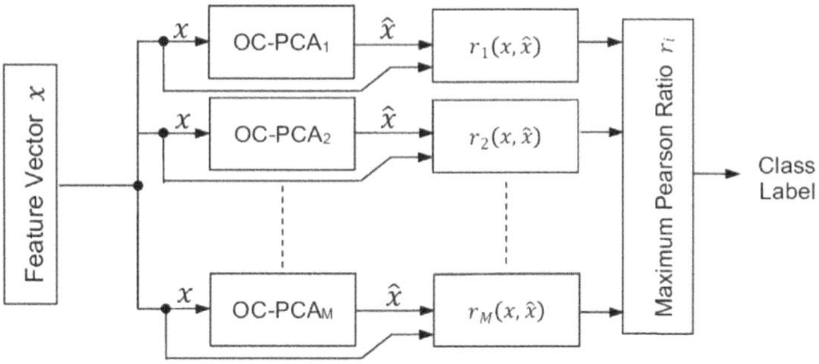

Fig. 2. Multi-class PCA classifier architecture.

4 Results

4.1 Data Description

PlantVillage (PV), Potato Leaf Diseases (PLD), and PlantDoc (PD) datasets are employed in the study (Table 1). PV and PLD images have a clear background and widely used in the state-of-the-art. While PD is collected from real conditions with complex backgrounds having two diseased potato classes. It is selected to justify the proposed approach's performance according to the paper's primary objective (Fig 3).

Table 1. Detailed Datasets Description (#C: Number of Classes)

Dataset	Ref	#C	Names of Classes	Image Size	#Images	Environnement
PV	[8]	3	Healthy, Early Blight, Late Blight	256×256	2152	Laboratory
PLD	[13]	3	Healthy, Early Blight, Late Blight	256×256	4161	Laboratory
PD	[18]	2	Early Blight, Late Blight	Variable	238	Field

(a) PlantVillage (b) Plant Leaf Disease (c) PlantDoc

Fig. 3. Sample Images from Each Dataset.

4.2 Evaluation Metrics

Various metrics are used for evaluating the classification performance. Accuracy is employed for assessing the proportion of correctly classified instances defined as:

$$Accuracy = \frac{\sum TP + TN}{\sum TP + TN + FP + FN} \tag{6}$$

such that TP and FP are instances correctly and incorrectly predicted as positive classes, respectively, while TN and FN are instances correctly and incorrectly predicted as negative classes, respectively. $F1 - Score$ is an additional metric used to evaluate class imbalance in classification tasks, defined as:

$$F1 - Score = \frac{2 \times Precision \times Recall}{Precision + Recall} \tag{7}$$

Precision and Recall are given as $Precision = \frac{\sum TP}{\sum TP+FP}$ and $Recall = \frac{\sum TP}{\sum TP+FN}$, respectively. Both $Accuracy$ and $F1 - Score$ metrics are ranged from

0 to 100%. An accuracy close to 100% indicates a high degree of precision in the model's predictions, while the F1-Score having a higher value indicating a more balanced and effective classification.

4.3 Experimental Protocol

The CNN is trained on a system with an Intel i7 CPU (Second generation) and 24 GB of memory. The training is conducted using the NVIDIA Quadro P1000 GPU, with 4 GB GDDR5 memory and CUDA support using PyTorch.

The first step tries to find the optimal CNN architecture using PV dataset. In this process, accuracy metric and cross-entropy loss are used in the CNN architecture design. Thereafter, the selected architecture is trained on the PLD and PD datasets. In this step, three CNNs are designed, each one is dedicated to a dataset namely CNN-PV, CNN-PLD and CNN-PD.

The second step involves feature extraction performed by each CNN model retrieved from the feature extractor of the CNN-FRB for feeding to the OC-PCA classifier. The main objective of these experiments is to evaluate more specifically the performance of the OC-PCA classifier when performed in a cross-domain.

4.4 CNN and OC-PCA System Co-Design

The CNN model is designed referring to the Cross Entropy Loss and the Accuracy metrics. Figure 4 demonstrates the performance achieved by the CNN model in terms of Loss and Accuracy on both the training and validation data. As can be shown, the training is attained within a short duration of only 70 epochs.

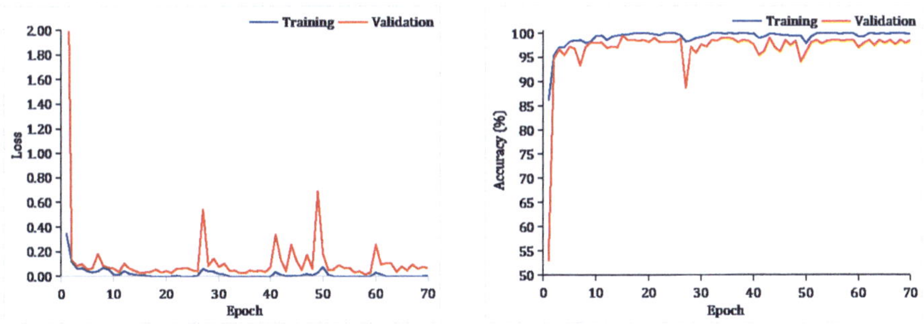

(a) Cross-Entropy Loss (b) Accuracy

Fig. 4. Loss and Accuracy versus Epoch for Training and Validation Datasets.

The OC-PCA design requires finding the optimal number of k PC for effectively separating data classes. Figure 5 provides an illustrative example for finding the optimal number of PC for OC-PCA classifiers when using CNN-PV for feature extraction and OC-PCA-PV for classification. As highlighted, the OC-PCA method exhibits high performance when the number of PC is around 48.

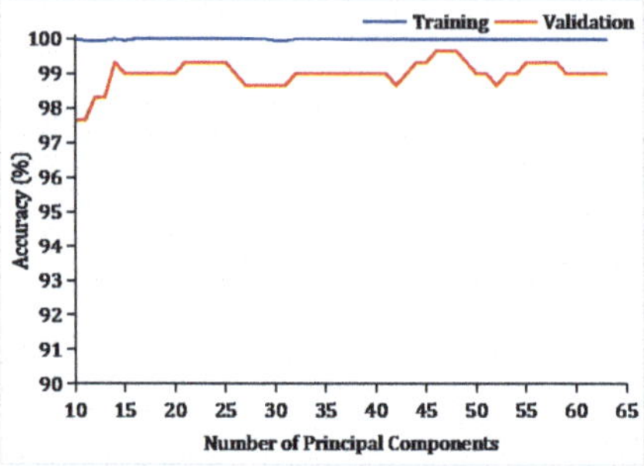

Fig. 5. Training and Validation Accuracies versus the Number of Principal Components for the PV dataset.

4.5 Experimental Evaluation

Intra Domain Evaluation: Each individual CNN is assessed on its dataset (intra-domain). The CNN achieves a training accuracy of 100% for PV and PLD, while 96% for PD. During testing, the offered accuracies are 99% for PV and PLD datasets while 100% for the PD. The lower accuracy offered by the PD during training can be justified by its variability since images are collected from real-world (Fig. 3.c). These results indicate that the CNN generalizes from the training data across datasets.

For OC-PCA, during training it achieves an accuracy of 100% for PV, 100% for PLD, and 97% for PD. In the testing phase, OC-PCA offers accuracies of 98% for PV, 98% for PLD, and 94% for PD. This highlight OC-PCA's strength in distinguishing classes. However, it is worth noting that OC-PCA's accuracy tends to be lower compared to the CNN, emphasizing the trade-off between OC-PCA's flexibility and simplicity of design versus its limitations when applied to the same dataset, compared to the CNN's performance.

Regarding class imbalance, PV is the most imbalanced because the Healthy class contains only 152 images compared to 1000 images for Early Blight and Late Blight. This is reflected in the degradation of the F1-Score, with a decrease to 95% for the CNN and 93% for OC-PCA, (Table 2). In contrast, PLD exhibits a more balanced distribution with 816, 1303, and 1142 images for Healthy, Early Blight, and Late Blight respectively. This balance contributes to an optimal F1-Scores of around 98%. In the case of PD, comprising only Early Blight and Late Blight, a constant F1-Score of 100% is achieved for CNN, while OC-PCA maintains a strong performance with an F1-Score of 94%. The regular distribution in PLD and PD shows the impact of class balance on the model performance.

Table 2. F1-Score Class Performance Evaluation Expressed in %

Dataset	PV			PLD			PD	
Classes	Healthy	Early Blight	Late Blight	Healthy	Early Blight	Late Blight	Early Blight	Late Blight
CNN	95	100	97	97	98	97	100	100
OC-PCA	93	100	98	98	97	98	94	94

Cross Domain Evaluation: A cross-domain evaluation is conducted to assess the system performance in handling data from different unseen domains. Table 3 depicts the cross domain testing using the CNN model. During the cross-domain evaluation, the CNN achieves not quite satisfactory testing accuracies across the three datasets. This indicates the CNN's limitation to generalize to different data domains. In addition, the absence of the accuracy values as can be seen in table 3 is due to the fixed nature of the CNN. Indeed, in the cases of PV and PLD, it is possible to apply cross-domain successfully. However, when performing to PD dataset, the implementation is not possible since this dataset contains only two classes. This limitation is the main drawback for CNNs when attempting to test a pre-trained model on a different dataset, as it lacks the necessary flexibility. In the other hand, the OC-PCA demonstrates promising accuracies on the testing subsets (Table 3) even when trained on a different dataset. The number of PCs varies across the training and testing dataset combinations. This indicates that different unseen datasets may behave differently from the training dataset and require a different number of components. Unlike the limitation with CNN, the application of OC-PCA eliminates this issue.

Table 3. Accuracies (%) of the cross domain evaluation on the testing dataset for CNN and OC-PCA performed on the PV, PLD and PD datasets with the corresponding optimal number of PC

Training dataset	Testing dataset								
	PV			PLD			PD		
	CNN	OC-PCA	#PC	CNN	OC-PCA	#PC	CNN	OC-PCA	#PC
PV	99	98	46	42	90	95	–	100	15
PLD	42	98	36	99	98	88	–	100	17
PD	–	95	15	–	82	18	100	94	56

4.6 Discussion

To contextualize the relevance of this study within the existing state of the art, two comparisons covering the intra-domain and the cross-domain datasets are conducted. It is important to note the absence of prior research using the PD dataset for comparison. When using the intra-domain datasets, as reported in Table 4, various advanced algorithms are harnessed incorporating DA, producing significant results. This suggests that the proposed approach demonstrates its competitiveness along state-of-the-art methods without DA. For both PV and PLD datasets, the CNN achieves accuracies of 99.0%, competing with deeper models such as VGG-19 (99.2%) and Depthwise-Separable CNN (99.7%) for PV dataset while VGG-16 (99.6%) and GoogleNet (99.3%) for PLD dataset. For the CNN-OC-PCA, PV dataset delivers an accuracy of 98.0% which is slightly comparable to the CNN's accuracy. However, the CNN-OC-PCA still surpasses many other models when performed in the same domain. For PLD , it achieves an accuracy of 98.0%, comparable to top-performing models like ResNet-50 (98.6%) without needing DA comparatively to other deep learning architectures.

When using the cross-domain datasets, the proposed combination of CNN-FEB-FRB and OC-PCA classifier highlights clearly its performance across different datasets. More precisely, when training the CNN-FEB-FRB on PV and testing on PLD, a prior study using PDCNN achieved an accuracy of 42% [12] while the proposed CNNFEB-FRB-FCB also reaches an accuracy of 42%, which demonstrates the challenge of cross-domain testing for standard CNNs. However, with OC-PCA, a remarkable accuracy of 98% is achieved, surpassing not only the standard CNN but also the PDCNN. This substantial improvement in cross-domain underscores the strength and adaptability of CNN-OC-PCA in handling variations between datasets.

For PD, the direct comparison with other datasets may not be possible, however, it is still possible to highlight its performance within its domain. Training and testing accuracies for CNN are impressive, with 100% accuracy during testing. This indicates that CNN excels in recognizing diseases within this dataset, despite potential variations in disease classes (Fig. 3.c). However, when considering the cross-domain datasets, as reported in Table 3, where gaps in accuracies are observed, it becomes obvious that the standard CNN exhibits inflexibility due to the different number of classes in PD dataset. Testing the CNN trained on PV and PLD datasets on the PD dataset yields no results due to the mismatch in the number of classes. This underscores the limitation of standard CNNs in adapting to datasets with different class structures.

Table 4. Comparison against the state of the art for potato leaf disease classification (DA means Data Augmentation)

Ref	Method	Dataset	DA	Accuracy (%)
[19]	VGG-19	PV	Yes	99.2
[15]	CNN	PV	No	97.0
[5]	Image processing+CNN	PV	Yes	96.8
[20]	Image processing+CNN	PV	Yes	97.0
[16]	VGG16/VGG19	PV	Yes	91.0/93.0
[1]	AlexNet/VGG-16/GoogleNet/ResNet-50	PLD	Yes	95.0/99.6/99.3/98.6
[13]	Depthwise-Separable CNN	PV/PLD	Yes	99.7/99.8
[14]	InceptionV3/ResNet-152V2/CNN	PV	Yes	84.0/97.5/94.0
[2]	DenseNet201	PLD	Yes	96.0
Proposed system				
Method 1	**CNN**	**PV**	**No**	**99.0**
		PLD	**No**	**99.0**
Method 2	**CNN-OC-PCA**	**PV**	**No**	**98.0**
		PLD	**No**	**98.0**

5 Conclusion

This study addresses the limitation of CNNs in handling various datasets for potato leaf disease identification. The proposed framework integrates OC-PCA into the identification system, offering a flexible and adaptable approach. Through the comparative analysis and cross-domain evaluation, the study demonstrates the coherence of combining CNN features with the OC-PCA classifiers, leading to significant improvements in classification performance, handling diverse classes with an absolute flexibility. Furthermore, this work can be extended to evaluate the importance of each module via an ablation study. Finally, this research opens the way for exploring the extention of the proposed system to other crops and datasets in order to confirm its robustness regardless of the conditions.

Disclosure of AI Tools Usage. The preparation of this manuscript involved the use of ChatGPT to correct the spelling, the grammar and for enhancing the language. Subsequently, the authors reviewed and edited the content as necessary, and take full responsibility for the paper's content.

References

1. Baihaki, R.I., Dafik, Agustin, I.H., Ridlo, Z.R., Kurniawati, E.Y.: The comparison of convolutional neural networks architectures on classification potato leaf diseases. In: Agustin, I.H. (ed.) ICONNSMAL 2022, pp. 125–145. Atlantis Press International BV, Dordrecht (2023). https://doi.org/10.2991/978-94-6463-174-6_12
2. Charisma, R.A., Adhinata, F.D.: Transfer learning with densenet201 architecture model for potato leaf disease classification. In: 2023 International Conference on Computer Science, Information Technology and Engineering (ICCoSITE), pp. 738–743. IEEE (2023)
3. Dhaka, V.S., Meena, S.V., Rani, G., Sinwar, D., Ijaz, M.F., Woniak, M.: A survey of deep convolutional neural networks applied for prediction of plant leaf diseases. Sensors **21**(14), 4749 (2021)
4. Hadjadji, B., Chibani, Y., Nemmour, H.: An efficient open system for offline handwritten signature identification based on curvelet transform and one-class principal component analysis. Neurocomputing **265**, 66–77 (2017)
5. Kasani, K., Yadla, S., Rachamalla, S., Hariharan, S., Devarajula, L., Andraju, B.P.: Potato crop disease prediction using deep learning. In: 2023 IEEE 12th International Conference on Communication Systems and Network Technologies (CSNT), pp. 231–235. IEEE (2023)
6. Khan, S.S., Madden, M.G.: One-class classification: taxonomy of study and review of techniques. Knowl. Eng. Rev. **29**(3), 345–374 (2014)
7. Khultsum, U., Taufik, G.: Komparasi kinerja densenet 121 dan mobilenet untuk klasifikasi citra penyakit daun kentang. JURIKOM (Jurnal Riset Komputer) **10**(2), 558–565 (2023)
8. Mohanty, S.P., Hughes, D.P., Salathé, M.: Using deep learning for image-based plant disease detection. Front. Plant Sci. **7**, 1419 (2016)

9. Nazir, T., Iqbal, M.M., Jabbar, S., Hussain, A., Albathan, M.: EfficientPNet: an optimized and efficient deep learning approach for classifying disease of potato plant leaves. Agriculture **13**(4), 841 (2023)
10. Nishad, M.A.R., Mitu, M.A., Jahan, N.: Potato leaf disease classification using k-means cluster segmentation and effective deep learning networks. In: Anwar, S., Ullah, A., Rocha, Á., Sousa, M.J. (eds.) ICITA 2022, pp. 35–45. Springer, Singapore (2023). https://doi.org/10.1007/978-981-19-9331-2_4
11. Pantazi, X., Moshou, D., Tamouridou, A.: Automated leaf disease detection in different crop species through image features analysis and one class classifiers. Comput. Elec. Agric. **156**, 96–104 (2019). https://doi.org/10.1016/j.compag.2018.11.005
12. Rashid, J., Khan, I., Ali, G., Almotiri, S.H., AlGhamdi, M.A., Masood, K.: Multilevel deep learning model for potato leaf disease recognition. Electronics **10**(17) (2021). https://doi.org/10.1109/10.3390/electronics10172064
13. Saeed, M.U., Ali, G., Rasheed, M., Mansoor, T., Abbas, A.W., Mansoor, Q.: DSCNN: a novel depthwise separable deep learning approach for automatic potato leaf disease classification. J. Jilin Univ. **42**(02) (2023)
14. Shewale, C., Sawadkar, A., Kadale, K., Sangle, P., Saswadkar, S.: Plant disease identification using convolutional neural network and transfer learning. IJRASET **11**(1) (2023)
15. Shirale, U.D., Sayyad, A.: Classification of potato plant leaf diseases using convolution neural networks. ESP JETA **3**(1), 98–106 (2023)
16. Sholihati, R., Sulistijono, I.A., Risnumawan, A., Kusumawati, E.: Potato leaf disease classification using deep learning approach. In: 2020 International Electronics Symposium (IES), pp. 392–397 (2020). IEEE. https://doi.org/10.1109/IES50839.2020.9231784
17. Singh, D., Jain, N., Jain, P., Kayal, P., Kumawat, S., Batra, N.: Plantdoc: A dataset for visual plant disease detection. In: Proceedings of the 7th ACM IKDD CoDS and 25th COMAD, pp. 249–253 (2020)
18. Singh, D., Jain, N., Jain, P., Kayal, P., Kumawat, S., Batra, N.: PlantDoc: a dataset for visual plant disease detection. arXiv preprint arXiv:1911.10317 (2019)
19. Upadhyay, N., Gupta, N.: Potato leaves disease detection with data augmentation using deep learning approach. In: Joshi, A., Mahmud, M., Ragel, R.G. (eds.) Information and Communication Technology for Competitive Strategies (ICTCS 2022): ICT: Applications and Social Interfaces, pp. 589–599. Springer, Singapore (2023). https://doi.org/10.1007/978-981-19-9638-2_51
20. Verma, R., Mishra, R., Gupta, P., Trivedi, S., et al.: CNN based leaves disease detection in potato plant. In: 2023 6th International Conference on Information Systems and Computer Networks (ISCON), pp. 1–5. IEEE (2023)

Computer Networks and Distributed Systems

Memory-Aware Scheduling for Time-Sensitive Many-Core Systems

Omar Kermia(✉), Ibrahim Mezzah, and Ramdane Kechad

Embedded System Design Team, Centre de Développement des Technologies Avancées (CDTA), Algiers, Algeria
okermia@cdta.dz

Abstract. As part of time-critical systems, designers of applications running upon multicore platforms face a complicated timing analysis because of inter-core interference. A real-time shared memory conflict-free schedule stands for an efficient alternative preventing interferences among I/O transactions from distributed applications. This paper introduces a task scheduling model to exploit multicore for real-time applications which is an extension of the 3-phase task model. A/R (Acquisition/Restitution) phases of tasks are non preemptive and strictly periodic whereas the execution phase is preemptive and simply periodic. In addition, we propose an efficient scheduling algorithm called SS-EDF and show its effectiveness throughout comparison tests. ...

Keywords: Hard Real-Time Systems · Shared Memory Multicore Platform · PREM · Schedulability Analysis

1 Introduction

Software-based solutions to enhance the predictability of real-time systems may be split mainly into two classes: (i) solutions with system-level control software and (ii) solutions where applications are modified [1]. Both solutions tend for ensuring predictability by achieving isolation among cores to ensure non-interference due to I/O. However, in the first approach, applications remain as they are whilst the system software is modified to handle memory access. When, in the second approach, applications themselves are modified to avoid interferences. In the present paper, we are interested in the second approach by acting on the application model to mitigate the system's unpredictability. Pellizzoni et al. proposed in [2] a Predictable Execution Model (or PREM) to ensure the timing predictability of multicore applications. The PREM concept assumes that a specific intrusive hardware bridge regulates the accesses of I/O peripherals to the DRAM subsystem. This concept has been extended by Durrieu et al. in [4] to port an avionic application on a multicore target. This concept, known as the 3-phase task model, assumes a three phases AER task model (Acquisition, Execution, Restitution) where execution and communications are decoupled. First, within the acquisition phase, the code and data are loaded, into the private L2

B. Djamaa et al. (Eds.): CSA 2024, LNNS 1145, pp. 169–176, 2025.
https://doi.org/10.1007/978-3-031-71848-9_16

cache of a core that has been previously configured in SRAM, to be used by the following phase. Second, the task is executed during the execution phase since all resources are available. Finally, data produced during the second phase is sent to the memory during the restitution phase. The key idea is to execute only one acquisition or restitution phase at a time in order to increase predictability. Still with the second approach solutions, Biondi et al. [5] discussed the possibility of using the Logical Execution Time (LET) execution model for controlling the accesses to global memory. The LET paradigm that was introduced with the time-triggered programing language Giotto [6], decouples communication from task execution.

In the present work, we propose to extend the 3-phase task model to a more predictable shared resources model where the memory access is strictly pre-defined thanks to the strict periodicity constraint. It consists of only two phases where the acquisition and the restitution have been merged. On the one hand, a two-phases model reduces the programming complexity and simplifies the resources management. On the other hand, strict periodicity constraint enhances the control of application behavior when accessing the memory as this constraint is linked to highly dependable hard real-time systems. In addition, the strict-periodicity access to the memory allows synchronizing the task scheduling on the cores through a global scheduling strategy. We assume that tasks are scheduled upon a multicore platform according to a partitioning algorithm. To deal with these features, we propose SS-EDF for Strict Start-times EDF which is a variation of the classical EDF scheduling. The proposed real-time tasks model as well as the scheduling algorithm may be of great help for decreasing the inter-core interference and enhancing the predictability of applications running on multicore chips without any isolation means between cores.

2 Memory Aware Task Model

As mentioned above, the 3-phase task model consists of three phases: (i) an acquisition phase in which the task fetches data and instructions from the main memory to the core's memory where this task is executed, (ii) an execution phase in which the task executes without needing any access to the shared memory and (iii) a restitution phase in which the resulting data are moved from the core's local memory to the main memory. This way, tasks never access the bus during their execution which minimize any possible interference. However, it is worth noting that such an execution model in a multicore architecture needs an appropriate global scheduling strategy to handle the accesses to the shared memory. This, necessarily, implies the add of an additional layer such as a hypervisor for example [1] to manage global resource accesses. In order to address this issue, we propose an extension of the 3-phase task model in which the memory accesses are performed in a strictly periodic time [7] which ensures a shared time synchronization upon all cores. Moreover, we propose to consider only two phases instead of three in order to reduce the application overhead and make the analysis easier. The key idea is as tasks are periodic then we assume that the restitution phase of the ith job and the acquisition phase of the $(i+1)^{th}$ job of the same task may be seen as a single phase performing the sending of data produced during the previous execution phase to the memory in a first step then loading data for the next execution phase in a second step. This way, the proposed task model just like the 3-phase task model tends to obtain an interference-free execution in a multicore system.

Thereby, a task $\tau_i \in \Gamma$ consists of a pair $(\tau_i^{AR}, \tau_i^{E})$ where τ_i^{AR} and τ_i^{E} are, respectively, the A/R (Acquisition/Restitution) phase and the execution phase of task τ_i. The computation time of τ_i which is C_i is equal to: $C_i = C_i^{AR} + C_i^{E}$ while the period of τ_i is noticed T_i where obviously: $T_i = T_i^{AR} = T_i^{E}$. The relative deadline of τ_i is noticed D_i and we assume that $D_i^{E} = D_i$. It is worth noting that unlike the 3-phase task model where all phases are non-preemptive, under the proposed model only acquisition/restitution phase is non-preemptive while the execution phase may be preempted, but not lead to any interference. Moreover, only the periodicity of the A/R phase is considered as strict while the execution phases periodicity is only periodic as the well-known classical period, which is a less restricted than the strict period.

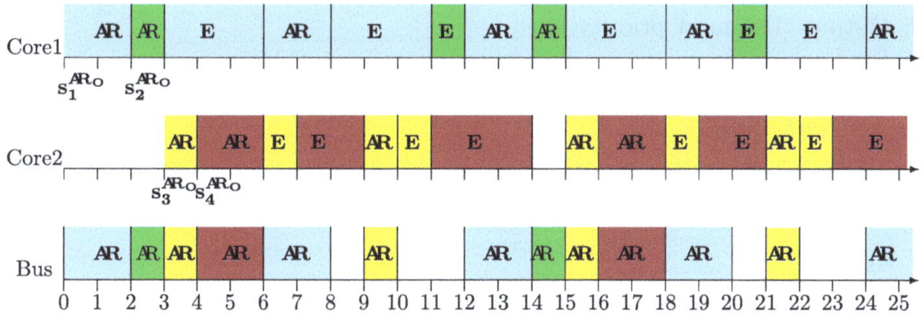

Fig. 1. Example of the proposed task model schedule

A/R phases, hence, are assumed to be non-preemptive and strictly periodic such that only a single A/R phase executes at a time across all cores as shown in Fig. 1. The highest priority is given to A/R phases over execution phases in order to prevent cross-core interference and to achieve isolation between cores. This way, $s_i^{AR_j}$ the start-time of the A/R phase of the jth job of τ_i is a function of the start-time of A/R phase of the first job $s_i^{AR_0}$ $(s_i^{AR_j} = s_i^{AR_0} + j \cdot T_i)$. The A/R phases of all tasks jobs share the bus and, hence, need to be scheduled on this bus. The strict start-times of these phases need to be defined off-line using an appropriate strategy (see Sect. 3) before launching the tasks scheduling. In Fig. 1, τ_1 and τ_2 (respectively the blue and the green tasks) are assigned to core 1 whilst τ_3 and τ_4 (respectively the yellow and the red tasks) are assigned to core 2. $T_1 = 6, T_2 = 12, T_3 = 6, T_4 = 12$ and $C_1 = 5, C_2 = 2, C_3 = 2, C_4 = 7$ are, respectively, their periods and computation times.

Computing times of A/R phases depend mainly on the exchanged data and consequently can be estimated. As well, their start-times are calculated beforehand after assessing their schedulability in such a way to ensure the isolation among the whole cores and makes the implementation easier. Thus, interferences emanating from data exchanged among cores are minimized without performing any change in the system structure. Moreover, we assume that the scheduling of tasks on the identical m cores is performed following a partitioned single-core scheduling algorithm, i.e. each task is assigned to a specific core at design time and each core schedules its subset of tasks at runtime by following this algorithm. In addition, strict periodicity of A/R phases necessarily impact tasks scheduling since it includes a static priority between tasks in addition to the earliest deadline priority. Indeed, a task execution phase τ_i^E may be suspended by a task A/R phase τ_j^{AR} even though $D_i \leq D_j$. We propose a strict start-time version of EDF, named SS-EDF (Algorithm 1), which handles single-core scheduling of the already defined tasks model.

Algorithm 1. SS-EDF

```
1: Input: Ready jobs set
2: Output: The most priority job
3: for Each job τ_i^j in the ready queue do
4:    if τ_i^j is a newly released job then
5:       SST_i ← s_i^{AR_0};
6:    else
7:       SST_i ← SST_i − 1;
8:    end if
9: end for
10: if The running job is not an A/R phase then
11:    if There exists a ready job τ_i^j such that SST_i = 0 then
12:       Return τ_i^j;
13:    else
14:       Return the highest priority job;
15:    end if
16: end if
```

As assumed before, the start-times of all A/R phases are known beforehand in addition to their feasibility. At every new task τ_i activation, the variable SST_i of the newly released job τ_i^j is initialized to $s_i^{AR_0}$ (lines 4 to 5) while SST_i of remaining jobs in the ready queue are decremented (line 7). Then, when a A/R phase is not running, if $SST_i = 0$ then τ_i is selected straight seeing that the current time t necessarily matches the start-time of the A/R phase of a given τ_i's job (lines 10 to 12). Otherwise, the scheduler selects from ready jobs the one with the earliest deadline (line 14). The complexity of SS-EDF is comparable to the one of EDF which is $O(n^2)$ (n is the number of tasks). On the other hand, every time the current time t matches one of the tasks strict start-times, SS-EDF complexity decreases to $O(n)$.

3 A/R Phases Schedulability Analysis

The main drawback of the strict-periodicity is its low schedulability ratio compared to the classical periodicity [7]. Thus, for that strictly periodicity of A/R phases does not degrade system schedulability, we propose to analyze the schedulability of A/R phases off-line in order to test the schedulability on the one hand and to have the start-times that fulfills both strict periodicities and deadlines constraints on the other hand.

Let us consider the case of of harmonic periods task set. Without any loss of generality, harmonic periodic tasks are assessed to be ordered in an increasing periods order (i.e. $T_1 \leq T_2 \leq ... \leq T_n$). Within harmonic periods every period divides the period of every longer period. For example, the periods $\{10, 20, 40\}$ are harmonic whereas $\{10, 20, 50\}$ are not harmonic. First, it is clear that:

$$\forall \tau_{i,i\neq 1},\ C_i^{AR} \leq T_1 - \sum_{\tau_q, T_q = T_1} C_q^{AR} \tag{1}$$

From this result, it is interesting to note that for the special case of harmonic periods tasks where $\forall i, j, T_i \neq T_j$ (each task has a distinct period) it is possible to infer a necessary and sufficient schedulability condition.

$$C_1^{AR} + \max_{T_i \neq T_1}(C_i^{AR}) \leq T_1 \tag{2}$$

In fact, A/R phases of harmonic tasks, where periods are two by two distinct, are schedulable if and only if condition 2 is met. The proof is quite elementary since each interval of $[(i-2) \cdot T_1 + C_1^{AR}, (i-1) \cdot T_1)$ allows to schedule each instance of A/R phase τ_i^{AR} $(i = 2, 3, ..., n)$ and for this largest C_i^{AR} value should be less or equal to the length of this interval.

The proposed schedulability condition is based on the calculation of a new parameter LS for Largest Sum that is given by Algorithm 2. The latter proceeds to split the task set Γ into a number of subsets Γ_k where $k \in \{\frac{T_2}{T_1}, ..., \frac{T_i}{T_1}, ..., \frac{T_n}{T_1}\}$ (duplicates are removed). Tasks within each subset Γ_k are sorted in a decreasing order of C_i^{AR}. Then, from the subset Γ_k every task τ_i^k such that its index id in this subset meets ($id \mod k \neq 1$) is removed. The value of LS is, hence, the sum of C_i^{AR} of the resulting subset. This way, tasks within each Γ_k are partitioned into k parts such that the sums of C_i^{AR} in each part are as similar as possible. Consequently, this minimizes the difference between the largest sums which are the values of ls_k (the only sums that are computed) and the smallest sums within this subsets of Γ_k.

Theorem 1. *Given a set of synchronous strictly periodic tasks such that ($T_1 \leq T_2 \leq ... \leq T_n$). If*

$$LS \leq T_1 - \sum_{T_i = T_1} C_i^{AR} \tag{3}$$

Algorithm 2. *LS* calculation

From Γ, build $\Gamma_k = \{\tau_1^k, \tau_2^k, ...\}$ subsets, $k = \frac{T_2}{T_1}, ..., \frac{T_i}{T_1}, ..., \frac{T_n}{T_1}$,
for each Γ_k **do**
    Tasks within each subset Γ_k are sorted in a decreasing order of C_i^{AR}
    In this order, if $(\forall \tau_i,\ id_i \mod k \neq 1)$ then τ_i^k is removed from Γ_k
    $ls_k = \sum_{\tau_i^k \in \Gamma_k} (C_i^{AR})$
end for
$LS = \sum_k ls_k$

then A/R phases are schedulable.

Sketch of proof: let us assume that the condition (3) is met for A/R phases but for all possible schedulings, an instance of at least one task τ_q is always scheduled after time T_q. Without any loss of generality we assume that it concerns the first instance. for every set $\{\tau_1, \tau_2, ..., \tau_{q-1}\}$ which is the set of tasks scheduled in one of intervals of length T_1, we have: $C_1^{AR} + C_2^{AR} + ... + C_{q-1}^{AR} + C_q^{AR} > T_1$. If we gather these tasks following their periods then the last equation becomes: $\sum_{T_i=T_1} C_i^{AR} + ... + \sum_{T_i=T_q} C_i^{AR} > T_1$ which contradicts 3.

The non harmonic A/R phases case is a more intractable problem than that of harmonic case because schedulability of non-harmonic tasks is related to their periods and computing times. This problem has already been solved in [7].

4 Evaluations

In this section, we conduct simulation experiments to compare the performance of SS-EDF under the proposed model with a global EDF approach with the 3-phase model that we denote AER-EDF. In these experiments, we evaluate the schedulability ratio criteria which is the ratio between schedulable task sets and the total amount of task sets tested. The UUnifast-Discard algorithm was used to generate the utilization u_i of each task τ_i ($u_i < 1$). Non-harmonic periods were chosen in the set $\{2^x 3^y 5^z,\ x, y, z \in [0, 4]\}$ whereas, harmonic periods are given by: $T_i = r_i \cdot T_{i-1}$ where $r_{i,(1 \leq i \leq n)}$ is a period ratio chosen randomly from the interval $[1, 6]$ (n is the number of tasks). The sizes of the A-phases and R-phases are decided by the A-ratio and R-ratio which in all experiments are set to 10% of the whole task. As these two phases are merged in the proposed model, C_i^{AR} stands for 20% of C_i meaning that the utilization of the memory will always be 20% of the task sets utilization. the AER model has been implemented according to [10] where both Acquisition and Restitution phases length are equal to 10% of the task. Each algorithm has been tested for different utilizations ranging from 0 to 4 with increments of 0.1. The task set cardinality is set to 20 and the number of cores is set to be as large as the schedulability issue will only result from the bus memory access.

Figures 2-a and 2-b depict the schedulability ratios of SS-EDF and AER-EDF with regards to the total utilization of the task set for non-harmonic and harmonic period tasks respectively. It appears, quickly, that SS-EDF outperforms

AER-EDF throughout the tests. However, disparities appear more clearly with the harmonic period tasks tests. Moreover, the schedulability ratio of AER-EDF had a quick initial drop off from utilization 0 to 1 whereas, for the same utilizations, the schedulability ratios of SS-EDF decrease more slowly. At around a utilization of 2 AER-EDF and SS-EDF schedulability ratios decrease similarly until they reach 0 around of an utilization of 4. Figure 2-c depict the obtained results for the special case of harmonic tasks where $\forall i, j, T_i \neq T_j$. This time, the improvement is definitely more impressive and this is due to the fact that the A/R phases schedulability assessment is based on a necessary and sufficient condition (condition 2).

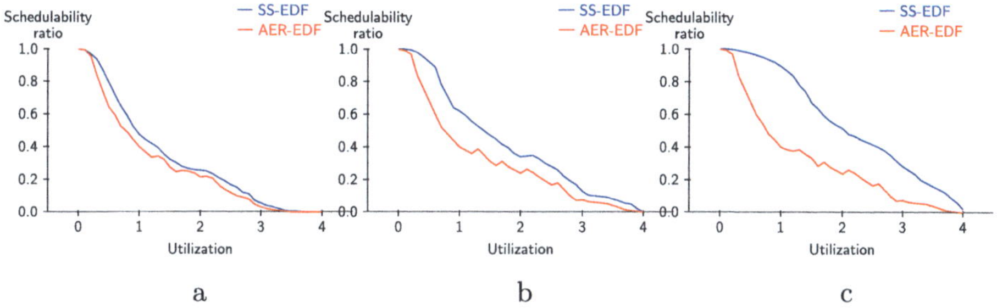

Fig. 2. Schedulability ratio versus utilization

5 Related Work

In [10] authors proposed a work close to ours. They targeted to improve results from an earlier paper by proposing an analysis for the 3-phase task model after reducing it to a single-core analysis. The LET model [6] assumes that a task is invoked and reads its input ports at the start of its LET interval, then performs a computation whose results are made available on its output ports exactly at the end of the LET, whereas in this paper, A/R phases are scheduled at different times as long as strict periodicity is fulfilled. Actually, the original LET model does not guarantee memory isolation and the recent works, which extend it to avoid memory contention, admit that this requires facing with several challenging design problems [5].

6 Conclusion

In this paper, we consider the problem of multicore real-time scheduling. We first present the scheduling model which achieves isolation among cores by ensuring non-interference due to I/O transactions. Tasks are seen as a junction of an A/R phase and an execution phase. The A/R phases within tasks are non preemptive and strictly periodic. Then, we propose to schedule tasks upon cores using a partitioned SS-EDF algorithm that has been introduced. Performed evaluations have shown the effectiveness of the proposed algorithm and its ability to provide a better memory access management.

References

1. Perret, Q., Maurère, P., Noulard, E., Pagetti, C., Sainrat, P., Triquet, B.: Temporal isolation of hard real-time applications on many-core processors. In: 2016 IEEE Real-Time and Embedded Technology and Applications Symposium (RTAS), Vienna, Austria, April 11–14, 2016, IEEE Computer Society, pp. 37–47 (2016)
2. Pellizzoni, R., et al.: A predictable execution model for cots-based embedded systems. In: 2011 17th IEEE Real-Time and Embedded Technology and Applications Symposium, pp. 269–279 (2011)
3. Betti, E., Bak, S., Pellizzoni, R., Caccamo, M., Sha, L.: Real-time I/O management system with COTS peripherals. IEEE Trans. Comput. **62**(1), 45–58 (2013)
4. Durrieu, G., Faugere, M., Girbal, S., Gracia Pérez, D., Pagetti, C., Puffitsch, W.: Predictable flight management system implementation on a multicore processor. In: ERTS 2014 (2014)
5. Biondi, A., Di Natale, M.: Achieving predictable multicore execution of automotive applications using the let paradigm. In: IEEE RTAS 2018, pp. 240–250 (2018)
6. Henzinger, T., Kirsch, C., Sanvido, M., Pree, W.: From control models to real-time code using Giotto. IEEE Control Syst. Mag. **23**, 50–64 (2003)
7. Kermia, O.: An efficient approach for the multiprocessor non-preemptive strictly periodic task scheduling problem. J. Syst. Archit. Embed. Syst. Des. **79**, 31–44 (2017)
8. Mok, A.K.L.: Fundamental design problems for the hard real-time environments, Ph.D. thesis, MIT (1983)
9. Davis, R.I., Burns, A.: Improved priority assignment for global fixed priority preemptive scheduling in multiprocessor real-time systems. Real-Time Syst. **47**(1), 1–40 (2011). https://doi.org/10.1007/s11241-010-9106-5
10. Maia, C., Nogueira, L., Pinho, L.M., Pérez, D.G.: A closer look into the AER model. In: 2016 IEEE 21st International Conference on Emerging Technologies and Factory Automation (ETFA), pp. 1–8. IEEE (2016)
11. Alhammad, A., Pellizzoni, R.: Schedulability analysis of global memory-predictable scheduling. In: International Conference on Embedded Software (EMSOFT), pp. 1–10 (2014)
12. Maia, C., Nelissen, G., Nogueira, L., Pinho, L.M., Pérez, D.G.: Schedulability analysis for global fixed-priority scheduling of the 3-phase task model. In: 2017 IEEE 23rd International Conference on Embedded and Real-Time Computing Systems and Applications (RTCSA), pp. 1–10 (2017)

Efficient Congestion and Popularity-Aware Caching Solution for Named Data Networking

Yakoub Mordjana(✉), Badis Djamaa, and Mustapha Reda Senouci

Ecole Militaire Polytechnique, BP 17, 16046 Bordj El Bahri, Algiers, Algeria
{yakoub.mordjana,badis.djamaa,mustaphareda.senouci}@emp.mdn.dz

Abstract. Named Data Networking (NDN) is a future Internet architecture that aims at the efficient utilization of network resources and improved data availability by enabling in-network caching. The efficiency of NDN depends, among others, on the performance of caching strategies, which depends on many parameters such as the caching decision policy, the replacement policy, and the forwarding algorithm. In this paper, we introduce a novel caching strategy, dubbed CPAC, that is both congestion- and popularity-aware. The aim is to store popular contents that are generally downloaded through congested and expensive links. This trade-off between the state of the link through which the content has been routed and the popularity of the content avoids caching unpopular contents that arrive from non-congested links. The efficiency of CPAC is demonstrated through comparative simulations on dedicated NDN platforms.

Keywords: In-Network Caching · Named Data Networking · Popularity-based Caching · Network Congestion.

1 Introduction

Named Data Networking (NDN) presents a prospective architectural approach for the future Internet, offering the capability to attain optimal content forwarding performance to end-users and ensuring robust data availability throughout the network. NDN constitutes a consumer-driven, data-centric architecture designed to supplant the classical IP host-centric architecture by directly naming content and using that name in the packet's forwarding process [2,15].

NDN routers communicate through two packet types: Interest and Data (Fig. 1), both encapsulating content names instead of source/destination addresses. Each NDN router manages three essential data structures [7], the Forwarding Information Base (FIB), the Pending Interest Table (PIT), and the Content Store (CS). The FIB acts as a routing table that maps name components to (inter)faces. The PIT retains all the Interests that a router has forwarded but not yet satisfied. The CS functions as a temporary cache of Data packets that the router has received and can be used to satisfy future interests.

B. Djamaa et al. (Eds.): CSA 2024, LNNS 1145, pp. 177–186, 2025.
https://doi.org/10.1007/978-3-031-71848-9_17

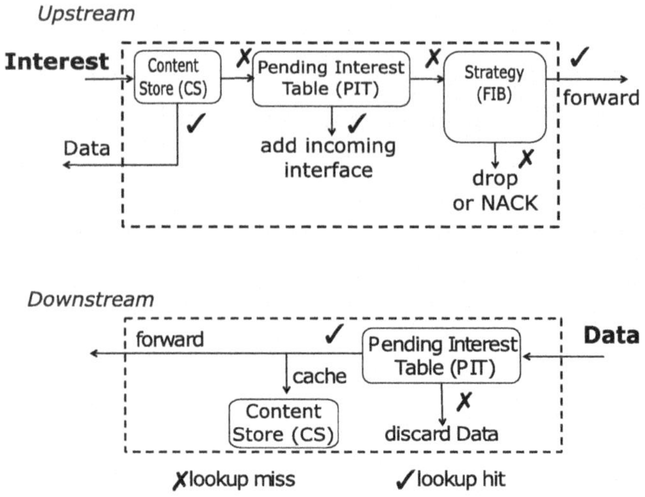

Fig. 1. Exchange of packets in an NDN node [15].

NDN ensures the high availability of content by placing the content near consumers, through a native in-network caching function. This process depends on several parameters, such as the caching decision strategy that decides which content to store and where it should be stored, the caching replacement policy that decides which content to evict if the cache is full, and the used forwarding strategy that helps meet the cached contents instead of reaching the producers.

In this regard, several issues arise such as (i) choosing the most suitable router to hold caches, (ii) choosing which content to cache, and (iii) forwarding interest packets to pass through a node storing a cached copy of matching content. Several solutions were proposed in the literature to define a caching strategy by considering several parameters that may intervene to make a caching decision such as the popularity of the content, the position of the node (at the edge or in-network), and the replacement policy when the cache is full. Another important issue in NDN is congestion which has been mainly treated as a separate problem. For instance, authors in [13] provided an analysis of NDN congestion control challenges without analyzing the effect of congestion on the caching strategy.

To deal with the aforementioned issues, we tackle in this paper, the problem of in-network caching while considering both the popularity of the content to be cached and the state of the links that provide such content. We devise a new Congestion- and Popularity-Aware Caching solution (CPAC). The main idea of CPAC is to store popular contents that are downloaded through congested links. This gives a trade-off between the state of the link through which the content has been routed and the popularity of the content. As a result, CPAC helps reduce caching operations by avoiding caching unpopular contents. Also, CPAC helps to decrease congestion and improve network performance by caching popular contents retrieved through congested links.

The remainder of this paper is organized as follows. An overview of caching strategies and replacement policies is given in Sect. 2. Related work are cited and discussed in Sect. 3. Section 4 describes CPACs design, while Sect. 5 evaluates its performance through simulations.

2 Overview on In-network Caching in NDN

When dealing with in-network caching, three main issues arise (Fig. 2):

- What and where to cache? This question is tackled by the *caching strategy*;
- Which data item to evict once the cache is full? This should be handled by the *replacement policy*;
- How to find cached items? This issue is related to the *forwarding strategy*.

In the sequel, we give a brief overview of the three aspects above-cited.

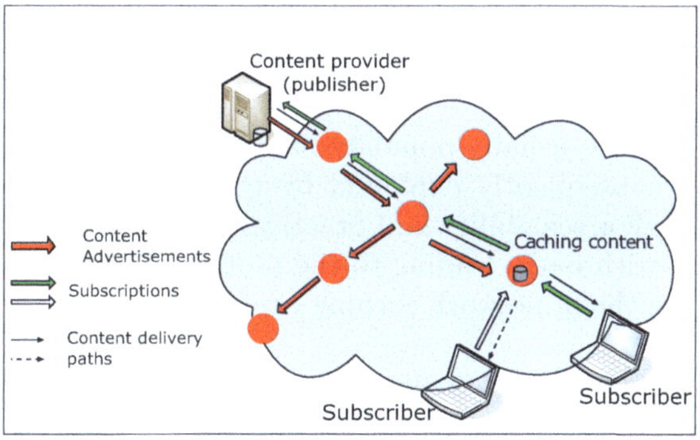

Fig. 2. In-network caching in NDN.

2.1 Caching Strategies

To answer the question of what and where to cache, the caching strategy defines the replication scheme used to spread copies of the content in the network. To decide on what, the characteristics of the content (e.g., popular or not) are usually considered. Whereas, the characteristics of the topology or node (e.g., the centrality of nodes) are considered to decide where.

Several caching strategies have been proposed in the literature, such as the Leave Copy Everywhere (LCE) [6] that involves caching a copy of every content requested and delivered to the consumer at each node along the response path. Clearly, this strategy provides rapid replication of all available Data packets through the network but causes high caching redundancy. To reduce such a redundancy, other strategies such as Probabilistic Caching (PROB) [10] adopt a strategy of caching contents probabilistically. This approach aims to ensure that distinct routers or paths store different Data packets.

2.2 Replacement Policies

Replacement or cache eviction policies play an important role in the performance of NDN. In fact, once the cache is full, keeping content forever without being requested does not offer any gain, and removing content that will be requested shortly is not profitable either.

There are several well-known replacement policies such as First In First Out (FIFO), Least Frequently Used (LFU), Least Recently Used (LRU), and Random Replacement (RR). LRU is the most well-known and widely used replacement policy [8,11,12,15], it evicts the least recently requested content to make room for new content.

Caching strategies and replacement policies improve efficiency in terms of content availability in the network, decrease consumers' access latency, and reduce the loads on the producers. Other parameters can affect the performance of the in-network caching process, such as the cache size, the content popularity distribution, and the size of the content.

2.3 Forwarding

In NDN, nodes' FIB are usually populated and updated by a name-based routing protocol, and subsequently exploited by the forwarding strategy to guide interest packets [3]. For scalability and practicality reasons, the routing protocol populates the FIB with paths leading to the producers, thus leaving the responsibility of exploiting the in-network caching capability, entirely to the forwarding strategy.

One of the most known forwarding strategies is *BestRoute* [14], which simply uses the next hop ranked highest by the routing protocol. More complex forwarding strategies undertake the challenge of efficiently coordinating the forwarding of interest packets with the volatile cache states at the nodes to fully take advantage of the in-network caching capability of the NDN architecture. The main idea of these strategies is to wisely guide interest packets to retrieve contents from the nearest caches.

3 Related Work

Caching techniques are continually evolving due to their pivotal role in emerging communication networks. These methods are anticipated to provide efficient data access by bringing it closer to end users. In this section, we delve into the relevant literature to explore the proposed caching solutions.

Leave copy everywhere (LCE) is the natively implemented strategy of caching in NDN. This powerful native caching mechanism offered by NDN [15] conducts to put every passenger content through a node without considering its popularity or the cost of its links of provenance. Authors in [1] propose a solution of congestion-aware caching (CAC) by assuring cooperation between neighbor nodes in the case of off-path caching by considering all links that rely on a concerned node. It presents a neighborhood cooperation solution by considering the state of all the links connected to the node concerned by making the caching decision. The disadvantage of CAC is the imprecise estimate of popularity [4].

The authors, in [9], propose a caching policy dubbed FGPC (Fine-grained popularity-based caching), which caches coming content when storage is available. Otherwise, it keeps only the most popular content. FGPC takes care of the threshold in a dynamic manner when selecting popular content, without caring about the state of the links or how their content is delivered.

Table 1 presents a synthesis of related work to situate our research within the existing literature. Differently from the above-cited work, we propose our CPAC solution, which considers a downstream link and takes into account the popularity of content before deciding whether or not to cache the concerned content. We also consider the case of on-path caching, where only the concerned links are located along the path from a Data packet to the requester. Given the favorable performance demonstrated by LRU in our previous work [8], we have opted to employ LRU as the replacement policy in this paper.

Table 1. Related work synthesis.

Caching Strategy	Description	Congestion	Popularity
LCE [5]	LCE involves caching a copy of every requested content at each node along the response path.	✗	✗
PROB [10]	PROB adopts a strategy of caching contents probabilistically to ensure that distinct routers or paths store different Data packets.	✗	✗
Mordjana et al. [8]	This study investigates and analyzes the performance of caching, replacement policy and forwarding strategies.	✗	✗
CAC [1]	CAC considers all links associated with a node to inform caching decisions and presents a neighborhood cooperation solution.	✓	✗
FGPC [9]	FGPC caches selectively incoming content based on availability, prioritizing the most popular content dynamically.	✗	✓
CPAC	CPAC evaluates downstream links, considering content popularity before deciding to cache, addressing both on-path caching and links along the Data packet path to the requester.	✓	✓

4 Our Proposal: Design Overview and Algorithm

This section presents our proposal Congestion- and Popularity-Aware Caching Mechanism (CPAC) for Named Data Networks. We aim to reduce the traffic load in congested NDN networks while efficiently satisfying user requests. CPAC is composed of three main steps that are triggered upon receiving a content 'x' as depicted in Fig. 3. The first step decides whether to forward the Data packet 'x' according to the PIT. If so, in the second step, CPAC retrieves a delay of the retrieved Data packet (T_D) and a hoped time (T_e) of the desired response time expressed by the user. This manner lets CPAC examine the state of congestion on the downstream link. This ratio is used to compute and update the cost of the retrieved content $Cost(x)$ (Eq. 1) which depends on the popularity of the concerned content (P_x) and the ratio between the packet size (S^p) and the cache size (S^c) in the router. The results of the first and second steps are exploited by the third step of CPAC, to decide whether to cache the received data. In this last step, CPAC will make a decision about caching based on the state of the CS whether it is full or not, and according to the threshold of popularity (P_{th}). All these steps are summarized in Algorithm 1.

It should be noted that if the CS is full, CPAC deploys an objective function to decide whether to replace cached content with the received data. In the current version, CPAC adopts the Least Recently Used (LRU) cache replacement strategy, thanks to its features [8]. The motivations and rationale behind CPAC's caching strategy are detailed below.

CPAC considers both the link status and content popularity when deciding whether to cache a specific piece of content. Our proposal primarily suggests caching content 'x' that originates from a congested (expensive) link and/or is popular and frequently requested. We compute popularity based on the number of requests of content (P_x), cache size (S^c), and size of packets (S^p) in the catalog and also when defining the cost threshold in Eq. 2. Indeed, CPAC is designed to optimize network performance by addressing the trade-off between network congestion and the popularity of content. To accomplish this task, Algorithm 1 is executed for every incoming Data packet. It makes a caching decision by determining whether to store the received packet data in each NDN router.

$$Cost(x) = P_x \times \frac{T_D}{T_e} \times \frac{S^p}{S^c} \tag{1}$$

$$Cost_{th} = n_{interest(th)} \times \frac{S^p}{S^c} \tag{2}$$

5 Performance Evaluation

This section is dedicated to assessing the effectiveness of our proposed solution. We outline the evaluation methodology, describe the different parameters involved, and detail the metrics measured in our simulation, which was carried out using the ndnSIM2.7 simulator [7]. Subsequently, we analyze and discuss the obtained results.

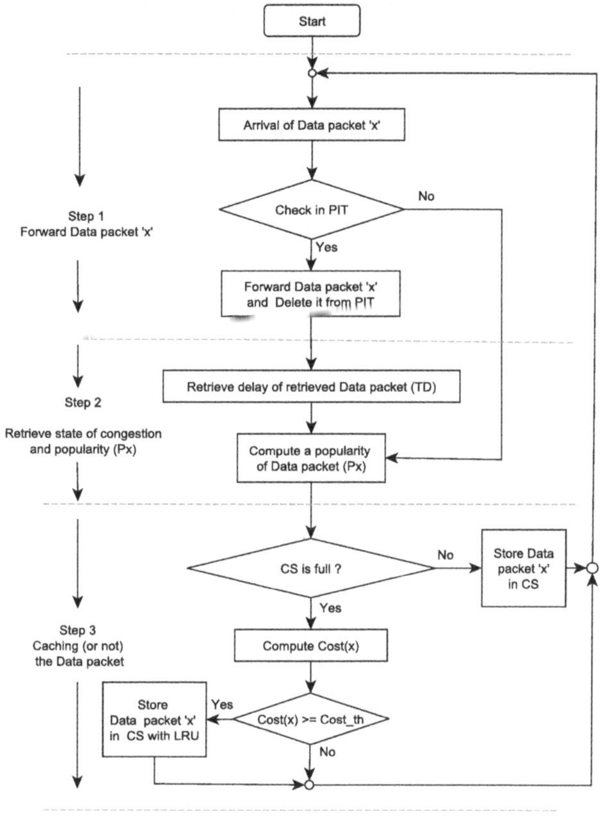

Fig. 3. Our CPAC solution for NDN.

5.1 Simulation Setup

We construct our simulation with a 3x3 grid topology, which gives 09 nodes, including one producer and one consumer, as depicted in Fig. 4. Each adjacent node is connected by a peer-to-peer link, and the congested link is between node 7 and the producer. For the forwarding algorithm, we adopted *BestRoute*, which forwards an Interest to the upstream having the lowest routing cost and LRU as a replacement policy when the cache is full. Besides, in this scenario, the consumer generates 100 Interest packets per second and asks for one of the three prefix names of the content. In our case, we respect the ratio between the cache size (S^c) and catalog size ($|S^p| \times S^p$), which typically spans from 10^{-5} to 10^{-1} [8]. Moreover, we make the assumption that the distribution of content popularity follows a Zipf distribution with a Zipf exponent of $\alpha = 1$. The parameters include S^p with a size of 64 Bytes, S^c with a size of 1024 Bytes, and T_e, which represent the product of the number of hops and the propagation delay.

To better evaluate our proposal, we use the following metrics:

- *Cache Hit:* the probability of finding a content in the network;

Algorithm 1: CPAC algorithm

```
for Every incoming Data packet 'x' do
    if 'x' exist in the PIT then
        Forward 'x' and delete the corresponding request from the PIT
        Retrieve a delay (T_D) of 'x'
    if CS is full then
        Compute the corresponding cost of retrieving 'x' (Eq. 1)
        if Cost(x) ≥ Cost_th then
            Store 'x' in the CS using LRU
    else
        Store 'x' in the CS
```

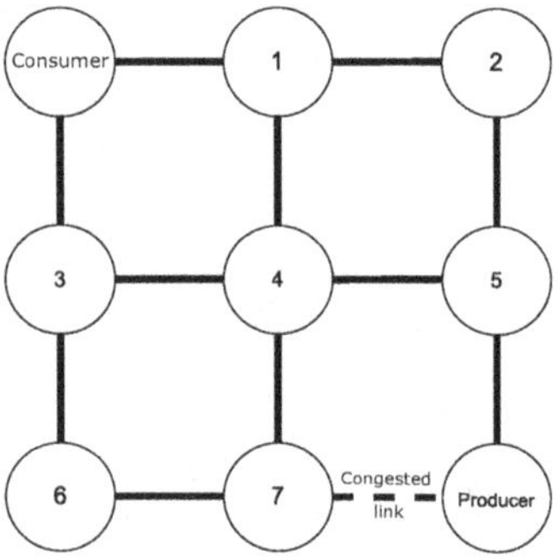

Fig. 4. 3x3 Grid topology.

- *Hop Count:* the number of hops taken to find a content;
- *Delay:* the time between issuing an Interest and receiving the corresponding Data packet.

We conducted each simulation scenario five times for 40 s. Unless explicitly stated, we report the mean value across all requests, accompanied by a 95% confidence interval. To provide a contextual understanding of the results, we compare our CPAC strategy with the LCE and PROB strategies, where PROB employs a fixed probability set at 0.3.

5.2 Obtained Results

Figure 5 illustrates the outcomes of the evaluation. Firstly, as depicted in Fig. 5a, CPAC demonstrates commendable performance in cache hit rate when compared to alternative strategies, specifically LCE and PROB. A higher cache hit rate signifies that a substantial number of data requests are fulfilled by the cached content, thereby diminishing the necessity to retrieve data directly from the original source.

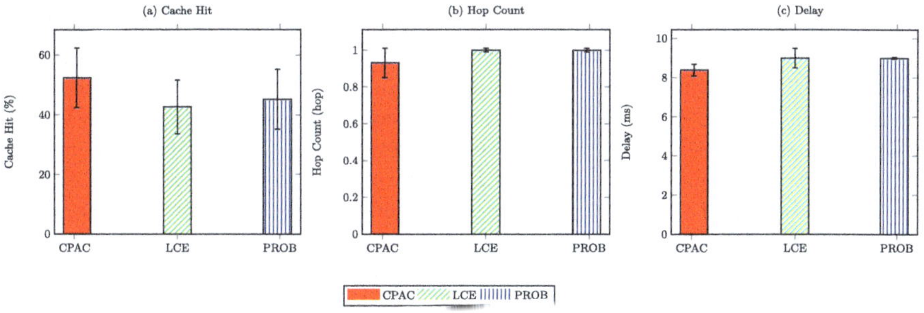

Fig. 5. Obtained results.

Secondly, Figs. 5b and 5c provide additional insights into the competitiveness of CPAC compared to LCE and PROB. Indeed, CPAC performs well in terms of both hop count and delay metrics. Lower delays in Fig.5c are typically favorable as they indicate faster data retrieval and response times. The fact that the mean number of hops (Hop Count) traveled by data packets fluctuates around one hop in all cases indicates efficient routing due to the nature of the 3x3 grid topology. A low mean hop count is generally desirable, as it implies more direct and efficient data packet routing. The mention of the mean number of hops traveling around a value of one hop in all cases is attributed to the 3x3 grid topology. This suggests that the network's structure, in this case, the grid topology plays a role in determining the efficiency of data packet routing.

To sum up, the results showcased in Fig. 5 highlight a notably superior performance of CPAC over alternative caching strategies regarding cache hit rate, hop count, and delay within the context of a 3x3 grid topology. These positive results underscore the effectiveness of CPAC in enhancing NDN caching performance within the specified experimental setting.

6 Conclusion and Future Work

In this paper, we introduced a new NDN caching strategy, dubbed CPAC, designed to be both congestion- and popularity-aware. Using the ndnSIM simulator, we quantitatively evaluated the performance of CPAC by inspecting the cache hit, hop count, and delay metrics. The obtained results show that our strategy outperforms other related schemes in terms of the cache hit, while slightly reducing both the hop count and the delay.

In our future work, we plan to explore how well CPAC performs in more complex scenarios. This exploration will involve examining adaptive cache replacement strategies and incorporating dynamic thresholds based on content popularity, all while considering larger network topologies.

References

1. Badov, M., Seetharam, A., Kurose, J., Firoiu, V., Nanda, S.: Congestion-aware caching and search in information-centric networks. In: Proceedings of the 1st international conference on Information-centric networking. pp. 37–46. ACM (2014)
2. Djama, A., Djamaa, B., Senouci, M.R.: Information-centric networking solutions for the internet of things: a systematic mapping review. Comput. Commun. **159**, 37–59 (2020)
3. Djama, A., Djamaa, B., Senouci, M.R., Khemache, N.: LAFS: a learning-based adaptive forwarding strategy for NDN-based IoT networks. Ann. Telecommun. **77**, 311–330 (2021). https://doi.org/10.1007/s12243-021-00850-2
4. Ioannou, A., Weber, S.: A survey of caching policies and forwarding mechanisms in information-centric networking. IEEE Commun. Surv. Tutorials **18**(4) (2016)
5. Jacobson, V., Smetters, D.K., Thornton, J.D., Plass, M.F., Briggs, N.H., Braynard, R.L.: Networking named content. In: Proceedings of the 5th International Conference on Emerging Networking Experiments and Technologies, pp. 1–12 (2009)
6. Laoutaris, N., Syntila, S., Stavrakakis, I.: Meta algorithms for hierarchical web caches. In: International Performance Computing and Communications Conference, pp. 445–452. IEEE (2004)
7. Mastorakis, S., Afanasyev, A., Zhang, L.: On the evolution of ndnSIM: an open-source simulator for NDN experimentation. ACM Comput. Commun. Rev. **47**(3), 19–33 (2017)
8. Mordjana, Y., Senouci, M.R., Mellouk, A.: Performance analysis of caching and forwarding strategies in content centric networking. In: GLOBECOM 2017-2017 IEEE Global Communications Conference, pp. 1–6. IEEE (2017)
9. Ong, M.D., Chen, M., Taleb, T., Wang, X., Leung, V.C.: FGPC: fine-grained popularity-based caching design for content centric networking. In: Proceedings of the 17th ACM International Conference on Modeling, Analysis and Simulation of Wireless and Mobile Systems, pp. 295–302 (2014)
10. Psaras, I., Chai, W.K., Pavlou, G.: Probabilistic in-network caching for information-centric networks. In: Proceedings of the Second Edition of the ICN Workshop on Information-centric Networking, pp. 55–60. ACM (2012)
11. Psaras, I., Clegg, R.G., Landa, R., Chai, W.K., Pavlou, G.: Modelling and evaluation of CCN-caching trees. In: Domingo-Pascual, J., Manzoni, P., Palazzo, S., Pont, A., Scoglio, C. (eds.) NETWORKING 2011. LNCS, vol. 6640, pp. 78–91. Springer, Heidelberg (2011). https://doi.org/10.1007/978-3-642-20757-0_7
12. Rossini, G., Rossi, D.: Coupling caching and forwarding: benefits, analysis, and implementation. In: Proceedings of the 1st ACM Conference on Information-Centric Networking, pp. 127–136 (2014)
13. Thibaud, A., Fasson, J., Arnal, F., Sallantin, R., Dubois, E., Chaput, E.: An analysis of NDN congestion control challenges. In: 2019 2nd International Conference on Hot Information-Centric Networking (HotICN), pp. 18–24 (2019)
14. Yi, C., Afanasyev, A., Wang, L., Zhang, B., Zhang, L.: Adaptive forwarding in named data networking. SIGCOMM Comput. Commun. Rev. **42**(3) (2012)
15. Zhang, L., et al.: Named data networking. ACM SIGCOMM Comput. Commun. Rev. **44**(3), 66–73 (2014)

Multicast Routing over Industrial Internet of Things

Issam Eddine Lakhlef(✉), Badis Djamaa, and Mustapha Reda Senouci

Distributed and Complex Systems Lab., Ecole Militaire Polytechnique, BP 17, 16046 Bordj El-Bahri, Algiers, Algeria
issameddinelakhlef@gmail.com

Abstract. In the industrial sector, we are currently witnessing the advent of the fourth industrial revolution, commonly known as the Industrial Internet of Things (IIoT). This cutting-edge digital technology relies heavily on connectivity, real-time processing, machine-to-machine communication, and machine learning. In this context, the imperatives of reliability, flexibility, mobility, and cost-effectiveness have led to a strong push towards Low-power and Lossy Networks (LLN) solutions for industrial systems. IP multicast presents a solution for a multitude of IIoT applications, ranging from over-the-air programming and information sharing to device configuration and resource discovery. It achieves this by enabling the transmission of packets from a single source to multiple destinations, thereby conserving resources such as bandwidth, energy, and time. The IPv6 over IEEE 802.15.4 TSCH (6TiSCH), as implemented, does not handle multicast traffic. To meet their needs, this document aims to propose an approach for implementing multicast in 6TiSCH networks via unicast without affecting MAC layer protocols. A series of extensive simulations were conducted to evaluate the performance of the multicast protocols under scrutiny. The obtained results not only affirm the viability of our proposed approach for implementing multicasting in 6TiSCH networks but also highlight the superior performance of ESMRF and BMFA when compared to MPL in terms of resource consumption and end-to-end delays. They also demonstrate the reliability of MPL.

Keywords: Industrial Internet of Things · Multicast · TSCH · 6TiSCH · Contiki-NG

1 Introduction

The Internet has now connected the whole world, from mainframes and servers to personal devices and objects. Today, billions of IoT devices are connected to the Internet worldwide. However, these objects generally operate with limited resources in terms of energy, computing, storage, and bandwidth. These constraints impose stringent challenges on all layers of the TCP/IP network stack. As a result, new standards and technologies for lightweight, low-energy communications have emerged to address the need for wired connectivity.

B. Djamaa et al. (Eds.): CSA 2024, LNNS 1145, pp. 187–199, 2025.
https://doi.org/10.1007/978-3-031-71848-9_18

Industrial IoT (IIoT) leans on industrial Low-power and Lossy Networks (LLNs), including IEEE 802.15.4-2015 [1], which are deployed in a factory and connect hundreds or thousands of devices. Although the data volume within industrial LLNs may not be extensive, ensuring reliability and minimizing latency are crucial factors. Multicasting is becoming increasingly important in scenarios such as information and command sharing, live programming, and device configuration. Despite the many protocols proposed in the literature for multicasting in LLNs, specific considerations for industrial LLNs remain little explored.

In IEEE 802.15.4 networks, we have three MAC layer communication modes: sender initiated, like ContikiMAC [2] and CSL [3], received initiated, which does not support multicast, and Time Slotted Channel Hopping (TSCH) [4]. TSCH allows nodes in an IIoT network to hop between the 16 channels of the IEEE 802.15.4 standard within specified time slots in order to meet the reliability and real-time requirements of industrial applications while saving nodes energy.

The MAC layer poses a particular problem for industrial LLNs, as messages transmitted in multicast are then distributed as broadcast packets at the MAC level. In the 6TiSCH (the IPv6 over IEEE 802.15.4 TSCH mode) paradigm, all these packets-whether data or control-are routed in the same time slot dedicated to this specific type of traffic. This operational characteristic generates high collision rates, reduced reliability, and prolonged latency, requiring tailor-made solutions for multicast management in industrial LLNs.

This paper introduces an approach to multicasting in 6TiSCH networks through the use of unicast communication. This choice was made to avoid having to modify the operation of the MAC layer and to take advantage of the benefits offered by unicast routing in 6TiSCH networks, particularly with regard to reliability and latency. This methodology has been seamlessly incorporated into various LLN multicast routing protocols, which have been thoroughly evaluated across a range of parameters, scenarios, and configurations using a comprehensive set of evaluation criteria.

The remainder of this paper is organized as follows: Sect. 2 furnishes a background on TSCH and the 6TiSCH architecture, while Sect. 3 provides an overview of related work, specifically focusing on multicast routing LLNs. In Sect. 4, we expound upon our proposal concerning multicast in IIoT. Following this, Sect. 5 presents the results of our performance evaluation conducted through realistic simulations. Ultimately, the paper ends in Sect. 6 with some concluding remarks and ideas for future work.

2 Background and Motivation

To provide context for this paper, this section introduces the basic concepts related to the IIoT. We will start with a concise overview of Industry 4.0. Next, we will introduce TSCH and the 6TiSCH architecture, followed by an overview of routing in 6TiSCH industrial networks.

2.1 Industrial IoT

Industry 4.0 represents the latest advance in manufacturing, aiming to collect, process, and share vast volumes of data in real time to dynamically adapt industrial processes. Industrial automation generally encompasses the automation of manufacturing, quality control, and material processing [5].

Numerous industrial applications rely on deterministic industrial wireless networks [5]. Wireless transmissions are widely recognized for their time-varying and lossy nature. Operating under such conditions presents substantial challenges in ensuring high reliability and low latency.

Deterministic Networking (DetNet) strives to significantly reduce packet loss while guaranteeing end-to-end delay for wired IP networks. Reliable and Available Wireless (RAW) extends DetNet to focus primarily on wireless links. RAW aims to offer specialized DetNet elements suitable for short-range radios. Specifically, RAW focuses on selected technologies that can be programmed in the lower layers, including IEEE Std. 802.15.4 TSCH, 3GPP 5G Ultra-Reliable Low Latency Communications (URLLC), and L-band Digital Aeronautical Communications System (LDACS) [6].

We chose TSCH because of its ability to meet the challenges prevalent in industrial environments, including node mobility, packet loss, and the stringent latency and data reliability requirements inherent in this type of environment. TSCH has been widely studied and adopted in IIoT network research. It stands out as the most practical MAC underlay protocol specifically crafted for critical applications, including areas such as process automation, equipment, and process monitoring.

2.2 TSCH and 6TiSCH Architecture

TSCH is a MAC protocol that divides time into fixed-length slices organized into a slotframe. Nodes remain tightly synchronized by periodically communicating with one another to compensate for clock drift. Many other synchronization methods have been proposed [7,8]. Frequency diversity is employed to mitigate the effects of multipath fading and enhance resilience to external interference (Fig. 1a). Channel hopping is achieved by transmitting packets one after another on different frequencies with a predefined and universally known sequence for all nodes. Each node can participate in transmission, reception, or sleep within a given cell that represents a time slot at a specific frequency. The scheduling function (SF) creates a communication schedule by assigning communication cells to different pairings of nodes that need to fulfill bandwidth, latency, and reliability requirements for specific applications. The SF aims to conserve energy efficiently by minimizing the number of schedule cells where nodes must transmit or listen.

Nevertheless, the TSCH specification never described the procedures for building and maintaining a schedule. However, many solutions were proposed later. They are classified into three categories: centralize [9–12], distributed

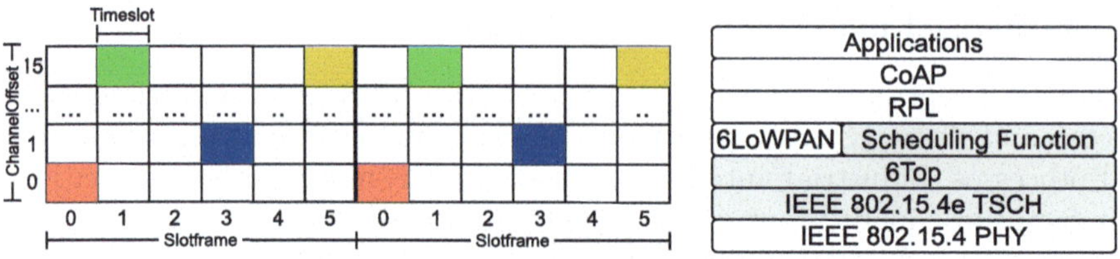

(a) TSCH schedule. (b) 6TiSCH protocol stack.

Fig. 1. 6TiSCH Scheduling.

(negotiation-based) [13–16], and autonomous [17]. Centralized SFs face scalability constraints due to high signaling overheads [4]. In negotiation-based scheduling, there is a recurring cost associated with negotiations between nodes and their neighbors to adapt to fluctuating traffic conditions. In addition, the risk of collisions cannot be neglected in densely occupied time slots.

In contrast, autonomous scheduling eliminates the need for negotiation by statically allocating active cells all the time. However, autonomous SF performs suboptimally when faced with changing traffic conditions in dynamic topologies, resulting in temporary traffic peaks. Clearly, the use of negotiation-based designs remains a viable consideration.

The formation of a TSCH network begins with the network coordinator broadcasting the existence of the network by transmitting announcement packets, known as Enhanced Beacons (EB). These beacons encapsulate information essential to the synchronization of a new node entering the network.

In [18], the IETF's 6TiSCH working group incorporated TSCH [19] into the IEEE 802.15.4 standard to mitigate the impact of external interference by leveraging its channel hopping capability. Figure 1b illustrates the 6TiSCH protocol stack. The 6Top layer [20] holds the central position, integrating the rest of the IPv6 stack with TSCH-MAC over IEEE 802.15.4.

2.3 Routing in Industrial 6TiSCH

The Routing Protocol for Low-Power and Lossy Networks (RPL) [21] is an IPv6 routing protocol designed by the IETF's ROLL working group. RPL employs the destination-directed acyclic graph (DODAG), anchored in the LLN Border Router (LBR), to establish parent-child relationships between network nodes.

After joining the TSCH network, a new node must join the RPL DODAG in order to reach other nodes and transmit data in a multi-hop topology. The formation of the DODAG is based on the RPL's information dissemination mechanism. DIO (DODAG Information Option) messages, which contain information on rank, objective function, IDs, etc., play a key role in this respect. They are multicast (periodically and locally) by each node to establish paths to the roots. Reached nodes will broadcast EB and DIO packets to extend the network. RPL control traffic and EBs responsible for synchronizing the network had higher priority than application messages.

3 Related Work

The requirements of real-world IIoT functionalities and applications, such as the dissemination of shared information to industrial equipment like sensors and actuators for distributed production organization, network configuration, resource discovery, and security management, necessitate the deployment of efficient multicast routing protocols. This mode of IIoT routing is still challenging, under active research, and is the main focus of this paper.

Numerous multicast protocols have been proposed for IoT, like ESMRF [22], BMRF [23], BMFA [24], M-LOADng [25], MPL [26], RECOUP [27], and MSDN-IoT [28], which usually follow one of the two basic modes: dense or sparse, with the dense mode being the most deployed one. However, studies of multicasting for the IIoT remain notably inadequate.

Stateless multicast RPL Forwarding (SMRF) is a lightweight stateless multicast forwarding algorithm introduced by Oikonomou *et al.* [29], utilizing RPL functionality for multicast traffic within the RPL's DODAG. To address SMRF limitations and enable bidirectional multicast communication, Qorany *et al.* proposed Enhanced SMRF (ESMRF) [22]. ESMRF supports both downstream and upstream multicast traffic flows by encapsulating packets at the source within ICMPv6 delegation packets, sent to the root of the RPL tree for forwarding.

Papadopoulos *et al.* [24] introduced the Bi-directional Multicast Forwarding Algorithm (BMFA) to facilitate bidirectional traffic and prevent routing loops in RPL-based networks. BMFA utilizes the flow label in the IPv6 header to manage incoming messages, leveraging the RPL group membership system.

The Multicast Protocol for Low Power and Lossy Networks (MPL), proposed by Hui *et al.* and standardized in 2016 through RFC 7731 [26], employs the Trickle algorithm to manage the transmission of both data and control messages. MPL employs a buffering system to retain the most recent data messages and retransmit them to neighboring nodes in the event of message loss.

Several SDN-based solutions for deploying multicast in the IIoT have been proposed. In [30], the authors introduce a novel SDN multicast approach based on Group Shared Tree (GST) and dynamic near receiver selection tailored for large-scale IIoT environments. An SDN controller gauges the delay disparity between the previous and updated GST and, based on this assessment, decides to switch GST to alleviate packet delay.

In [31], the authors proposed Mobile Multicast Forwarding with Software Defined Network (MMF-SDN) as an SDN-based alternative that enables nodes to enhance determinism and ensure QoS parameters in IIoT. This is achieved by leveraging the global knowledge provided by the SDN and the synchronization achieved with TSCH, facilitating the implementation of a centralized TSCH scheduler that performs a cumulative allocation of multicast resources.

All the protocols outlined in this section are specifically designed for 802.15.4 networks using sender-initiated RDC protocols. However, none of these protocols address the multi-channel aspect of TSCH. As a result, multicast packets are broadcasted at the MAC layer level, leading to their transmission concurrently with other RPL protocol or scheduling function control packets. Consequently, these protocols require heightened attention to ensure their seamless operation within industrial LLNs.

Following an exhaustive analysis of the implementation of TSCH and scheduling functions such as the MSF and Orchestra [17] in Contiki-NG, it was found that there are no rules dedicated to the efficient handling of broadcast and multicast traffic. In addition, in the default TSCH configuration, we observed that the multicast data packets frequently collided with control packets such as those from RPL. This is due to the fact that only one scheduled cell is allocated for the transmission of broadcast or multicast messages. Given our objective of evaluating multicast traffic in LLN networks, we found it necessary to handle this kind of traffic at the network level without affecting the TSCH and SFs rules.

In [32], the authors analyze the performance of BMRF [23] on TSCH and ContikiMAC. They introduced new rules to manage multicasting, ensuring that each node has one broadcast-related transmit slot and as many broadcast-related listen slots as the number of children. The results indicate that BMRF broadcast over TSCH can lead to increased energy consumption in high-interference environments; however, it significantly improves packet delivery rate and reduces delay compared to ContikiMAC.

4 Multicast Routing over IIoT

Multicast traffic is generally less reliable than unicast traffic. Given its point-to-multipoint nature, ensuring reception by all recipients requires multiple acknowledgments. However, as multicast packets do not involve acknowledgments, senders have no means of ensuring whether retransmissions are necessary.

Figure 2 provides an overview of how our approach operates. Upon the establishment of the DODAG tree and the allocation of TSCH cells by the scheduling functions, the multicast routing protocol transmits the unicast packets containing the multicast message. The general diffusion scheme is depicted in the final topology by the bold red arrows.

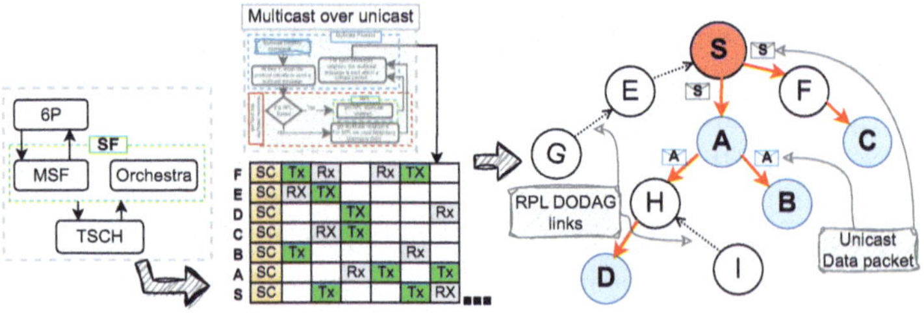

Fig. 2. Multicast over Unicast data diffusion overview.

We have chosen to employ unicast transmission for multicast messages to ensure the fulfillment of reliability and end-to-end delay requirements within the IIoT environment. In this paradigm, each node intending to send a multicast packet transmits it individually to all its relevant neighbors via unicast communication (Fig. 2). This strategic adjustment is designed to guarantee a higher rate of reliability without incurring additional resource consumption, all without violating the established rules of the MAC protocol.

The diagram in Fig. 3 illustrates our approach. Following the processing of the multicast protocol, at send time "t," we must determine the subsequent nodes to which the multicast message needs to be sent. Two approaches exist for this step: either the multicast protocol is RPL-based (which is the case for most protocols proposed in the literature), in which case this information is provided by the RPL, or the protocol must have a mechanism for identifying the next-hop nodes. In the case of MPL, the Neighbors Discovery protocol can retrieve all neighbor addresses. Subsequently, the multicast protocol sends unicast packets to each interested neighbor, containing the multicast message to be transmitted.

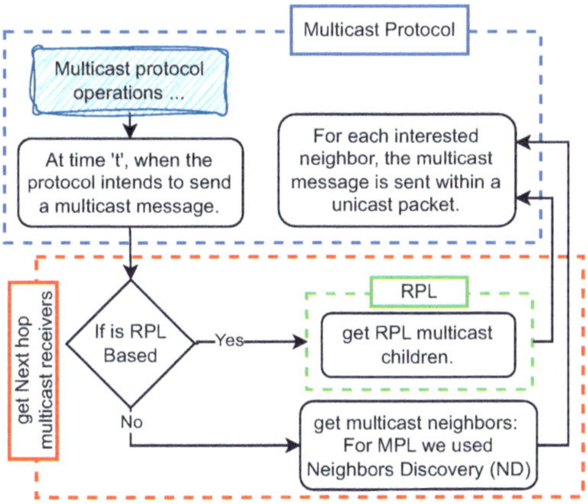

Fig. 3. Multicast over Unicast diagram.

5 Performance Evaluation

To achieve the objectives of this study, we evaluate, present, and compare the performance of the studied multicast routing protocols using the Contiki-NG operating system. We perform the simulations using the Cooja simulator.

5.1 Evaluation Methodology and Metrics

To conduct our study, we have selected the following comparative criteria:

- Packet Delivery Ratio (PDR): this metric holds significant importance in evaluating the reliability and efficiency of multicast communication within LLNs. It is defined by Eq. 1 as the ratio of the total number of unique multicast messages received by members R to the product of the number of unique multicast messages sent by traffic sources to the group's IPv6 address D and the number of multicast group members M.

$$PDR = \frac{R}{D \times M} \tag{1}$$

- End-to-End delay: defined as the average time it takes for a multicast data message to be received by member nodes in the network. Equation 2 defines the end-to-end delay of a message as the difference between the time at which it was last received by the members of the group and the time of its transmission from the multicast source node.

$$\frac{1}{D}\sum_{i=1}^{D} t\_last\_recive(d_i) - t\_send(d_i) \tag{2}$$

- Number of packets sent (per message): defined as the average number of multicast packets sent by each node for the same message generated by the source. In Eq. 3, $Packets\_sent(d_i)$ is the number of packets sent for the same datagram d_i.

$$\frac{1}{D}\sum_{i=1}^{D} Packets\_sent(d_i) \tag{3}$$

- Radio Duty Cycle (RDC): refers to the ratio of time a device spends actively transmitting or receiving data (i.e., its active state) compared to its total operational time.

We use the radio duty cycle as an indicator of network energy consumption since the energy consumption of the radio transceiver is linear and depends on its duration of activity.

5.2 Simulation Results and Discussion

We selected three multicast protocols-MPL, ESMRF, and BMFA-and incorporated our approach into each protocol. Table 1 presents the detailed parameters and their respective values utilized for configuring the envisioned IIoT network scenarios in the Cooja simulator.

Variation of the reception probability: As an initial step, we assessed the impact of variations in reception probability on the studied multicast protocols. This parameter signifies the network's condition, specifically the probability of

Table 1. Simulation setup.

Parameter	Value
Duration	10 min
Number of repetitions (RandomSeed)	5 times
Transmission rate of the multicast sources	1 pkt / 10 sec
SF - MAC - PHY	Orchestra - TSCH - 802.15.4
MPL, ESMRF and BMFA	Default parameters in Contiki-NG

Fig. 4. Simulation topologies.

message reception (at the physical level) by the destination node. To conduct this evaluation, we employed a topology encompassing 36 nodes (Fig. 4a). The corresponding results are illustrated in Fig. 5.

Observing the results, it becomes evident that the probability of reception exerts a direct influence on the PDR. MPL exhibits greater reliability compared to ESMRF and BMFA. Furthermore, MPL demonstrates a significantly higher end-to-end delay and message transmission count in comparison to the other two protocols, with this gap widening as the reception probability decreases.

ESMRF exhibits greater reliability than BMFA, with very similar outcomes observed in terms of end-to-end delay and the number of packets sent.

Regarding the RDC, the results indicate that MPL yielded the least favorable outcomes, exhibiting higher energy consumption compared to the other two protocols. ESMRF and BMFA consistently produced similar results, with ESMRF showcasing superiority in scenarios where the reception probability was 0.7 and 0.8. This disparity can be attributed to the fact that ESMRF employs multicasting solely for downstream traffic, whereas BMFA utilizes it in both upstream and downstream directions.

These findings align with those presented in [33], where MPL is noted for its high reliability attributed to the implementation of a retransmission mechanism for lost messages. However, a noteworthy flaw of MPL lies in its substantial consumption of resources, particularly energy and bandwidth.

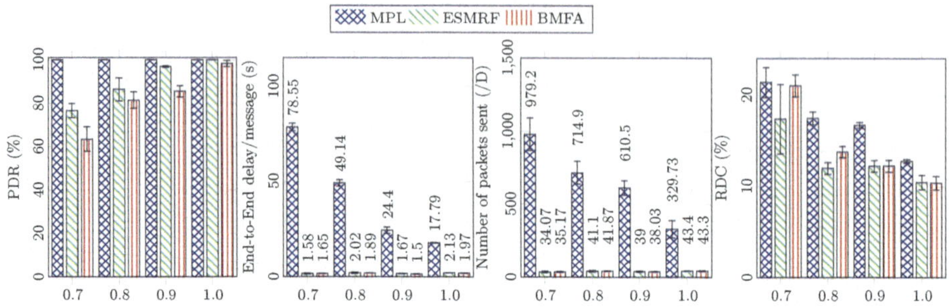

Fig. 5. Variation of the reception probability results.

Variation of the Network Topology: In this section, we have evaluated the performance of the protocols studied in other topologies. First, we increased the number of network nodes to 64, then we changed the network topology to a tree topology (see Fig. 4).

In the case of the 8×8 network, MPL maintains its reliability by achieving a PDR of 100%, affirming the robustness of this protocol. Conversely, ESMRF and BMFA exhibit a decline in PDR, attributed to the increase in collisions and subsequently lost messages as the number of nodes and network density increase. For both end-to-end delay and the number of packets sent, MPL exhibits a significant increase, while the other two protocols demonstrate nearly identical results for the first topology (6×6) (Fig. 6).

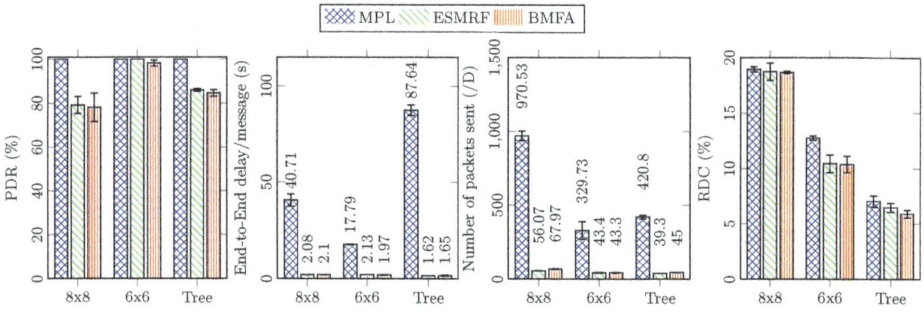

Fig. 6. Variation of the network topology results.

In the scenario of the tree topology comprising only 30 nodes, it becomes evident that the end-to-end delay for MPL is less favorable compared to the results demonstrated in larger grid topologies with 36 and 64 nodes. On the other hand, the number of packets sent is lower in the 64-node topology, which shows the influence of the number of nodes on this parameter. However, it is essential to note that in industrial environments, machines are typically arranged in grid topologies, thereby justifying our selection of the test topology. For ESMRF and BMFA, a decrease in PDR is observed, while the other parameters remain unaffected. This phenomenon underscores the influence of topology on the performance of these two protocols.

It's evident that the number of nodes and network density have a direct impact on the RDC, as higher density leads to a greater allocation of new cells for packet transmission. In each topology, the MPL protocol consistently exhibits the highest RDC, while ESMRF and BMFA yield nearly identical results.

6 Conclusion and Future Work

In conclusion, this article has introduced an approach to the implementation of multicast routing in industrial LLNs. Based on an exhaustive analysis of the most studied multicast protocols in the literature, taking into account various factors influencing performance, the approach we propose has demonstrated its effectiveness. MPL has demonstrated superior reliability to RPL-based protocols, such as ESMRF and BMFA, under favorable network conditions. Whereas MPL exhibits significant resource consumption and suboptimal end-to-end delay. In addition, the protocols examined performed better in grid topologies than in tree topologies. However, this discrepancy is inconsequential, given that industrial networks mainly adopt grid structures. Our future research efforts will focus on the development of a generic multicast routing protocol suitable for industrial LLNs. The main objective of this protocol is to guarantee a robust PDR with reasonable resource consumption.

References

1. IEEE Standard Association: IEEE Standard for Low-Rate Wireless Networks: IEEE 802.15. 4-2015 (Revision of IEEE Std 802.15. 4-2011) (2016)
2. Dunkels, A.: The ContikiMAC Radio Duty Cycling Protocol (2011)
3. IEEE Standard for Low-rate Wireless Networks: IEEE Std 802.15.4-2020 (Revision of IEEE Std 802.15.4-2015), pp. 1–800 (2020)
4. Hermeto, R.T., Gallais, A., Theoleyre, F.: Scheduling for IEEE802. 15.4-TSCH and slow channel hopping MAC in low power industrial wireless networks: a survey. Comput. Commun. **114**, 84–105 (2017)
5. Grossman, E.: Deterministic networking use cases. Technical report (2019)
6. Thubert, P., Cavalcanti, D., Vilajosana, X., Schmitt, C., Farkas, J.: Reliable and available wireless technologies, ID draft-thubert-rawtechnologies-03. IETF (2019)
7. Chefrour, D.: Evolution of network time synchronization towards nanoseconds accuracy: a survey. Comput. Commun. **191**, 26–35 (2022)
8. Chang, T., Watteyne, T., Pister, K., Wang, Q.: Adaptive synchronization in multi-hop TSCH networks. Comput. Netw. **76**, 165–176 (2015)
9. Palattella, M.R., Accettura, N., Dohler, M.: Traffic aware scheduling algorithm for reliable low-power multi-hop IEEE802.15.4e networks. In: IEEE 23rd International Symposium on Personal Indoor and Mobile Radio Communications, pp. 327–332 (2012)

10. Palattella, M.R., Accettura, N., Grieco, L.A.: On optimal scheduling in duty-cycled industrial IoT applications using IEEE802. 15.4 e TSCH. IEEE Sens. J. **13**(10), 3655–3666 (2013)
11. Soua, R., Minet, P., Livolant, E.: MODESA: an optimized multichannel slot assignment for raw data convergecast in wireless sensor networks. In: IEEE IPCCC, pp. 91–100. IEEE (2012)
12. Soua, R., Livolant, E., Minet, P.: MUSIKA: a multichannel multi-sink data gathering algorithm in wireless sensor networks. In: IEEE IWCMC, pp. 1370–1375 (2013)
13. Antoni, M., Xavier, V., López, V.J., Thomas, W.: Label switching over IEEE802. 15.4 e networks. Trans. Emerg. Telecommun. Technol. **24**(5), 458–475 (2013)
14. Soua, R., Minet, P., Livolant, E.: Wave: a distributed scheduling algorithm for convergecast in IEEE 802.15. 4e TSCH networks. Trans. Emerg. Telecommun. Technol. **27**(4), 557–575 (2015)
15. Soua, R., Minet, P.: DiSCA: a distributed scheduling for convergecast in multichannel wireless sensor networks. In: IFIP/IEEE Integrated Network Management Symposium, pp. 156–164 (2015)
16. Accettura, N., Palattella, M.R., Boggia, G., Grieco, L.A., Dohler, M.: Decentralized traffic aware scheduling for multi-hop low power lossy networks in the internet of things. In: IEEE WoWMoM, pp. 1–6 (2013)
17. Duquennoy, S., Al Nahas, S.B., Landsiedel, O., Watteyne, T.: Orchestra: robust mesh networks through autonomously scheduled TSCH. In: Proceedings of SenSys, pp. 337–350. ACM (2015)
18. Dujovne, D., Watteyne, T., Vilajosana, X., Thubert, P.: 6TiSCH: deterministic IP-enabled industrial internet (of things). IEEE Commun. Mag. **52**(12), 36–41 (2014)
19. Watteyne,T., Palattella, M., Grieco, L.: Using IEEE 802.15. 4e time-slotted channel hopping (TSCH) in the internet of things (IoT): problem statement. Technical report (2015)
20. Wang, Q., Vilajosana, X., Watteyne, T.: 6TiSCH operation sublayer (6top). IETF Standard draft-wang-6tisch-6top-sublayer, vol. 1 (2014)
21. Winter, T., et al.: RPL: IPv6 routing protocol for low-power and lossy networks. In: RFC, vol. 6550, pp. 1–157 (2012)
22. Abdel Fadeel, K.Q., El Sayed, K.: ESMRF: enhanced stateless multicast rpl forwarding for IPv6-based low-power and lossy networks. In: Proceedings of the Workshop on IoT challenges in Mobile and Industrial Systems, pp. 19–24 (2015)
23. Lorente, G.G., Lemmens, B., Carlier, M., Braeken, A., Steenhaut, K.: BMRF: bidirectional multicast RPL forwarding. Ad Hoc Netw. **54**, 69–84 (2017)
24. Papadopoulos, G.Z., Georgallides, A., Tryfonas, T., Oikonomou, G.: BMFA: bidirectional multicast forwarding algorithm for RPL-based 6LoWPANs. In: Mitton, N., Chaouchi, H., Noel, T., Watteyne, T., Gabillon, A., Capolsini, P. (eds.) InterIoT/SaSeIoT -2016. LNICST, vol. 190, pp. 18–25. Springer, Cham (2017). https://doi.org/10.1007/978-3-319-52727-7_3
25. Sobral, J.V., Rodrigues, J.J., Rabelo, R.A., Al-Muhtadi, J.: Multicast improvement for LOADng in internet of things networks. Measurement **148**, 106931 (2019)
26. Hui, J., Kelsey, R.: Multicast protocol for low-power and lossy networks (MPL). Technical report (2016)
27. Conti, M., Kaliyar, P., Lal, C.: A robust multicast communication protocol for low power and lossy networks. J. Netw. Comput. Appl. **164**, 102675 (2020)

28. Baddi, Y., Sebbar, A., Zkik, K., Maleh, Y., Bensalah, F., Boulmalf, M.: MSDN-IoT multicast group communication in IoT based on software defined networking. J. Reliable Intell. Environ. 10, 93–104 (2024). https://doi.org/10.1007/s40860-023-00203-x
29. Oikonomou, G., Phillips, I.: Stateless multicast forwarding with RPL in 6LowPAN sensor networks. In: 2012 IEEE International Conference on Pervasive Computing and Communications Workshops, pp. 272–277. IEEE (2012)
30. Kim, H.S., Yun, S., Kim, H., Shin, H., Kim, W.T.: An efficient SDN multicast architecture for dynamic industrial IoT environments. Mobile Inf. Syst. **2018**(1) (2018)
31. Orozco-Santos, F., Sempere-Payá, V., Silvestre-Blanes, J., Albero-Albero, T.: Multicast scheduling in SDN wise to support mobile nodes in industrial wireless sensor networks. IEEE Access **9**, 141651–141666 (2021)
32. Teshome, E., et al.: Time slotted channel hopping and contikiMAC for ipv6 multicast-enabled wireless sensor networks. Sensors **21**(5), 1771 (2021)
33. Lakhlef, I.E., Djamaa, B., Senouci, M.R.: A comprehensive study of multicast routing protocols in the internet of things. In: Lejdel, B., Clementini, E., Alarabi, L. (eds.) Artificial Intelligence and Its Applications, pp. 325–335. Springer, Cham (2022). https://doi.org/10.1007/978-3-030-96311-8_30

When Curriculum Learning Meets Multi-Agent DRL in Connected Autonomous Vehicles

Mohamed El Amine Ameur[1,2](✉), Habiba Drias[1], and Bouziane Brik[3]

[1] Laboratory of Research in Artificial Intelligence (LRIA), USTHB, BP32, EL Alia, 16111 Bab Ezzouar, Algiers, Algeria
{m_ameur,hdrias}@usthb.dz

[2] Department of Computer Science, Laghouat University, BP G37, Mkam, Laghouat, 3049 Laghouat, Algeria
me.ameur@lagh-univ.dz

[3] Computer Science Department, College of Computing and Informatics, University of Sharjah, Sharjah, UAE
bbrik@sharjah.ac.ae

Abstract. Efficient lane changing in Connected Autonomous Vehicles (CAVs) plays a pivotal role in congestion reduction by optimizing traffic flow, minimizing bottlenecks, and enhancing overall road capacity. This paper introduces an innovative automated lane-changing strategy based on Curriculum Learning, offering notable advantages in learning efficiency while maintaining stable performance. The trained agent acquires the capability to develop a safe and time-efficient driving policy for lane-change decisions. Numerical results provide a comprehensive perspective on the strengths of the proposed framework, opening avenues for future advancements in autonomous driving systems and intelligent transportation networks.

Keywords: Autonomous Vehicles · Lane Changing · Multi-Agent Deep Reinforcement Learning · Curriculum Learning

1 Introduction

The emergence of Connected Autonomous Vehicles (CAVs) represents a transformative paradigm shift in the landscape of transportation, promising profound impacts on various facets of society [1]. As self-driving technology advances, it offers numerous benefits, particularly in safety, by reducing human errors, a major cause of road accidents.

In the realm of Autonomous Vehicles (AVs), achieving proficient motion planning is paramount for ensuring safe and efficient navigation through dynamic environments. The intricate nature of real-world scenarios demands intelligent decision-making mechanisms capable of adapting to unforeseen challenges. Moreover, the advent of autonomous driving holds the promise of optimizing traffic flow, reducing congestion, and enhancing overall transportation efficiency.

B. Djamaa et al. (Eds.): CSA 2024, LNNS 1145, pp. 200–210, 2025.
https://doi.org/10.1007/978-3-031-71848-9_19

As the integration of autonomous driving technologies accelerates, its societal, economic, and environmental impacts are poised to be both substantial and far-reaching [2]. However, realizing the benefits of autonomous driving hinges on addressing significant challenges, notably effective lane changing. Nuanced solutions are essential to seamlessly integrate this aspect into dynamic traffic environments, as shown in Fig 1.

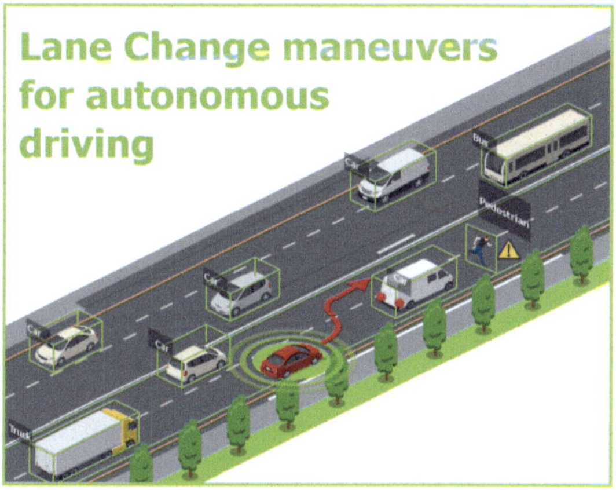

Fig. 1. Lane Changing Scenario in CAV

Lane changing is a complex maneuver that requires a delicate balance between safety, efficiency, and adaptability. In the context of AVs, this maneuver becomes particularly challenging due to the need for real-time decision-making in response to dynamic traffic conditions [3]. The successful implementation of robust lane-changing strategies is essential for realizing the full potential of autonomous driving systems and ensuring a harmonious coexistence with traditional human-driven vehicles. At the heart of lane-change behavior lies the lane-change decision model, a pivotal element that dictates whether a vehicle should switch lanes and, if so, whether it should move to the left or the right to enhance driving conditions, such as achieving the desired speed or improving visibility [4].

Even though, Multi-Agent RL algorithms have found application in autonomous driving, particularly in highway lane change tasks, displaying promising and scalable performance [6]. In these scenarios, AVs learn collaboratively by sharing a common objective, typically defined by a reward/cost function that takes into account safety and efficiency considerations. Nevertheless, these reward designs often neglect passengers' comfort, potentially resulting in abrupt acceleration and deceleration that can cause discomfort during the ride. Furthermore, these models assume that human-driven vehicles adhere to unaltered, universally consistent behaviors, a simplification that proves impractical in the diverse and nuanced real-world driving environment where human drivers exhibit varying behaviors. Additionally, this type of modeling frequently encounters challenges related to poor learning rates [5], demanding extensive time and resources for training, prompting the need for innovative solutions to enhance the learning phase and address this research gap.

In this study, our focus centers on addressing the issue of slow learning rates by implementing Curriculum Learning alongside multi-agent deep reinforcement learning. This combination proves to be particularly effective in navigating lane-changing scenarios for AVs. This approach draws inspiration from educational curricula, where students progressively tackle more challenging topics as they master foundational concepts. A curriculum is designed to guide the agent through a sequence of learning stages, starting from simpler scenarios and gradually introducing more complex ones. This sequential exposure helps the agent develop a robust and generalized policy by first mastering basic skills and then incrementally building on them.

We employ a Multi-Agent Advantage Actor-Critic network (MA2C) for decision making in AV lane-changing scenarios, incorporating Curriculum Learning techniques. This involves elevating the learning environment's complexity by introducing a higher number of vehicles. The model includes an innovative local reward design that integrates safety, efficiency, and passenger comfort, along with a parameter-sharing scheme to encourage collaborative interactions among the agents.

The primary contributions and technological advancements presented in this paper are outlined as follows:

1. Our investigation delves into AV highway lane changes within mixed traffic, scrutinizing the current state-of-the-art to glean insights into existing methodologies and pinpoint the gaps and challenges requiring attention.
2. Formulating the lane-changing problem as a Partially Observable Markov Decision Process (POMDP), our approach involves cooperative learning among agents to derive a driving policy that prioritizes safety and efficiency.
3. In our study, we introduce an innovative Curriculum Learning framework grounded in Multi-Agent DRL, incorporating a parameter-sharing mechanism and a well-crafted reward function design.
4. To provide a comprehensive understanding of the efficacy of our framework, we propose a comparative performance evaluation, with other DRL algorithms (DQN, A2C) within the context of AV highway lane changes in mixed traffic.

2 System Model

2.1 Curriculum Learning

Curriculum Learning is a machine learning strategy inspired by the educational concept of curricula, which involves organizing learning tasks in a structured and progressive manner [10]. In the context of machine learning, Curriculum Learning involves presenting the model with training examples in a deliberately sequenced order, starting with simpler instances and gradually introducing more complex ones. This approach helps the model to learn incrementally, building upon previously acquired skills and knowledge. By exposing the model to easier tasks initially, Curriculum Learning facilitates a smoother convergence and enables the model to generalize more effectively. This training methodology has shown success in various domains, from natural language processing to reinforcement learning, providing a mechanism for improved model robustness and performance on challenging tasks[10].

In our novel curriculum A2C algorithm for autonomous lane changing, we incorporate a dynamic learning rate (γ) schedule, beginning with simple tasks and gradually decreasing γ as the algorithm encounters more advanced scenarios. This novel curriculum A2C algorithm is designed to expedite convergence and optimize the learning process, ultimately enhancing the adaptability and decision-making capabilities of AVs in highway mixed-traffic environments.

2.2 Problem Formulation

Discontinuous evaluation is a prevalent approach in autonomous driving, extensively employed in numerous papers [8,9]. Specifically, we conceptualize the mixed-traffic lane-changing scenario as a multi-agent network: $\mathcal{G} = (\mathcal{V}, \mathcal{E})$, where each agent (i.e., ego vehicle) $i \in \mathcal{V}$ communicates with its neighbors $\mathcal{N}_i$ through the communication link $\mathcal{E}_{i,j} \in \mathcal{E}$. Within a multi-agent POMDP, each agent i adheres to a decentralized policy $\pi_i : \mathcal{O}_i\mathcal{S}_i \rightarrow [0, 1]$ to make decisions at time step t. The described POMDP is formally defined as:

1. **State Space:** Agent i's state space, denoted as $\mathcal{O}_i$, is characterized by a matrix with dimensions $\mathcal{N}_{\mathcal{N}i} \times \mathcal{F}$, where $\mathcal{N}_{\mathcal{N}i}$ is the count of detected vehicles, and $\mathcal{F}$ is the number of features utilized to depict the present state of vehicles. This representation encompasses the longitudinal position (x), lateral position (y) of the observed vehicle in relation to the ego vehicle, longitudinal speed (v_x), and lateral speed (v_y) of the observed vehicle relative to the ego vehicle.
2. **Action Space:** The collection of high-level control decisions accessible to agent i is represented as the action space $\mathcal{A}_i$. This collection includes actions like speeding up, slowing down, cruising, turning left, and turning right. The combined action space for AVs is defined as $\mathcal{A} = \mathcal{A}_1 \times \mathcal{A}_2 \times \ldots \times \mathcal{A}_N$, where N denotes the total number of vehicles in the scene.
3. **Reward Function:** The design of the reward function incorporates various metrics, including safety, traffic efficiency, and passenger comfort. The specific evaluations within the reward function are as follows:
   - **Safety Evaluation R_{safety}:** Ensuring the vehicle operates without collisions.

$$R_{safety} = \begin{cases} -10 & \text{, if Collision occurs} \\ 0 & \text{, Otherwise} \end{cases}$$

   - **Effectiveness Evaluation R_{eff}:** Encouragement to uphold a consistent speed v_t, preventing the agent from decelerating to consistently avoid collisions.

$$R_{eff} = \begin{cases} \dfrac{v_t - v_{\min}}{v_{\max} - v_{\min}} & \text{, if AV speed } v_t < \text{min speed limit} \\ \\ \dfrac{v_{\max} - v_t}{v_{\max} - v_{\min}} & \text{, if AV speed } v_t > \text{min speed limit} \\ \\ \dfrac{v_t}{v_{\max}} & \text{, Otherwise} \end{cases}$$

- **Lane Change Penalty** $R_{LaneChange}$**:** The avoidance of frequent lane changes is encouraged.

$$R_{LaneChange} = \begin{cases} -1 & \text{, if Change Lane} \\ 0 & \text{, Otherwise} \end{cases}$$

- **Driving Comfort** $R_{confort}$**:** Achieving smooth acceleration a and deceleration d to ensure safety and comfort, prioritizing smooth deceleration. c_a and c_d are weighing coefficients to balance the importance of acceleration and deceleration.

$$R_{confort} = c_a \cdot a - c_d \cdot d$$

The probability of transition, denoted as $\mathcal{T}(s', |s, a)$, delineates the shift from one state to another. Given that our MARL algorithm is model-free in design, no assumptions are made regarding prior knowledge of transition probability.

2.3 Curriculum Multi-Agent Advantage Actor-Critic (C-MA2C) for Lane Changing

In this research, we enhance the actor-critic network to operate in a multi-agent context, specifically as a multi-agent Actor-critic network. The incorporation of A2C enhances the stability and scalability of the learning process by facilitating communication among agents [7]. To capitalize on the uniformity of agents within the considered MARL framework, we assume a shared network structure and parameters among all agents. Nevertheless, each agent retains the ability to execute distinct maneuvers based on diverse input states. In the cooperative MARL scenario, the overarching objective is to maximize the collective reward for all agents. To address challenges such as communication overhead and the credit assignment problem [11], expressed by the A2C loss function:

$$\mathcal{L}_t = \sum_{k=0}^{\infty} \gamma^k r_{t+k} + \gamma^{T-t} V(s_T) \tag{1}$$

where r_{t+k} represents the immediate reward at time step t. γ is the discount factor, where $0 < \gamma < 1$, and $V(s_T)$ is the value function representing the expected cumulative future rewards from state S_t. The final time step in an episode is denoted as T. This formula combines the immediate rewards with the discounted future rewards and the value function estimate. It captures the notion of assigning value to each action based on both its immediate consequences and its potential for long-term gain.

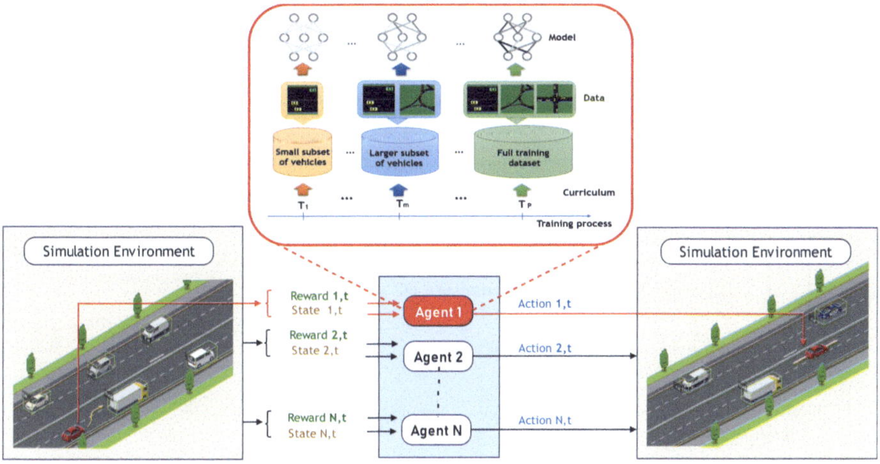

Fig. 2. The envisioned Curriculum Lane Changing Framework

Curriculum learning offers a diverse range of techniques for effective implementation, including adaptive policies, transfer learning, and more [10]. In our specific context, we employ two key strategies to enhance learning dynamics. Firstly, dynamic parameter adjustment, particularly in the form of adapting the γ, allows for fine-tuning the model's responsiveness to different tasks. Secondly, we leverage the task sequencing technique by progressively introducing complexity to the training environment, specifically by augmenting the scenario with additional vehicles. As illustrated in Fig. 2, this tailored approach facilitates a structured learning progression, enabling the model to gradually master increasingly intricate aspects of the given task.

As presented in the pseudo-code 1, our algorithm functions by iteratively training individual agents to make optimized lane-changing decisions. In each iteration, agents observe their local environment, including factors like current lane, neighboring vehicles, and speed. The agents then select actions using personalized policies, considering their specific states. The algorithm computes advantages, reflecting the impact of chosen actions on immediate rewards and future state values. This collaborative learning process involves communication among agents and parameter synchronization to enhance overall decision-making. Over multiple iterations, the multi-agent A2C algorithm adapts agent behaviors, aiming to improve the efficiency and safety of AVs during lane changes in mixed traffic scenarios.

Algorithm 1: Curriculum A2C for Lane-Changing (pseudo)

for *each episode* **do**
  Initialize/update learning rate (γ);
  for *each training step* **do**
    for *each AV* **do**
      Observe local state $\mathbf{s}_i$, including current lane, surrounding vehicles, and speed;
      # **Dynamic Parameter Adjustment:** Adjust γ based on task difficulty;
      Select lane-changing action $\mathbf{a}_i$ using policy $\pi_{\boldsymbol{\theta}_i}$;
      Execute lane-changing action $\mathbf{a}_i$ and observe local reward r_i and next state $\mathbf{s}_{i+1}$;
      # **Advantage Calculation:** Calculate advantage δ_i using local value function V_{ϕ_i}:

$$\delta_i = r_i + \gamma V_{\phi_i}(\mathbf{s}_{i+1}) - V_{\phi_i}(\mathbf{s}_i)$$

      # **Policy and Value Function Updates:** Update policy parameters $\boldsymbol{\theta}_i$ using gradient ascent:

$$\boldsymbol{\theta}_i \leftarrow \boldsymbol{\theta}_i + \alpha_{\theta_i} \nabla_{\boldsymbol{\theta}_i} \log \pi_{\boldsymbol{\theta}_i}(\mathbf{a}_i|\mathbf{s}_i)\delta_i$$

      Update value function parameters $\boldsymbol{\phi}_i$ using TD error:

$$\boldsymbol{\phi}_i \leftarrow \boldsymbol{\phi}_i + \alpha_{\phi_i} \nabla_{\boldsymbol{\phi}_i} \delta_i^2$$

    end
  end
  Communicate information among AVs (Agents) ;
  Update network parameters;
end

3 Performance Evaluation

3.1 Experiment Setup

The integration of Simulation of Urban MObility (SUMO) [12] stands as a widely adopted approach for simulating traffic scenarios with mixed vehicle types. In our particular study, we harnessed SUMO to emulate the behavior of five ego vehicles and a fleet of AVs (35 vehicles) while training three distinct reinforcement learning models, namely, Deep Q-Network (DQN), A2C, and our customized Curriculum-based Multi-agent A2C (C-MA2C) algorithm. The workflow involved constructing the highway layout using NetEdit, extracting simulation parameters through TraCi, establishing a tailored Gym environment, and subsequently training the neural networks. TraCi is a Python library employed to extract simulation parameters of SUMO. This entails connecting a Python script to SUMO through a TCP connection. To facilitate RL training, a custom Gym environment was crafted, providing a structured definition of the state space, rewards, and observations based on information gathered through TraCi.

3.2 Training Details

The experiment was conducted utilizing an Ubuntu virtual machine possessing the following specifications: an 8-core CPU with a clock speed of 3.2 GHz, a 16-gigabyte memory capacity, and a graphics processing unit of NVIDIA GeForce

RTX 3060. The training process integrates the Multi-Agent A2C algorithm with Curriculum Learning techniques, focusing on the learning rate parameter γ.

The neural network architecture comprises three layers with 64, 128, and 64 neurons in each layer, As depicted in Table 1. This tailored training regimen aims to enhance convergence, improve performance, and promote adaptability in AV lane-changing tasks, providing a comprehensive learning experience for the multi-agent system.

Table 1. Hyper parameter setup for the experiment

Hyper-parameter	Value
Learning Rate (γ)	0.001
Number of Vehicles (total)	35
Neural Network Architecture	3 layers (64-128-64)
Activation Function	ReLU
Optimizer	Adam
Framework	Pytorch 1.4

3.3 Experiment Results and Discussion

In this section, we elucidate the outcomes achieved in the course of this study. The obtained results from the experimental evaluation are presented in Table 2, detailing the performance metrics for three different reinforcement learning methods: DQN, A2C, and Curriculum-MA2C. In terms of success rates, C-MA2C outperformed both DQN and A2C, achieving an impressive 99.7%. A2C followed closely with a success rate of 98.77%, while DQN exhibited a slightly lower success rate at 97.02%.

Table 2. Performance Metrics

Method	Success (%)	Collision (%)	Avg. Brake time (s)
DQN	97.02	0.12	0.55
A2C	98.77	0.085	0.45
C-MA2C	**99.7**	**0.0039**	**0.31**

When considering collision rates, it is noteworthy that C-MA2C demonstrated flawless performance with a 0.0039% collision rate. A2C exhibited a low collision rate of 0.085%, indicating a robust avoidance of collisions, while DQN had a slightly higher collision rate at 0.12%. Additionally, analyzing the average brake duration provides insights into the efficiency of the methods in controlling the simulated vehicle. C-MA2C showcased the shortest average brake duration, requiring only 0.31 s. A2C followed with 0.45 s, and DQN had the longest brake duration at 0.55 s. These findings suggest that C-MA2C not only excels in success and collision rates but also demonstrates superior control efficiency through minimized braking actions. Figures 3a and 3b provide visual insights into the convergence behavior. Notably, the C-MA2C method exhibits superior performance compared to both A2C and DQN. The convergence of C-MA2C occurs at a faster rate, reaching stability around timestep 110,000 in the loss function and 145,000 in cumulative reward. In contrast, A2C follows, with DQN converging last. This observed pattern strongly suggests the effectiveness of Curriculum Learning in expediting the learning process. By decomposing the learning tasks into smaller and simpler components, represented in our case by initiating training with a lower number of vehicles, C-MA2C achieves a more rapid convergence. The gradual increment in the complexity of the environment, coupled with the strategic adjustment of the learning rate parameter γ, contributes to this accelerated learning phenomenon, resulting in improved learning efficiency and faster convergence.

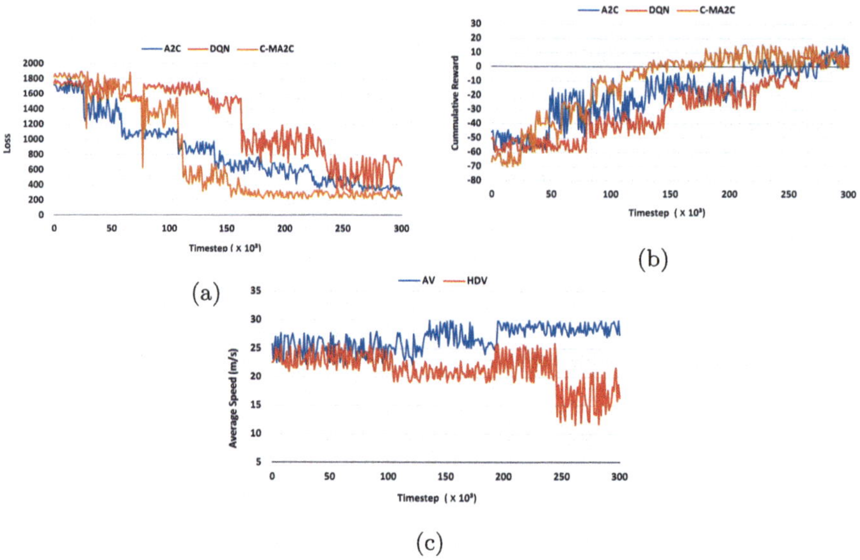

Fig. 3. Different performance metrics. (a) Loss function graph, (b) Cumulative reward graph, (c) AV Average Speed graph.

As illustrated in Fig. 3c, our findings reveal a noteworthy disparity in the average speed of AVs when executing lane changes compared to human-driven vehicles. The data unequivocally demonstrates that AV maneuvers exhibit greater efficiency, characterized by swifter lane changes. This disparity can be logically attributed to the inherent predictive capabilities of AVs in anticipating and responding to traffic conditions. Unlike human drivers who rely on real-time observations, AVs leverage advanced sensors and algorithms to foresee potential congestion, enabling them to make informed decisions for optimal lane changes. The expedited lane-changing capabilities of AVs contribute to a reduction in overall traffic time, providing a tangible benefit for commuters and the environment.

4 Conclusion

This study introduced an innovative automated lane-changing strategy based on Curriculum Learning, addressing the limitations of existing models. The proposed framework offered notable advantages in learning efficiency while maintaining stable performance, as demonstrated by numerical results compared to two other DRL solutions. By emphasizing the significance of multi-agent dynamics and introducing a more efficient learning approach, this research set the stage for further innovations in ensuring the safety, congestion management, and overall efficiency of highway traffic in the era of Connected AVs. For future works, our focus will be on exploring the proposed Curriculum framework in conjunction with other traffic scenarios to validate its effectiveness and assess its adaptability across diverse environments (various complexities).

References

1. Yue, W., Li, C., Wang, S., Xue, N., Wu, J.: Cooperative incident management in mixed traffic of CAVs and human-driven vehicles. IEEE Trans. Intell. Transp. Syst. **24**, 12462–12476 (2023). https://doi.org/10.1109/tits.2023.3289983
2. Hegde, B., Bouroche, M.: Multi-agent reinforcement learning for safe lane changes by connected and autonomous vehicles: a survey. AI Commun. **37**(2), 203–222 (2023). https://doi.org/10.3233/aic-220316
3. Duan, X.: Cooperative lane-change motion planning for connected and automated vehicle platoons in multi-lane scenarios. IEEE Trans. Intell. Transp. Syst. **24**(7), 7073–7091 (2023)
4. Wang, Y.: Ego-efficient lane changes of connected and automated vehicles with impacts on traffic flow. Transp. Transp. Res. C Emerg. Technol. **138**, 103478 (2022)
5. Mirchevska, B., Pek, C., Werling, M., Althoff, M., Boedecker, J.: High-level decision making for safe and reasonable autonomous Lane changing using reinforcement learning. In: 2018 21st International Conference on Intelligent Transportation Systems (ITSC). IEEE (2018)
6. Ha, P.Y.J., Chen, S., Dong, J., Du, R., Li, Y., Labi, S.: Leveraging the capabilities of connected and autonomous vehicles and multi-agent reinforcement learning to mitigate highway bottleneck congestion. arXiv preprint arXiv:2010.05436 (2020)
7. Du, W., Ding, S.: A survey on multi-agent deep reinforcement learning: from the perspective of challenges and applications. Artif. Intell. Rev. **54**, 3215–3238 (2021). https://doi.org/10.1007/s10462-020-09938-y
8. Schester, L., Ortiz, L.E.: Longitudinal position control for highway on-ramp merging: a multi-agent approach to automated driving. In: 2019 IEEE Intelligent Transportation Systems Conference (ITSC). IEEE (2019)

9. Palanisamy, P.: Multi-agent connected autonomous driving using deep reinforcement learning. In: 2020 International Joint Conference on Neural Networks (IJCNN). IEEE (2020)
10. Narvekar, S., Stone, P.: Learning curriculum policies for reinforcement learning. arXiv preprint arXiv:1812.00285 (2018)
11. Sutton, R.S., Barto, A.G.: Reinforcement learning: an introduction. IEEE Trans. Neural Netw. **9**, 1054–1054 (1998). https://doi.org/10.1109/tnn.1998.712192
12. Lopez, P.A., et al.: Microscopic traffic simulation using SUMO. In: 2018 21st International Conference on Intelligent Transportation Systems (ITSC). IEEE (2018)

Fault Tolerant Aware Scheduling on IoT Systems: A Comparative Study

Amel Sekkiou[1,3(✉)], Ouassila Hioual[2,3], and Ouided Hioual[3]

[1] ICOSI Laboratory of Khenchela, khenchela, Algeria
{amel.sekkiou,hioual.ouided}@univ-khenchela.dz
[2] LIRE Laboratory of Constantine, Constantine, Algeria
[3] Abbes Laghrour University Khenchela, Khenchela, Algeria

Abstract. In recent years, the rise of IoT technology has led to a significant increase in the demand for real-time applications that require execution with prioritization. The challenge of scheduling fault-tolerant tasks in fog and cloud environments has become particularly crucial, especially when considering the need to minimize energy consumption in the nodes. This paper presents a comprehensive comparative study on fault-aware scheduling in IoT (Internet of Things) systems. The aim is to evaluate different scheduling approaches that improve fault tolerance in the dynamic and resource-constrained environment of IoT systems (especially at a fog or cloud level). We analyse and compare various related works that consider different constraints such as system reliability, resource utilization, and responsiveness to failures. The study aims to provide an overview of the strengths and weaknesses of existing fault-tolerant scheduling techniques in IoT systems, thereby facilitating a better understanding of their practical implications. Thus, this study contributes to the advancement of robust and reliable scheduling strategies for IoT applications.

Keywords: Fault-tolerance · Scheduling · Energy Consumption · Fog Computing · IoT

1 Introduction

The Internet of Things (IoT) application has become an integral part of our daily lives, and as a result, the number of connected devices has increased significantly. This growth has led to the generation of enormous amounts of data. Initially, Cloud computing was considered the primary solution for processing this vast amount of data generated by IoT devices. However, due to the increasing demand for processing time-sensitive tasks and the fact that the cloud is distant from the devices, there is a delay in response.

Consequently, there is a pressing need to find a new technology that is closer to the IoT devices where the data is generated. In response to this, Cisco proposed a new computing paradigm called Fog computing, which brings computation capabilities closer to the end user at the edge of the network. This solution

B. Djamaa et al. (Eds.): CSA 2024, LNNS 1145, pp. 211–222, 2025.
https://doi.org/10.1007/978-3-031-71848-9_20

aims to address the challenge of latency by providing computing resources in proximity to the IoT devices themselves [1].

Fog computing is an enhanced version of Cloud computing that aims to bring computation, processing, storage, and networking closer to end users at the edge of the network [2]. This approach is implemented to achieve a faster response time and avoid delays. Cisco predicts that Fog computing can effectively address the challenges faced by IoT applications or systems, such as low latency, high response time, cost, and support for mobility [3,4]. The goal is to enhance the Quality of Service (QoS) for users.

However, there are challenges associated with providing services through both Fog and cloud environments. These challenges include issues with software, human errors, and hardware failures, which can significantly impact the user's QoS [5]. Therefore, it is crucial to implement fault-tolerant scheduling strategies to improve the reliability and availability of the system in such environments.

This paper focuses on the problem of fault-tolerant scheduling in fog and cloud environments. Our objective is to address this issue by conducting a comprehensive survey of existing approaches found in the literature. This comparative study holds paramount significance within the existing literature for several reasons. At first, as IoT environments are dynamic and resource-constrained, the need for robust fault-tolerant scheduling mechanisms is crucial to ensure the reliability of applications. Our study addresses this issue by examining and comparing various scheduling approaches, providing explicit insights into their strengths and weaknesses.

Secondly, the comparative analysis carried out in this paper delves into several aspects such as resource utilization and responsiveness to failures. Thus, it offers a comprehensive assessment of the practical implications of different scheduling methodologies. Furthermore, this contribution is vital for advancing the state-of-the-art in IoT systems, ultimately leading to more resilient and dependable deployments in real-world scenarios.

The remainder of this paper is structured as follows: Section 2 provides an overview of the fundamental concepts related to fault-tolerant scheduling tasks in Cloud and Fog environments. Section 3 presents a comprehensive review of the existing related work in this field. In Sect. 4, we present a comparative study of various methods and provide a detailed discussion of their strengths and limitations.Finally, in Sect. 5, we conclude the survey study.

2 Context

In today's era, the advancement of IoT technology has resulted in a substantial increase in the number of connected devices. This, in turn, has led to the generation of massive amounts of data and an increased demand for task execution and quick response times. Initially, Cloud computing emerged as the primary computing paradigm providing support for IoT acrossdivers scenarios. It offered a diverse range of services to end users in domains like healthcare, smart cities, and scientific computing.

However, one of the challenges faced by Cloud computing is the geographical distance between the cloud infrastructure and end users. This geographical separation often results in delays and suboptimal response times.

The delay in response, especially for time-sensitive requirements, becomes a significant issue in Cloud computing. To address this challenge, fog computing emerges as a solution by bringing computation closer to the end-users, right where the data is generated. Fog computing significantly improves IoT performance by reducing latency, delay, and response time.

However, fog computing introduces its own set of challenges due to the heterogeneous nature of fog nodes, which are susceptible to failures. In such environments, resource management becomes a crucial concern. To optimize the utilization of fog nodes and ensure reliability and availability, fault-tolerance techniques or paradigms need to be incorporated. This includes the implementation of scheduling mechanisms that effectively manage resources and nodes, thereby enhancing reliability, availability, and resource management in fog computing environments.

Scheduling. In IoT systems, the scheduling problem can be divided into five distinct categories (see Fig. 1):

- Resource Allocation: Achieving faster response times, improved resource utilization, reduced energy consumption, minimized makespan, and cost optimization are key objectives in IoT systems. To accomplish these goals, it is crucial to optimize the allocation of a set of tasks {T1, T2... Tn} with varying Quality of Service (QoS) requirements to a set of densely-distributed heterogeneous fog nodes {F1, F2... Fm}.
- Task scheduling: In order to optimize resource usage and minimize delay, cost, and energy consumption, the tasks within the IoT system have varying lengths and must be executed based on their priorities.
- Workflow scheduling: In the context of IoT systems, one important category of scheduling is workflow scheduling. In this scenario, jobs generated by Internet of Everything (IoE) devices are composed of multiple tasks. These tasks can be divided into independent and dependent tasks. Independent tasks are not influenced by the execution order of other tasks, while dependent tasks rely on the completion of their parent tasks. To efficiently manage and sequence these tasks, workflows are used, often represented as Directed Acyclic Graphs (DAGs). The objective of workflow scheduling in distributed systems is to assign tasks to heterogeneous fog nodes while minimizing the make span (total execution time) and respecting task dependencies. However, due to the computational complexity of this task, workflow scheduling in distributed environments is considered an NP-complete problem that has received significant attention from researchers [6].
- Job scheduling: The concept here is to allocate a group of jobs to the fewest possible fog resources, ensuring minimal CPU execution time.
- Resource scheduling: The objective is to identify the optimal allocation of resources that aligns with the client's requirements. This goal aims to improve

resource utilization, minimize processing delays, and enhance the Quality of Service (QoS).

Fault Tolerance Aware Task Scheduling. Fault-tolerance task scheduling refers to the assignment of tasks to resources, such as VMs or fog nodes, in a manner that ensures the successful completion of those tasks. This scheduling process aims to optimize Quality of Service (QoS) requirements while maintaining the desired level of system utilization, even in the presence of resource failures.

Task scheduling, in general, is a computationally challenging problem and is known to be NP-hard [7,8]. Introducing fault tolerance into the scheduling process further increases its complexity and difficulty [9].

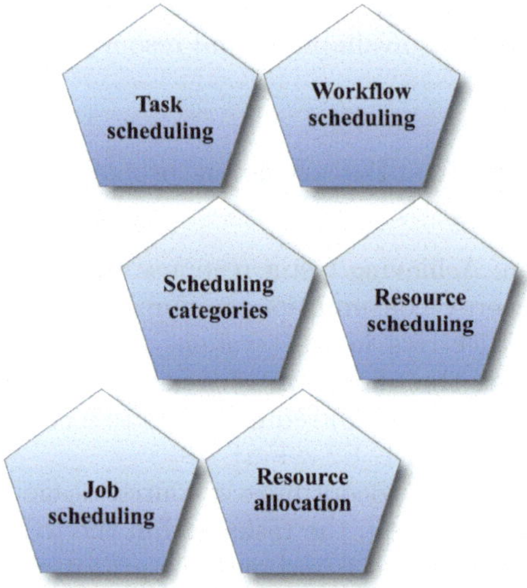

Fig. 1. Scheduling categories on IoT systems

Fault tolerance involves designing strategies to ensure the continuous operation of tasks and programs, even when multiple components experience failures. These algorithms aim to distribute the system load evenly, enabling task requests and execution to adapt to different situations, even in the presence of failures. It is important to note that fault tolerance proposals do not guarantee 100% task execution or uninterrupted services. However, they strive to maintain system usability at a satisfactory level of performance [10].

Fault tolerance, in essence, refers to the capability of a system (such as a computer, network, cloud cluster, or IoT system) to operate seamlessly without any interruptions in the event of a failure of one or more components. The primary objective of a fault-tolerant system is to prevent disruptions caused by component failures and ensure the availability and reliability of the overall system [11].

To achieve fault tolerance, various types have been explored for different distributed systems. They can be categorized into two models (see Fig. 2): proactive and reactive.

Proactive Approaches, also known as active fault tolerance, refer to strategies aimed at preventing errors before they occur. These later's involve the utilization of techniques such as monitoring, prediction, and preemption to anticipate and avoid potential errors or failures.

Reactive Approaches, also known as passive fault tolerance, involve measures taken after a failure has occurred to mitigate its impact. These measures include replication, check pointing, restarting, detection, and recovery. A commonly used technique in fault-tolerant systems is replication, especially primary-backup replication. Replication means that some system components are copied and deployed simultaneously in different resources. The purpose is to form a robust system and ensure the execution of jobs [12]. Whereas, primary-backup replication means that a primary task can be supported by zero, one, or multiple backup tasks. This replication mechanism plays a crucial role in enhancing system reliability by providing redundancy and ensuring continuity in case of failures.

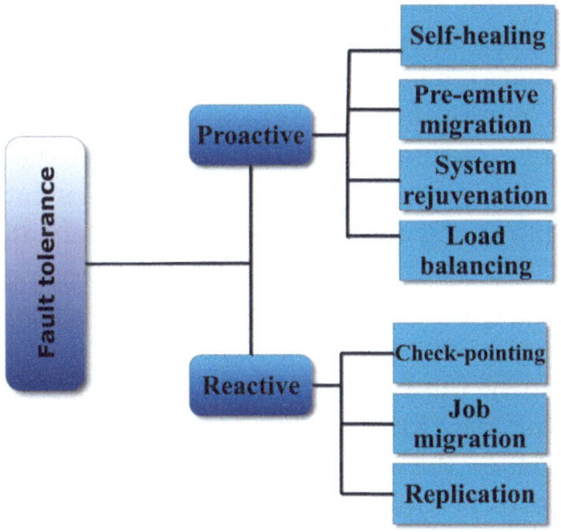

Fig. 2. Fault tolerance methods Taxonomy

3 Related Work

In this section, we present a literature review that focuses on the key fault-tolerant scheduling methods in the IoT environment. The aim is to provide a comprehensive overview of the most significant research conducted in this field.

In [13], Gupta et al. proposed a real-time fault-tolerant scheduling approach for scientific workflows in the IoT environment. They extended the traditional

Primary Backup (BP) method and incorporated cloud characteristics, which they referred to as FASTER. The approach includes features such as task allocation and message transmission, aiming to make full use of idle resources. Additionally, the authors employed a backward shifting method to optimize resource utilization. They, also, used VM migration techniques to, further, enhance resource utilization in the system.

The authors in [14] presented a scheduling algorithm for allocating services requests at the cloud-fog environment. The method classifies the fog devices into three classes according to time sensitivity of the requests, in goal to reduce the delay and increase the reliability and the capacity of cloud by providing the fault tolerance using three techniques: replication, check pointing, and resubmission.

In [15], A. Sharif et al. investigate the fault-tolerant allocation of resources in the fog computing environment. They proposed a novel fault-tolerant scheduling algorithm for modules within the fog. The main idea revolves around classifying the resources and different modules and calculating their energy consumption. The objective is to minimize energy consumption by efficiently distributing the modules among fog resources. To achieve this, the authors employed an energy-efficient check pointing and load balancing approach, augmented by Bayesian classification techniques. The proposed algorithm aims to optimize resource utilization while ensuring fault tolerance in the fog computing environment.

In [16], a novel approach, referred to as DFTLA (Dynamic Fault Tolerant using Learning Automata) has been developed. This method focuses on task scheduling for the fog computing platform and aims to achieve fault tolerance while keeping costs low, optimizing quality of service (QoS), and minimizing energy consumption. The proposed approach involves integrating a variable learning automaton into each fog node, which actively monitors the process and makes decisions.

The researchers of [17] focused on fault tolerance model for real time tasks, when the uncertainty of task runtime is considered in the cloud environment. They proposed an efficient fault tolerant algorithm (DEFT): Dynamic fault tolerance elastic scheduling algorithm. The strategy based on two methods: replication and resumption, where the overlapping mechanism is applies. The objective of this work is to achieve the tradeoff between reliability and resource optimization.

Authors of [18] examined the task scheduling in fog computing and put forth two heuristic algorithms, namely Priority-aware semi-greedy (PSG) and PSG with multi-start procedure (PSG-M). These algorithms aim to address the task scheduling challenges in volunteer computing systems (VCSs) by tackling deadline violations and minimizing the overall energy consumption and communication costs.

In [19], J. Xuet al. tackled the scheduling problem of complex tasks with priority constraints in IoT applications within a Cloud-Fog environment. They introduced an algorithm called LBP-ACS (Laxity and Ant Colony System), which takes into account task priorities while minimizing the overall energy consumption. The algorithm utilizes an ant colony strategy to approximate the scheduling scheme and obtain optimal results.

A novel algorithm called EFTR (Energy Efficient Fault-Tolerant Scheduling)has been developed in [20]. Their research focused on green computing within a cloud environment, aiming to reduce energy consumption and improve system processing capacity. The proposed algorithm employed a primary-backup method for fault tolerance. Specifically, the baseband processing was centralized in the cloud DCNs through C-RAN. The results were compared with the FESTAL algorithm, revealing that EFTR outperformed FESTAL in terms of energy conservation.

Tang et al. explored the fault-tolerant scheduling problem for task allocation on the Cloud in their work [21]. They presented the APSDQN (self-adaptive active-passive scheme with deep reinforcement learning) algorithm, which enabled the selection of an optimal fault-tolerant scheme (active or passive) for a given task based on the system's real-time information at that time. By transforming the fault-tolerant process of faulty tasks into a Markov Decision Process (MDP), the APSDQN algorithm significantly enhanced resource utilization and increased the success rate of tasks.

In their research [22], Ranjan et al. addressed the challenge of managing application failures in a fog computing environment. They proposed a solution called FLBFH (Fuzzy Logic-Based Failure Handling) that leveraged fuzzy logic techniques. The FLBFH approach involved predicting and handling both predicted and unpredictable failures intime sensitive applications. For reactive failures, check pointing and replication techniques were employed, while proactive failures were addressed through migration, based on a failure score. The primary objective of the study was to reduce processing time and delay in handling application failures.

Zulfiqar et al. [23] investigated the issue of fault-tolerant data management in a fog computing environment for healthcare IoT. They proposed a novel scheme called FTDM (Fault Tolerant Data Management), which incorporated two fault tolerance mechanisms. The first mechanism was task-based fault tolerance, utilizing a resubmission technique that re-executed tasks on newly available fog nodes in case of failure. The second mechanism was node-based fault tolerance, employing replication to execute one of the task copies if a node failure occurred. By implementing these mechanisms, the study aimed to minimize energy consumption, execution time, and cost while addressing network delays and ensuring efficient task execution.

In [24], Zhang et al. tackled the fault-tolerant scheduling problem for workflows in the cloud environment. They introduced the EFFT (Efficiency-First Fault-Tolerant)algorithm, which aimed to achieve a cost-optimal resource scheduling strategy that balanced cost and reliability considerations. The algorithm determined the most suitable migration plan by evaluating the efficiency ratio of the initial allocation schemes, with the goal of minimizing execution costs. Additionally, replication was employed to ensure reliability in the event of task execution failures.

Xue et al. focused on enhancing reliability while considering energy consumption in the Edge Cloud system in their research [25]. They proposed an energy-aware fault-tolerant resource scheduling algorithm to improve system reliability. This was achieved through a VM migration mechanism aimed at optimizing energy consumption. Additionally, the study incorporated CPU temperature prediction and time between failure (TBF) prediction methods as part of the fault tolerance strategy. By integrating these techniques, both reliability and energy efficiency were improved in the Edge Cloud system.

The next section will delve into the discussion of Table 1, which presents a comprehensive summary and comparison of various existing works.

Table 1. Some existing works: summarize and comparison

Ref	Problem	Method	Algorithm	Objective	Challenges	Environment
13	Migration	-Primary backup -Migration	Faster	-Resource utilization -Tolerating hardware failures	Considering extra constraints	Cloud/ Real
14	Requests scheduling	-Replication -Check pointing -Resubmission	FTSM	-Reduce latency -Reliability	-Quality of service -Cost	Cloud-fog/ iFog-Sim
15	Modules scheduling	-Primary backup -Check pointing -Load balancing	ECLB	Resource allocation	Energy consumption	Fog/ iFog-Sim
16	Task scheduling	-Backup	DFTLA	-Reliability -Optimize the response time	-Quality of service -Cost -Energy consumption	Fog
17	Task scheduling	-Backup -Resubmission	DEFT	-Reliability -Resource optimization	-Volatility -Runtime estimation	Cloud/ CloudSim
18	Complex task scheduling	-Not supporting fault tolerence	PSG/ PSG-M	-Minimization of deadline violation time	-Quality of service -Energy consumption	Fog
19	Complex task scheduling	-Not supporting fault tolerence	LBP-ACS	-Satisfying the mix deadline	-Task delay -Energy consumption	Cloud-fog
20	Task scheduling	-Primary backup	EFTR	-Increase system processing capacity	Energy consumption	Cloud/ CloudSim
21	Task allocation	-Primary backup	Deep Q-Network (APSDQN)	-Resource utilization	success rate	Cloud/ CloudSim
22	Modules scheduling	-Replication check-pointing Migration	FLBFH	-Resource management -Failure prediction	-Cost -Delay	Fog/ CloudSim
23	Task scheduling	-Replication -Resubmission	FTDM	-Energy consumption -Makespan	-Cost -Delay	Fog/ IFOGSIM
24	Workflow scheduling	-Migration -Replication	EFFT	-Reliability	-Cost	Cloud
25	Task scheduling	-Migration	TEM/TBF	-Reliability	-Energy consumption	Edge Cloud/ CloudSim

4 Discussion

To ensure the reliability of a fault-tolerant system in fog computing, it is essential to consider the diverse characteristics of fog nodes, including limitations in terms of processor power, memory capacity, bandwidth, and energy resources. However, there have been limited efforts in this field to address fault tolerance in such environments and effectively handle the demanding requirements of real-time applications.

Hence, optimizing the energy consumption of nodes involves effectively prioritizing task execution based on factors such as delay, deadline, and time sensitive requirements. This becomes crucial as cloud computing has limitations in supporting real-time applications and handling time sensitive requests due to the considerable distances between end users and data centers. Consequently, numerous fault tolerance scheduling methods have been proposed in both fog and cloud computing to ensure the optimal utilization of these technologies.

In [15], a scheduling approach for fault tolerance modules was proposed, utilizing a backup method. The main challenge addressed in this approach was to create backup copies for each model and execute them only in case of primary failures, thereby conserving energy. However, one limitation of this approach is that it does not take into account sensitive-time requirements.

However, in [16], the consideration of sensitive time tasks is addressed through the proposal of automata based on mathematical equations. These automata are designed to monitor the process and make decisions accordingly. The approach focuses on setting up backups specifically for the sensitive tasks, ensuring their execution within the specified deadlines. By doing so, the approach guarantees the timely execution of these tasks without any deadline violations.

In [14], a fault-tolerant scheduling system was proposed to achieve the desired objectives. The authors employed three fault tolerance techniques, namely replication, check pointing, and resubmission. However, it is worth noting that the approach does not specifically address energy consumption considerations.

The heuristic algorithms proposed in [18,19] aim to address the scheduling problem in fog computing. However, these proposals do not specifically support fault tolerance mechanisms.

In [21], the authors utilized Q-learning-based task scheduling for fault-tolerant systems. By employing a Markov decision process, they selected the most optimal scheme, either the active scheme or the passive scheme, for handling faulty tasks based on the system's state. This approach improved resource utilization in the cloud environment. However, one limitation of this work is that it is based on a first-come-first-serve principle, which overlooks the priority of tasks as an important constraint.

Our review of related work reveals that the literature on fault -tolerance-aware scheduling can be classified based on the study environment and the fault-tolerance technique used, such as shown in Fig. 3.

After reviewing and discussing some related works, we noticed that there is a notable lack of initiatives to address fault tolerance in such environments that need to be able to effectively handle the requirements of real-time applications.

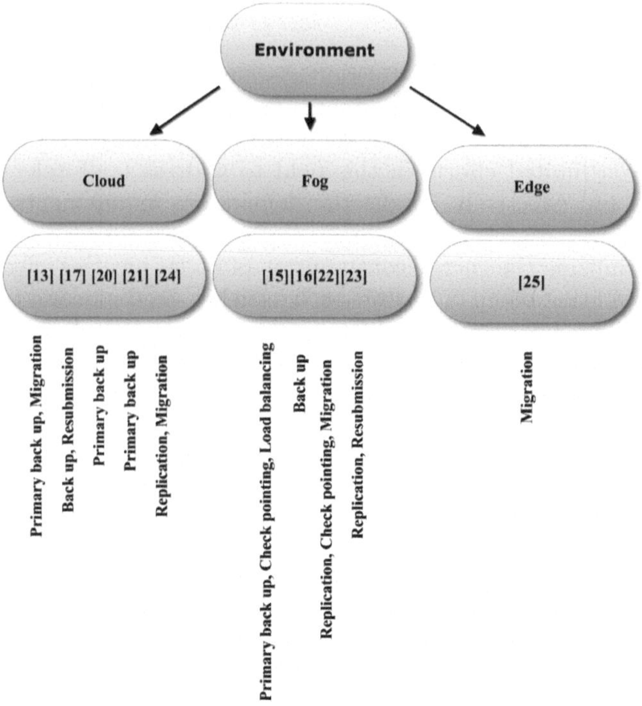

Fig. 3. Literature approach's Taxonomy

5 Conclusion

In this paper, we have presented a comparative study of various fault tolerance scheduling proposals in fog and cloud environments. The comparison was based on the fault tolerance methods employed, the study environments, and the proposed algorithms. Our investigation revealed that key quality of service (QoS) attributes such as response time, delay, and reliability were of utmost importance.

Additionally, energy consumption emerged as a significant challenge in this domain, emphasizing the need for its consideration in future research. Haut du formulaireMoreover, the study highlighted the lack of works addressing fault tolerance-aware scheduling specifically in fog computing, indicating an area that warrants further attention and exploration in our future works. Furthermore, by explicitly describing the results of the comparative study, it helps to make informed decisions regarding the selection and implementation of fault-tolerant planning approaches tailored to the specific needs of IoT applications.

References

1. Bonomi, F., Milito, R., Zhu, J., Addepalli, S.: Fog computing and its role in the internet of things. In: Proceedings of the First Edition of the MCC Workshop on Mobile Cloud Computing, pp. 13–16 (2012)
2. Hao, Z., Novak, E., Yi, S., Li, Q.: Challenges and software architecture for fog computing. IEEE Internet Comput. **21**(2), 44–53 (2017)
3. Mahmud, R., Kotagiri, R., Buyya, R.: Fog computing: a taxonomy, survey and future directions. In: Internet of Everything, pp. 103–130. Singapore, Springer Nature (2018)
4. Gao, L., Luan, T.H., Yu, S., Zhou, W., Liu, B.: FogRoute: DTN-based data dissemination model in fog computing. IEEE Internet Things J. **4**(1), 225–235 (2017)
5. Kalra, M., Singh, S.: Multi-objective energy aware scheduling of deadline constrained workflows in clouds using hybrid approach. Wireless Pers. Commun. **116**, 1743–1764 (2021)
6. Xu, R., et al.: Improved particle swarm optimization based workflow scheduling in cloud-fog environment. In: Business Process Management Workshops: BPM 2018 International Workshops, Sydney, NSW, Australia, September 9–14, 2018, Revised Papers 16, pp. 337–347 (2019)
7. Abawajy, J.H.: An efficient adaptive scheduling policy for high-performance computing. Futur. Gener. Comput. Syst. **25**(3), 364–370 (2009)
8. Bjerkevik, H.B., Botnan, M.B., Kerber, M.: Computing the interleaving distance is np-hard. Found. Comput. Math. **20**, 1237–1271 (2020)
9. Xiao, F., Zhang, Z., Abawajy, J.: Workflow scheduling in distributed systems under fuzzy environment. J. Intell. Fuzzy Syst. **37**(4), 5323–5333 (2019)
10. Churi, P., Rao, N.: Teaching cyber security course in the classrooms of nmims university. Int. J. Modern Educ. Comput. Sci. (IJMECS) **13**(4), 1–15 (2021)
11. Marahatta, A., Wang, Y., Zhang, F., Sangaiah, A.K., Tyagi, S.K.S., Liu, Z.: Energy-aware fault-tolerantdynamic task scheduling scheme for virtualized cloud data centers. Mobile Netw. Appl. **24**, 1063–1077 (2019)
12. Karthikeyan, L., Vijayakumaran, C., Chitra, S., Arumugam, S.: Saldeft: self adaptive learning differential evolution based optimal physical machine selection for fault tolerance problem in cloud. Wireless Pers. Commun. **118**, 1453–1480 (2021)
13. Gupta, V., Kaur, B.P., Jangra, S.: An efficient method for fault tolerance in cloud environment using encryption and classification. Soft. Comput. **23**, 13591–13602 (2019)
14. Alarifi, A., Abdelsamie, F., Amoon, M.: A fault-tolerant aware scheduling method for fog-cloud environments. PLoS ONE **14**(10), 0223902 (2019)
15. Sharif, A., Nickray, M., Shahidinejad, A.: Fault-tolerant with load balancing scheduling in a fog-based iot application. IET Commun. **14**(16), 2646–2657 (2020)
16. Ghanavati, S., Abawajy, J., Izadi, D.: Automata-based dynamic fault tolerant task scheduling approach in fog computing. IEEE Trans. Emerg. Top. Comput. **10**(1), 488–499 (2020)
17. Yan, H., Zhu, X., Chen, H., Guo, H., Zhou, W., Bao, W.: Deft: Dynamic faulttolerant elastic scheduling for tasks with uncertain runtime in cloud. Inf. Sci. **477**, 30–46 (2019)
18. Azizi, S., Shojafar, M., Abawajy, J., Buyya, R.: Deadline-aware and energyefficient iot task scheduling in fog computing systems: A semi-greedy approach. J. Netw. Comput. Appl. **201**, 103333 (2022)

19. Xu, J., Hao, Z., Zhang, R., Sun, X.: A method based on the combination of laxity and ant colony system for cloud-fog task scheduling. IEEE Access **7**, 116218–116226 (2019)
20. Cai, M., Zhou, P., Lourenço, L.C., Zhang, D.: Endoscopic full-thickness resection (eftr) for gastrointestinal subepithelial tumors. Gastrointestinal Endoscopy Clin. **26**(2), 283–295 (2016)
21. Tang, H., Tang, Z., Dong, T., Hai, Q., Xue, F.: Fault-tolerant scheme of cloud task allocation based on deep reinforcement learning. In: Pan, L., Cui, Z., Cai, J., Li, L. (eds.) Bio-Inspired Computing: Theories and Applications: 16th International Conference, BIC-TA 2021, Taiyuan, China, December 17–19, 2021, Revised Selected Papers, Part II, pp. 54–66. Springer Singapore, Singapore (2022). https://doi.org/10.1007/978-981-19-1253-5_5
22. Tuli, S., et al.: AI augmented edge and fog computing: trends and challenges. J. Netw. Comput. Appl. **216** 103648 (2023)
23. Saeed, W., Ahmad, Z., Jehangiri, A.I., Mohamed, N., Umar, A.I., Ahmad, J.: A fault tolerant data management scheme for healthcare internet of things in fog computing. KSII Trans. Internet Inform. Syst. (TIIS) **15**(1), 35–57 (2021)
24. Zhang, Y., Fan, G., Yu, H., Chen, X.: Efficiency-first fault-tolerant replica scheduling strategy for reliability constrained cloud application. In: Cérin, C., Qian, D., Gaudiot, J.-L., Tan, G., Zuckerman, S. (eds.) Network and Parallel Computing: 18th IFIP WG 10.3 International Conference, NPC 2021, Paris, France, November 3–5, 2021, Proceedings, pp. 131–143. Springer International Publishing, Cham (2022). https://doi.org/10.1007/978-3-030-93571-9_11
25. Xue, Y., Fan, G., Yu, H., Sun, H.: Energy-aware resource scheduling with fault tolerance in edge computing. In: Network and Parallel Computing: 16th IFIP WG 10.3 International Conference, NPC 2019, Hohhot, China, August 23–24, 2019, Proceedings 16, pp. 327–332 (2019)

Multi-agent Deep Reinforcement Learning for Self-organized Aggregation of a Swarm of Robots

Ahmed Abdelli(✉), Ali Yachir, and Abdenour Amamra

Ecole Militaire Polytechnique, BP 17, Bordj El-Bahri, Algiers, Algeria
ahmedabdelli@kaist.ac.kr

Abstract. Self-organized aggregation is a fundamental behavior for a swarm of robots wherein the robots are required to form one aggregate. It allows interaction between the swarm members and it is considered as a prerequisite for other complex behaviors. Designing automatically such behavior is attractive and promising, yet challenging, particularly in a homogeneous environment and without communication. In this paper, we tackle this problem using the vanilla policy gradient (VPG) algorithm following the centralized training decentralized execution scheme (CTDE). We assess its performance in our implemented physics-based simulator following the OpenAI Gym framework. Our results demonstrate the viability of this approach for designing automatically a scalable self-organized aggregation.

Keywords: Swarm robotics · Self-organized aggregation · Automatic design · Multi-agent deep reinforcement learning · policy gradient methods

1 Introduction

Accomplishing tasks that are beyond the capacity of a single robot requires the cooperation of a group of robots. When this cooperation is decentralized and self-organized using only local sensing and communication, it produces a swarm of robots that are robust, flexible, and scalable. This is the ultimate goal of swarm robotics which is a sub-field of multi-robot systems. It takes inspiration from natural swarms such as ant colonies, fish schools, and flocks of birds which exhibit a sort of swarm intelligence through their behavior as a collective [1,19].

A swarm of robots exhibits a collective behavior that emerges from local interactions between individual robots composing the swarm without any central entity of control. One fundamental behavior is collective decision-making which has many forms [2]. One of particular interest is aggregation where the robots are required to gather into one cluster in the environment and form a single aggregate. Despite its simplicity, it represents a building block that lets the swarm members interact with each other. Also, it is considered as a prerequisite for other forms of cooperation such as self-assembly [11].

B. Djamaa et al. (Eds.): CSA 2024, LNNS 1145, pp. 223–232, 2025.
https://doi.org/10.1007/978-3-031-71848-9_21

The aggregation behavior in the literature is either cue-based [18] or self-organized [4]. In cue-based aggregation, the robots exploit cues/features of the environment such as light or temperature. For example, the robots look for the brightest spot in the environment to aggregate. In the self-organized aggregation, no cues are available to the swarm, thus the robots should perform a collective decision-making behavior to reach a consensus on the aggregation location. In contrast to [6,8] where the problem is formulated as the best-of-n problem [23], in this paper, we are interested in self-organized aggregation in a homogeneous environment without any sites or shelters. This setting is more challenging since the number of alternatives becomes infinite.

Methods for designing self-organized aggregation are either manual[14] or automatic[12]. Manual design is difficult and it relies heavily on the intelligence of the designer and requires manual tuning, whereas automatic methods are promising and attractive[7]. The most used paradigm for automatic design in swarm robotics is artificial evolution[12]. The only work, that applied recent achievements in deep reinforcement learning to tackle the self-organized aggregation is the work of Hüttenrauch et al.[13]. However, this work relied on communication and its performance has not been well evaluated for scalability with local observability. In addition, their experimental setup is a non-physics-based simulator where robots are considered as point masses.

In this work, we investigate the viability of the vanilla policy gradient algorithm [20,21] without communication to synthesize decentralized robot controllers for self-organized aggregation. Moreover, we evaluate its performance in a physics-based simulator that accounts for the environment dynamics [15].

This paper is organized as follows: Sect. 2 gives a brief overview of the related works on the automatic design of self-organized aggregation. Section 3 describes the vanilla policy gradient algorithm and its mathematical background. In Sect. 4, we present and discuss the obtained results of our experiments. Finally, Sect. 5 summarizes our work and gives some future directions.

2 Related Works

Various approaches have been proposed for self-organized aggregation. We focus only on automatic methods for designing swarm controllers in homogeneous environments. The reviewed literature can be classified according to its theoretical background as follows:

2.1 Artificial Evolution

Artificial evolution is based on the principle of the survival of the fittest. An initial population of artificial genotypes, each representing a robot controller, is randomly generated in the environment. This population is tested to select the fittest for reproduction of the next generation. This process is repeated until a robot controller meets the desired performance defined by a fitness function.

One of the first works that used artificial evolution for synthesizing the self-organized aggregation behavior is the work of Trianni et al. [22]. The robot controller is a simple perceptron of 16 sensory inputs representing proximity sensors, microphones, and gripper sensors mapped to 3 motor outputs controlling the gripper and the speed of the two wheels. The evolutionary process resulted in the emergence of two types of aggregation strategies: static and dynamic. In the former, the robots are compact and stable aggregates. In the latter, the aggregates are loose and moving. The dynamic aggregation behavior was shown to be relatively scalable compared to the static one which is tuned for a specific swarm size. In the same vein, Dorigo et al. [5] improved the scalability of the dynamic aggregation by varying the number of robots during evaluation.

Gaucci et al. [9] used a fully recurrent neural network with only one input that represents a single binary range sensor mapped into two outputs to control the left and the right wheels. They succeeded in aggregating robots in a single location. However, a long-range binary sensor is required to achieve good performance.

2.2 Reinforcement Learning

Reinforcement learning relies on interactions of agent(s) with the environment through a trial-and-error process while getting a reward after each performed action. This approach is a goal-driven approach compared to other machine learning approaches. The objective is to make a decision about the suitable action for the current situation in order to maximize the reward [21].

Hüttenrauch et al. [13] applied recent advances in deep reinforcement learning to the rendez-vous task using the mean embeddings of distributions for state representation. To reduce the curse of dimensionality of the observation space, a neural network is used to define the feature space. The authors employed trust region policy optimization (TRPO) [10] for producing a policy that is trained and executed by 20 agents. The simulated environment is of size 1 m by 1 m and the communication range is 40 cm. Results showed that performance drops as we change the swarm size to 10. Even though additional information is transmitted using local communication (i.e., the number of observed neighbors), the policy seemed to be tuned to the number of agents involved in the learning process.

3 Methods

3.1 Preliminaries

The problem of self-organized aggregation is formulated as a decentralized partially-observable Markov decision process (Dec-POMDP) [17], which is defined by the tuple $\langle \mathcal{N}, \mathcal{S}, \{\mathcal{A}\}, \mathcal{R}, \mathcal{T}, \{\mathcal{O}\} \rangle$, where $\mathcal{N}$ is the set of agents, $\mathcal{S}$ is the set of states of the environment, $\mathcal{A}$ is the set of joint actions, $\mathcal{R}$ is the reward function, $\mathcal{T}$ is the transition probability function, $\mathcal{O}$ is the set of joint observations. This model describes well sequential decision-making made by a group of agents interacting with the environment.

3.2 Policy Gradient Methods and the Actor-Critic Scheme

Since our action space is continuous, policy gradient methods [21,24] are the most suitable for learning a control policy for our robots. In these methods, the goal is to adjust the policy's parameters θ in order to maximize the objective $J(\pi_\theta) = \mathbb{E}_{\tau\sim\pi_\theta}[R(\tau)]$ which represents the expected return of the trajectory τ using the policy function π (i.e., the actor neural network). To optimize the policy, the policy gradient is defined as follows:

$$\nabla_\theta J(\pi_\theta) = \mathbb{E}_{\tau\sim\pi_\theta}\left[\sum_{t=0}^{T} \nabla_\theta \log \pi_\theta(a_t|s_t)\hat{A}^{\pi_\theta}(s_t, a_t)\right] \quad (1)$$

where $\hat{A}(s_t, a_t) = \sum_{t'=t}^{T} R(s_{t'}, a_{t'}, s_{t'+1}) - V_\phi(s_t)$ is the advantage function [20]. The on-policy value function $V_\phi(s_t)$ which acts as the critic is approximated using another neural network of parameters ϕ. Both actor and critic networks are optimized via a stochastic gradient ascent algorithm such as Adam.

Since our swarm consists of homogeneous robots, a scalable approach for training the swarm is parameter sharing used by Gupta et al. [10] where only one policy is trained using the experiences of all robots. Differently from their work, we omit the indexing of the agents making our robots homogeneous and indistinguishable. Also, we follow the centralized training with a decentralized execution (CTDE) scheme. In this scheme, the robots have access to the observations of the whole swarm during training, while they have access only to their own observation during execution. Algorithm 1 describes the vanilla policy gradient algorithm. Even though Proximal Policy Optimization (PPO) represents the state-of-the-art policy gradient methods, the Vanilla Policy Gradient (VPG) method still gives the state-of-the-art performance, however, a larger number of samples in training is needed [16].

In the following, we describe the observation space and the action space of our robots and the reward obtained by each robot after taking an action in the environment.

Observation Space. A simulated range-and-bearing sensor (RAB) that provides the distance and orientation to the neighboring robots with a maximum of 10 robots, its detection limit is set to 50 cm. Also, eight proximity sensors of range 30 cm.

Action Space. It consists of two actions of real values. The linear velocity $v_t \in [0, v_{max}]$ cm/s where v_{max} is set to 16 cm/s and the angular velocity $\omega_t \in [-2\pi, +2\pi]$ rad/s.

Reward. The proposed reward function for a robot i is defined as follows:

$$R_t(s, a) = w_1 d_{[t-1,t]} - c_t - w_2\omega_t \quad (2)$$

Algorithm 1: Vanilla Policy Gradient Algorithm

Input: initial policy parameters θ_0, initial value function parameters ϕ_0

for $k = 0, 1, 2, ...$ **do**

Collect the set of trajectories $D_k = \tau_i$ by running policy $\pi_k = \pi(\theta_k)$

Compute rewards-to-go $\hat{R}_t$

Compute advantage estimates, $\hat{A}_t$ based on the current value function V_{ϕ_k}

Estimate policy gradient as

$$\hat{g}_k = \frac{1}{|D_k|} \sum_{\tau \in D_k} \sum_{t=0}^{T} \nabla_\theta \log \pi_\theta(a_t|s_t)|_{\theta_k} \hat{A}_t$$

Compute policy update, using Adam

$$\theta_{k+1} = \theta_k + \alpha_k \hat{g}_k$$

Fit value function by regression on mean-squared error:

$$\phi_{k+1} = \arg\min_{\phi} \frac{1}{|D_k|T} \sum_{\tau \in D_k} \sum_{t=0}^{T} (V_\phi(s_t) - \hat{R}_t)^2$$

via Adam,

end

where $d_{[t-1,t]}$ is the displacement from time $t-1$ to t, w_t is the angular velocity and c_t is distance of the robot i from the swarm's center of mass.

This reward function promotes exploration by maximizing $d_{[t-1,t]}$ and penalizes both moving away from the swarm's center of mass c_t and rotation on the spot w_t. The weights w_1 and w_2 serve for normalization and they are set to 50 and 10^{-1}, respectively.

Policy Architecture. Figure 1 illustrates our policy (i.e., robot controller) is a feed-forward neural network. It consists of three layers, two hidden layers of size 64, and an output layer of two nodes, controlling the linear and angular velocities. The input to this network represents the range and bearing values of the neighboring robots and the proximity sensors values. For the activation function, the hyperbolic tangent (tanh) function is used.

3.3 Simulation Setup

We implemented our physics-based simulator using the 2D physics engine (Box2D) that accounts for collisions and PyGame for the graphics. It follows the OpenAI Gym framework [3] for episodic settings of reinforcement learning. The environment is represented by a bounded arena of size 2 by 2 m. Our

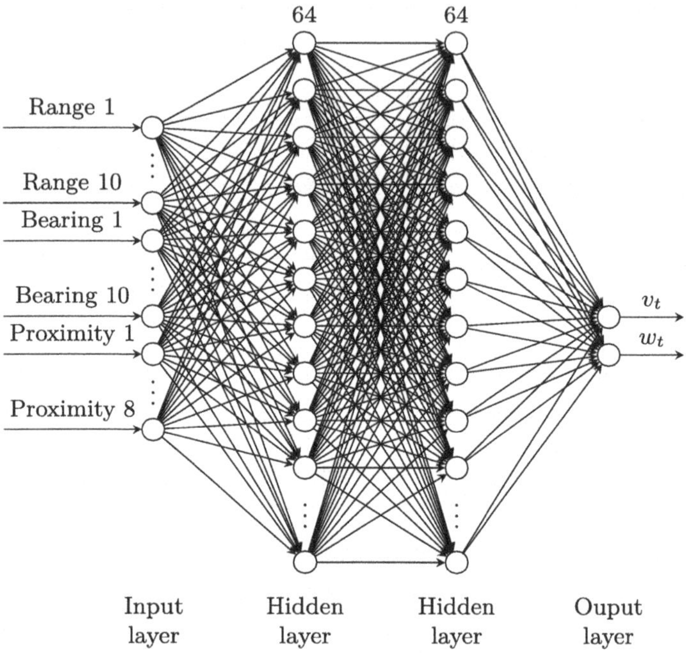

Fig. 1. Each robot multi-layer perception (policy function π).

simulated robot radius is 3.5 cm. We used the unicycle model to describe its kinematics and it is controlled through linear and angular velocity denoted v and ω, respectively. The swarm robots are spawned randomly in the arena following a uniform distribution at the initialization of the environment.

4 Results and Discussion

For the purpose of assessing the viability of the vanilla policy gradient algorithm for self-organized aggregation, we adopt the dispersion metric used by [9], for assessing the aggregation performance. This metric is formulated as follows:

$$u^{(t)} = \frac{1}{4r^2} \sum_{i=1}^{n} ||p_i^{(t)} - \bar{p}^{(t)}||^2 \tag{3}$$

where r is the robot radius, $p_i^{(t)}$ is the position of the robot i at time t, and $\bar{p}^{(t)} = \frac{1}{n} \sum_{i=1}^{n} p_i^{(t)}$ is the center of mass of the swarm.

Another metric is the average distance (AD) to the center of mass, which is defined as follows:

$$AD^{(t)} = \frac{1}{n} \sum_{i=1}^{n} \left(1 - \frac{d_i(t)}{1.4}\right) \tag{4}$$

where $d_i(t)$ is the distance of the robot i from the center of mass, which is normalized to obtain values in the interval $[0, 1]$.

4.1 A Dynamic Aggregate

The snapshots from Fig. 2 show that at t=0 the robots are positioned randomly in the arena exploring the environment. After 200 s, the robots successfully aggregate forming one aggregate. This aggregate is moving dynamically in the environment exhibiting a sort of flocking behavior. This is beneficial for scalability as sub-groups of isolated robots tend to move to each other forming a larger aggregate over time. We argue that this is due to the interplay of exploration and attraction to the center of mass in the reward function. This finding is in accordance with [5,22] using evolutionary robotics. Furthermore, compared to Hüttenrauch et al. [13], our settings are more challenging as our robots have no communication means and our arena size is larger twofold which reduces the chance for the robots to see each other.

Also, we assessed the performance of the synthesized controller over 10 trials using the dispersion metric. As depicted in Fig. 3, our controller generated one aggregate in all trials. Even though some of them have taken a longer time to converge, all of them succeeded in forming one aggregate in less than 200 s.

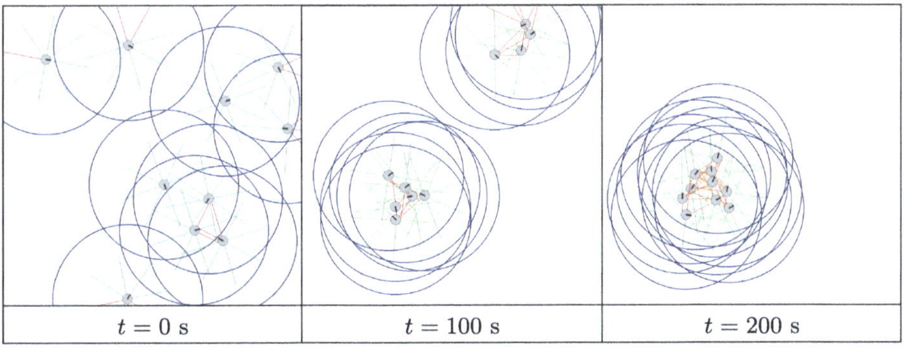

Fig. 2. Evolution of self-organized aggregation over time.

4.2 Scalability

We have shown that our learned policy using 10 agents works well when executed by a swarm of 10 robots. Now, we will explore its scalability with different swarm sizes. For this, we perform 30 trials, each trial duration is 200 s. To measure the performance of our learned controller, we use the average distance (AD) metric at the end of the trial. As shown in Fig. 4, the boxplots represent the performance of our controller for different swarm sizes. As you can see, it maintains relatively

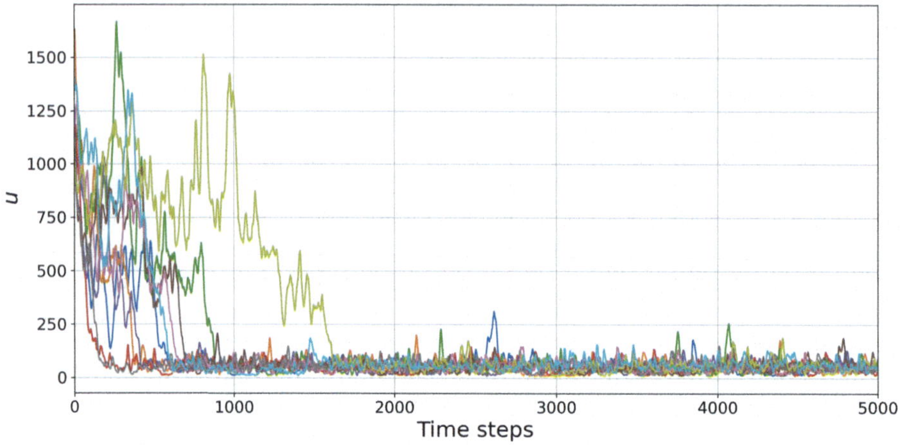

Fig. 3. Dispersion metric of 10 trials for a swarm of 10 robots.

high performance as we increase the swarm size demonstrating its scalability. The outliers represent trials where small aggregates did not have enough time to join each other and form a single aggregate within the trial time limit, we believe that giving more time will let them find each other and form one aggregate.

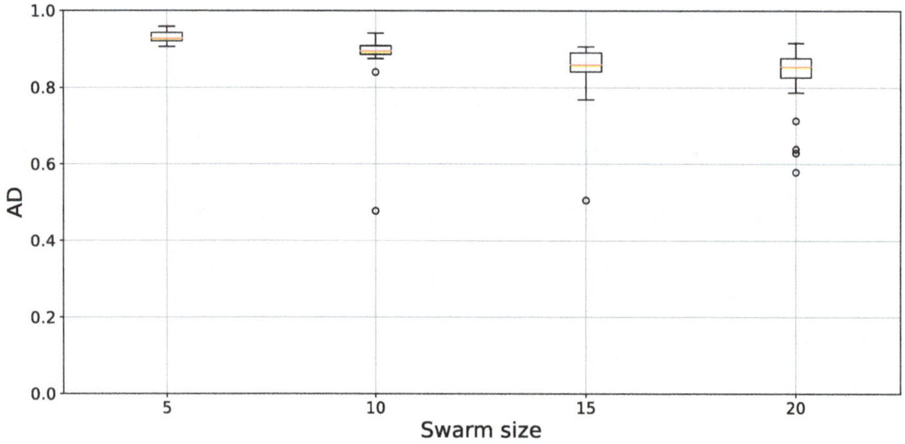

Fig. 4. Performance of the synthesized controller over different swarm sizes.

5 Conclusion

In this paper, we investigated deep reinforcement learning as an automatic method for the emergence of self-organized aggregation. This task is challenging due to local sensing and the absence of communication in our problem setting. Nonetheless, our results showed that deep reinforcement learning using the vanilla policy gradient (VPG) algorithm is a viable approach for this task. The synthesized controller produced a dynamic aggregate which promotes scalability as the aggregate moves around the arena. As a future work, we intend to explore more complex behaviors such as flocking.

References

1. Abdelli, A., Amamra, A., Yachir, A.: Swarm robotics: a survey. In: Senouci, M.R., Boulahia, S.Y., Benatia, M.A. (eds.) Advances in Computing Systems and Applications. CSA 2022. LNNS, vol. 513, pp. 153–164. Springer, Cham (2022). https://doi.org/10.1007/978-3-031-12097-8_14
2. Abdelli, A., Yachir, A., Amamra, A., Khaldi, B.: Maximum likelihood estimate sharing for collective perception in static environments for swarm robotics. Robotica **41**(9), 2754–2773 (2023). https://doi.org/10.1017/S0263574723000668
3. Brockman, G., et al.: OpenAI Gym. Tech. rep. (2016). https://doi.org/10.48550/arXiv.1606.01540, arXiv:1606.01540 [cs] type: article
4. Cambier, N., Albani, D., Frémont, V., Trianni, V., Ferrante, E.: Cultural evolution of probabilistic aggregation in synthetic swarms. Appl. Soft Comput. **113**, 108010 (2021). https://doi.org/10.1016/j.asoc.2021.108010
5. Dorigo, M., et al.: Evolving self-organizing behaviors for a swarm-bot. Auton. Robot. **17**(2), 223–245 (2004). https://doi.org/10.1023/B:AURO.0000033973.24945.f3
6. Firat, Z., Ferrante, E., Gillet, Y., Tuci, E.: On self-organised aggregation dynamics in swarms of robots with informed robots. Neural Comput. Appl. **32**(17), 13825–13841 (2020). https://doi.org/10.1007/s00521-020-04791-0
7. Francesca, G., Birattari, M.: Automatic design of robot swarms: achievements and challenges. Front. Robot. AI **3**, 29 (2016). https://doi.org/10.3389/frobt.2016.00029
8. Francesca, G., et al.: Automode-chocolate: automatic design of control software for robot swarms. Swarm Intell. **9**(2), 125–152 (2015)
9. Gauci, M., Chen, J., Dodd, T.J., Groß, R.: Evolving aggregation behaviors in multi-robot systems with binary sensors. In: Ani Hsieh, M., Chirikjian, G. (eds.) Distributed Autonomous Robotic Systems. STAR, vol. 104, pp. 355–367. Springer, Heidelberg (2014). https://doi.org/10.1007/978-3-642-55146-8_25
10. Gupta, J.K., Egorov, M., Kochenderfer, M.: Cooperative multi-agent control using deep reinforcement learning. In: Sukthankar, G., Rodriguez-Aguilar, J.A. (eds.) AAMAS 2017. LNCS (LNAI), vol. 10642, pp. 66–83. Springer, Cham (2017). https://doi.org/10.1007/978-3-319-71682-4_5
11. Hamann, H.: Swarm robotics: a formal approach. Springer International Publishing (2018). https://doi.org/10.1007/978-3-319-74528-2
12. Hasselmann, K., Ligot, A., Ruddick, J., Birattari, M.: Empirical assessment and comparison of neuro-evolutionary methods for the automatic off-line design of robot swarms. Nat. Commun. **12**(1), 4345 (2021). https://doi.org/10.1038/s41467-021-24642-3
13. Hüttenrauch, M., Šošić, A., Neumann, G.: Deep reinforcement learning for swarm systems. J. Mach. Learn. Res. **20**(1), 1966–1996 (2019)

14. Khaldi, B., Harrou, F., Cherif, F., Sun, Y.: Self-organization in aggregating robot swarms: A DW-KNN topological approach. Biosystems **165**, 106–121 (2018). https://doi.org/10.1016/j.biosystems.2018.01.005
15. Liu, C.K., Negrut, D.: The role of physics-based simulators in robotics. Ann. Rev. Contr., Robot., Autonom. Syst. **4**(1), 35–58 (2021). https://doi.org/10.1146/annurev-control-072220-093055
16. Mnih, V., et al.: Asynchronous methods for deep reinforcement learning. In: Balcan, M.F., Weinberger, K.Q. (eds.) Proceedings of The 33rd International Conference on Machine Learning. Proceedings of Machine Learning Research, vol. 48, pp. 1928–1937. PMLR, New York, New York, USA (2016)
17. Oliehoek, F.A., Amato, C.: A Concise Introduction to Decentralized POMDPs, 1st edn. Springer Publishing Company, Incorporated (2016)
18. Sadeghi Amjadi, A., et al.: Reinforcement learning-based aggregation for robot swarms. Adaptive Behavior p. 10597123231202593 (2023). https://doi.org/10.1177/10597123231202593
19. Schranz, M., Umlauft, M., Sende, M., Elmenreich, W.: Swarm robotic behaviors and current applications. Front. Robot. AI **7**, 36 (2020). https://doi.org/10.3389/frobt.2020.00036
20. Schulman, J., Moritz, P., Levine, S., Jordan, M.I., Abbeel, P.: High-dimensional continuous control using generalized advantage estimation. In: Bengio, Y., LeCun, Y. (eds.) 4th International Conference on Learning Representations, ICLR 2016, San Juan, Puerto Rico, May 2-4, 2016, Conference Track Proceedings (2016)
21. Sutton, R.S., Barto, A.G.: Reinforcement Learning: An Introduction. A Bradford Book, Cambridge, MA, USA (2018)
22. Trianni, V., Groß, R., Labella, T.H., Şahin, E., Dorigo, M.: Evolving aggregation behaviors in a swarm of robots. In: Banzhaf, W., Ziegler, J., Christaller, T., Dittrich, P., Kim, J.T. (eds.) Advances in Artificial Life, pp. 865–874. Springer Berlin Heidelberg, Berlin, Heidelberg (2003). https://doi.org/10.1007/978-3-540-39432-7_93
23. Valentini, G., Ferrante, E., Dorigo, M.: The best-of-n problem in robot swarms: formalization, state of the art, and novel perspectives. Front. Robot. AI **4**, 9 (2017). https://doi.org/10.3389/frobt.2017.00009
24. Williams, R.J.: Simple statistical gradient-following algorithms for connectionist reinforcement learning. Mach. Learn. **8**(3), 229–256 (1992). https://doi.org/10.1007/BF00992696

Enhanced Navigation Based Obstacle Avoidance and Target Tracking with Raspberry-Pi for Mobile Robots

Hadjira Belaidi[1(✉)], Fethi Demim[2], Adel Mezhary[1], Abdenebi Rouigueb[3], Sofiane Yahoui[1], and Abdelkrim Nemra[2]

[1] Signals and Systems Laboratory, Institute of Electrical Electronic Engineering University M'Hamed Bougara of Boumerdes, Boumerdes, Algeria
ha.belaidi@univ-boumerdes.dz, adel.mezheri@aiesec.net
[2] Laboratory of Guidance and Navigation, Ecole Militaire Polytechnique, Bordj El Bahri, Algiers, Algeria
[3] Laboratory of Artificial Intelligence and Virtual Reality, Ecole Militaire Polytechnique, Bordj El Bahri, Algiers, Algeria

Abstract. Our study focuses on developing a self-navigating mobile robot using advanced computer vision techniques and a microprocessor. The innovative integration of these technologies results in an advanced robotic system. Our main objective is to create a seamless navigation system from a starting point to a predetermined target, with real-time obstacle avoidance and dynamic decision-making capabilities. The robot efficiently detects and adjusts its trajectory to avoid obstacles by continuously scanning its surroundings with computer vision. Precise obstacle identification is ensured through intricate embedded algorithms, enabling secure and collision-free navigation. The robot autonomously halts upon reaching the target, and if the target is hidden, it initiates comprehensive environmental exploration. Powered by the Raspberry Pi, our system offers a robust and cost-effective solution for mobile robot navigation. This strategic amalgamation of technologies greatly enhances the robot's operational prowess, enabling adept navigation and focused target pursuit in complex environments.

Keywords: Autonomous navigation · mobile robot · computer vision system · collision avoidance · image processing and sensor fusion

1 Introduction

Robotics is increasingly vital in today's tech landscape, driving progress across industries like manufacturing, hazardous materials handling, and planetary exploration. Vision sensors are key here, providing a visual map for robots to navigate environments, detect obstacles, and prevent collisions.

Additionally, vision holds immense significance in target tracking tasks. It is worth highlighting that object tracking stands as a cornerstone technique in

B. Djamaa et al. (Eds.): CSA 2024, LNNS 1145, pp. 233–244, 2025.
https://doi.org/10.1007/978-3-031-71848-9_22

various robot vision applications. Examples of its utilization encompass surveillance [1], tracking moving objects [2,3], monitoring extended moving entities [4], and facilitating real-time motion tracking and human following [5].

In the dynamic realm of robotics, the quest for precise target tracking has spurred numerous inventive methods. A prominent approach involves integrating diverse sensor types beyond traditional vision-based systems. Laser range scanners, for example, are utilized to identify unique features of a target individual, like their shins [6]. Similarly, microphone arrays are employed not only for tracking but also for detecting sound emitted by moving objects, thus augmenting the tracking process with an auditory dimension [2]. Sound sources are emerging as alternatives to cameras, particularly in low-light scenarios. Recognizing the limitations of cameras in certain conditions, researchers are exploring sound as a tracking tool [7,8]. Additionally, there's growing interest in guiding mobile robots towards targets using range sensors exclusively, showcasing diverse tracking strategies [9].

Numerous studies have delved into target tracking, providing insights into moving targets, and path planning approaches for mobile robots [10–13]. Although vision systems provide robust tracking, their computational demands may not be practical for microcontroller based applications [14]. Thus, advanced microprocessor based platforms are essential to integrate advanced tracking techniques efficiently. This highlights the necessity for powerful processing units to support complex computations in target tracking applications.

Image processing algorithms perform well on conventional computers [11,15], but deploying them on microprocessor based robot systems presents challenges [16]. Microprocessors may struggle to handle the computational demands, affecting algorithm execution and robot control. Previous work used a fuzzy inference engine for object tracking, but its computation time was lengthy, highlighting the need for faster methods. This study aims to develop simpler algorithms balancing efficiency and accuracy for real-time tracking on microprocessor based robots. Specifically, it focuses on enhancing obstacle avoidance and target tracking using computer vision on limited-memory microprocessors like the Raspberry Pi. The Raspberry Pi is celebrated for its versatility and cost-effectiveness in robotics [17], making it an ideal platform to address the limitations of onboard computation in autonomous navigation and object tracking.

The algorithm targets real-time computer vision for obstacle avoidance and target tracking in autonomous mobile robots, addressing microprocessor resource limitations. It balances precision and computational efficiency for seamless navigation. This research advances algorithmic design and explores the broader impact of computer vision techniques in enhancing microprocessor-based autonomous robots. By optimizing resource use, it paves the way for these robots to perform complex tasks within their constrained computational environments, signaling potential future advancements in the field.

The paper's structure is as follows: Sect. 2 discusses the interplay between the robot's behavior and its environment recognition, detailing environment processing and hardware. Section 3 examines background anti-matching and covariance matching, crucial for object recognition and tracking accuracy. In Sect. 4, global motion compensation is discussed, seamlessly integrated into an object tracking mean shift algorithm to enhance effectiveness. Section 5 evaluates the algorithms using real video sequences to assess performance and practicality. Finally, Sect. 6 concludes by summarizing contributions, insights, and implications, underlining the approach's significance and suggesting future research directions in this evolving field.

2 Mobile Robot Platform Building

This section provides a comprehensive overview of the mobile robot platform crucial for obstacle avoidance and target tracking. It serves as the physical foundation for smooth movement and task execution, integrating components like the Raspberry Pi microprocessor, motors, wheels, and sensors for motion and perception. Designed for stability, maneuverability, and scalability, it features a robust chassis for steady movement, precise motor-wheel coordination for agile motion, and various sensors such as proximity sensors or cameras for safe navigation. To enhance communication and control, the platform includes wireless connectivity options like $Wi-Fi$ or Bluetooth, enabling remote operation, data transmission, and real-time interaction, thus expanding its versatility and potential applications [9,18]. This holistic view emphasizes the platform's pivotal role in implementing our approach effectively, paving the way for subsequent implementation and evaluation.

2.1 The Platform Design

The mobile robot platform is based on the *Raspberry Pi* 2 *Model B*, featuring the powerful *Broadcom BCM*2836 *ARMv*7 *Quad Core* 32 – *bit* Processor operating at 900 MHz with 1 GB RAM. It offers connectivity with 4 *USB* ports, 40 *GPIO* pins, HDMI, and Ethernet ports, as illustrated in Fig. 1. Originating from the UK, Raspberry Pi aims to teach computer science in educational settings. The platform runs on Linux-based operating systems, primarily *Raspbian*, known for robustness and compatibility. This chosen hardware aligns with our commitment to enhancing obstacle avoidance and target tracking capabilities.

The robot's motion is driven by two geared DC motors, as depicted in Fig. 2(a). Each motor features a gear assembly for torque amplification and speed control. Rotation precision is managed by a servomotor. Circuitry design, detailed in Fig. 1(a), ensures seamless integration of components, enhancing mobility and enabling navigation and perception functions. The circuit follows the diagram in Fig. 1(b), facilitating:

- The 2 *DC* motors are connected to 4 output ports of the *L*293*D* *H*-Bridge driver;
- The 4 input ports of *L*293*D* are connected to the *GPIO* of the Pi through the ports (*GPIO* 22 & 23 and *GPIO* 12 & 6);

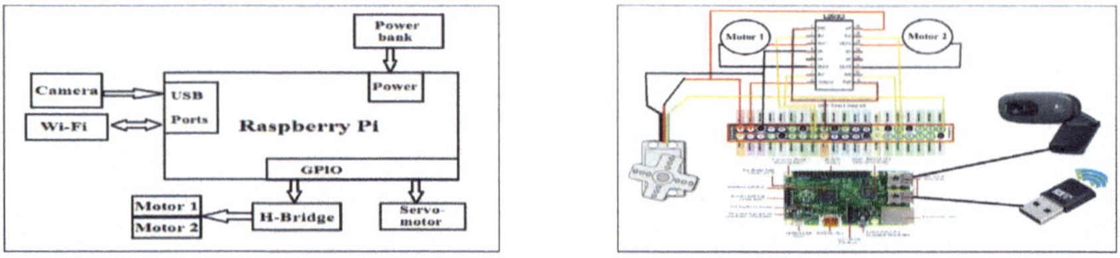

(a) Platform diagram. (b) Platform circuit.

Fig. 1. The built robot platform.

- The $L293D$ is powered throw its port 16 for internal ship and 8 for the motors using the 5 V pins of the Pi;
- The servo motor is powered throw its 2 cables Vcc and Gnd and the control cable is connected to pin 40 of the $GPIO$ of the Pi;
- A $Wi - Fi$ 802.11n dongle is used to connect Pi to the PC remotely. The $Wi - Fi$ key and the camera are connected to the USB ports of Pi;
- A power bank of 16000 mAh that delivers 5V 2 mA is used, which is corresponding to the power needed by the Pi card to comfortably meet the needs of the whole system. The raspberry pi is powered by the power bank using $USB - miniUSB$ cable. The choice of the power bank is decided after some calculations given in Table 1, where:

$$Power = \frac{Current \times voltage}{1000} \tag{1}$$

The Raspberry Pi serves as the central processing unit, coordinating all system components including motors, sensors, and wireless modules. Its robust processing power enables real-time decision-making for intricate navigational tasks. Integration of a vision sensor enhances the robot's perception of its environment, with cameras acting as its "eyes". Visual data is processed by the Raspberry Pi to identify obstacles, track targets, and make navigational decisions. Wireless communication via $Wi - Fi$ or Bluetooth enables remote operation and data exchange with external devices, facilitating real-time monitoring and control adjustments, thus enhancing adaptability and practicality. The entire system circuitry is shown in Fig. 2(b).

2.2 Remote-Controlling the Raspberry

A Virtual Network Computing (VNC) Server was meticulously set up for remote control of the Raspberry Pi's (GUI) via $Wi{-}Fi$. VNC allows control of one computer's interface from another, transmitting keyboard and mouse inputs remotely and displaying real-time screen updates. This integration enables immersive access to the Raspberry Pi's desktop interface on an external computer, providing full control akin to direct operation. After installing VNC on both devices and configuring the Raspberry Pi's unique IP address, comprehensive real-time

Table 1. The system power consumption calculations.

Components	Intensity consumption (mA)	Power consumption (W)
Raspberry pi	400	2
USB camera	500	2.5
Servo motor	200	1
DC motors	200	1
Wi-Fi key	250	1.25
SUM	1550	7.75

control over its interface is achieved remotely. This bridging of physical presence and remote operation enhances operational fluidity and accessibility. The *VNC*-enabled interface extends beyond local environments, enabling remote operation, troubleshooting, and real-time monitoring, thereby broadening the robot platform's versatility and aligning with the goal of creating a sophisticated and accessible system.

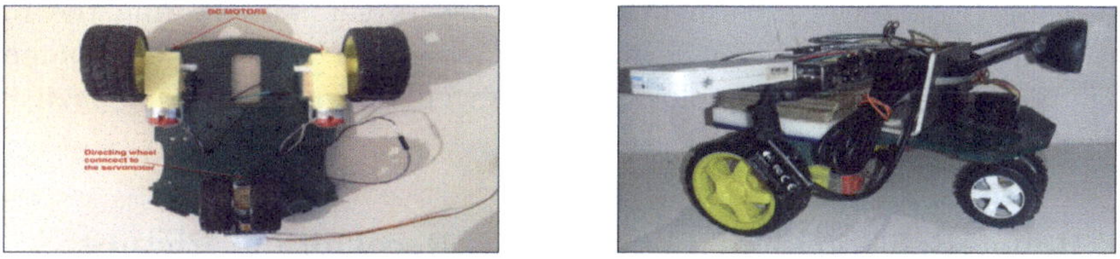

(a) The position of the motors. (b) The entire robot platform.

Fig. 2. A picture of the robot platform.

3 Obstacle Detection and Avoidance

The robot's autonomous navigation relies on analyzing sequential snapshots from its onboard camera to traverse its environment independently. This involves complex coordination between visual perception, decision-making, and motor control to circumvent obstacles and reach a predefined target. Our approach centers on intricate image processing steps to enhance the robot's navigation. It begins with capturing an image, which is then converted to grayscale to streamline processing and improve edge detection efficiency. The grayscale image undergoes subtle blurring to reduce noise and enhance the clarity of detectable edges, revealing environmental features [9].

The use of the Canny edge detection algorithm transforms images by identifying pixel intensity transitions, outlining contours, and providing the robot

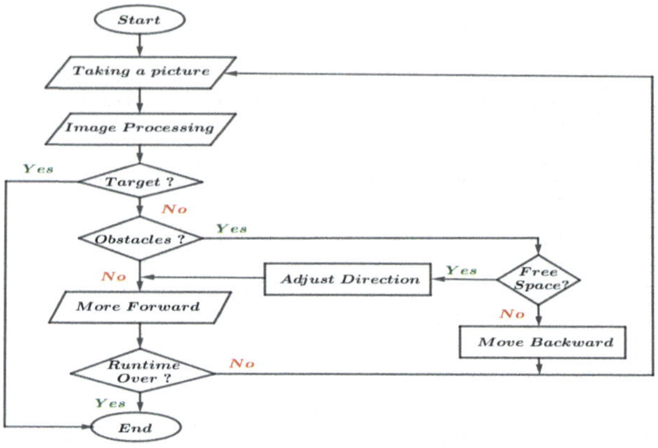

Fig. 3. Flowchart summarizing the proposed approach.

with a dynamic map of its surroundings. This visual perception enables real-time decision-making, distinguishing obstacles from navigable pathways. Integrating image processing with autonomous mobility enables obstacle avoidance and precise trajectory adjustments toward the target, empowering the robot with cognitive capabilities for seamless navigation in changing environments. This approach signifies a technological synergy enhancing the robot's navigation precision and adaptability.

3.1 Robot Behavior and Environment Recognition: Picture to Processing

The robot employs systematic edge detection by scanning images from the left-most edge, moving upwards in defined intervals. Upon detecting a white pixel, it captures edge coordinates for analysis, guiding navigation decisions to avoid obstacles. The upper image half undergoes less meticulous analysis. Figure 3 illustrates this process, showcasing the fusion of technology and cognition for autonomous navigation. A function facilitates visual data acquisition, providing insights into image dimensions and characteristics. Figure 4(a) depicts this process, capturing visual perception as the robot's camera takes a picture. To enhance CPU processing on the Raspberry Pi, images measuring 640×480 pixels are resized to 80×60 pixels using "$Cv2.resize(.)$" (refer to Fig. 4(b)). Edge extraction starts with converting the image to grayscale as shown in Fig. 5(a), smoothing the way for further analytical tasks.

After grayscale transformation, the Canny filter highlights edges with intensities from 0 to 255, revealing contours with precision (Fig. 5(b)). A meticulously crafted function generates a matrix mirroring the image, with white pixels marked as 1 and black pixels as 0. This binary array guides an algorithm to navigate around obstacles by discerning the spatial distribution of these pixels, ensuring elegant and cautious movement.

(a) Robot's camera picture. (b) Resize the 640×480 image to 80×60.

Fig. 4. Illustration example of taking a picture from the robot camera.

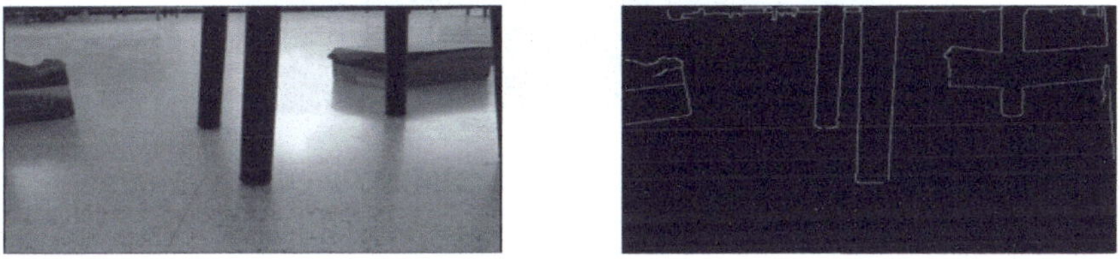

(a) The result of the grayscale image. (b) The canny edge transformation.

Fig. 5. Image processing.

3.2 Obstacle Avoidance

The obstacle avoidance algorithm begins with a systematic scan from the bottom row upwards, examining the binary matrix bidirectionally for white pixels, indicating edges. When detected, their coordinates are recorded. The image is divided into four columns A, B, C, and D, each playing a crucial role. The algorithm carefully examines rows with white pixels, probing five subsequent rows with precision, covering 10% of the image's height [9,19]. It identifies open passages amid rows, marking lanes for unhindered passage. Figure 6 visually illustrates this process: upon reaching the first row of white pixels, the algorithm scans the following five rows, calculates the environment's free space, and determines the next action accordingly.

- B & C columns are empty then go forward;
- A & B columns are empty then turn left with 30°;
- C & D columns are empty then turn right with 30°;
- B & C columns not free :
  - The D column is empty then turns right with 45°;
  - The A column is empty then turn left with 45°;
  - A & D columns are not free then go back and turn right with 60° then move forward.

The strategic choice of angles 30° and 45° guides the robot's obstacle evasion. These angles are meticulously calibrated to ensure a balanced approach, enabling the robot to elegantly sidestep obstacles while minimizing deviations. This calculated selection safeguards the delicate equilibrium between avoidance and deviation, ensuring effective obstacle evasion while maintaining proximity. The chosen angles are particularly crucial when the robot's forward motion is limited to 30 cm within a one-second interval, underscoring the thoughtful calibration required to optimize evasion strategies.

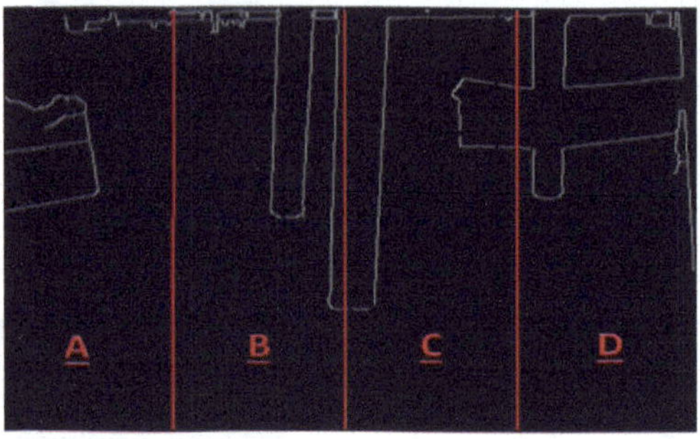

Fig. 6. Dividing the image into four columns and looking for the empty space.

3.3 Gaussian Smoothing for Error Minimization

In images, lines often appear in the trapeze region due to ground markings, surface imperfections, or light reflections (Fig. 7). To address this, Gaussian blur from (*OpenCV*) is utilized to reduce unwanted details. Gaussian blur, also called Gaussian smoothing, applies a Gaussian function to blur an image, reducing noise and intricate details [20]. Unlike the Bokeh effect caused by out-of-focus lenses, Gaussian smoothing produces a smooth, translucent effect. It serves as a pre-processing step in computer vision, enhancing image structures and improving subsequent algorithm efficacy [11,20]. Gaussian blur is achieved by convolving an image with a Gaussian kernel, where kernel size determines smoothing strength. Its use notably improved our approach by up to 60%. To enhance effectiveness further, the algorithm now scans only the lower image portion, disregarding distant objects in the upper region. This adjustment focuses solely on nearby obstacles, optimizing computational resources. Figure 8 visually illustrates this modification, marking objects beyond the red line as distant. This optimization prioritizes proximity-based obstacle detection. Concrete testing and experimentation verified a significant time delay of approximately 2 min on the Raspberry Pi during image scanning for edges. While obstacle avoidance remained effective, the robot's responsiveness between operational cycles suffered. To overcome this bottleneck, a novel approach dynamically reduces captured image dimensions to optimize processing time. Illustrated in Fig. 4(b), this resizing occurs immediately post-capture, minimizing computational load for subsequent stages. Crucially, the Canny edge detection algorithm, vital for obstacle avoidance, maintains efficacy on resized images, ensuring uninterrupted functionality. This strategic resizing not only addresses processing delays but also preserves obstacle avoidance integrity, enhancing overall robot performance and responsiveness.

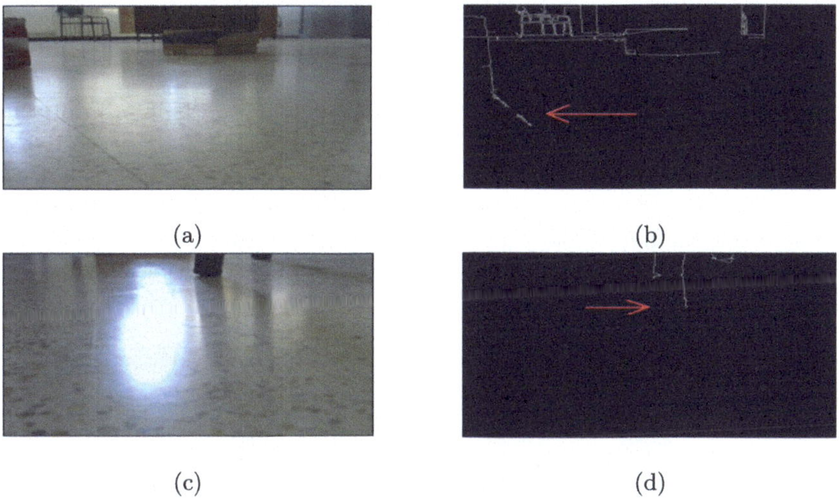

Fig. 7. The problem of presence of lines (edges) even if there is no obstacle in the real picture.

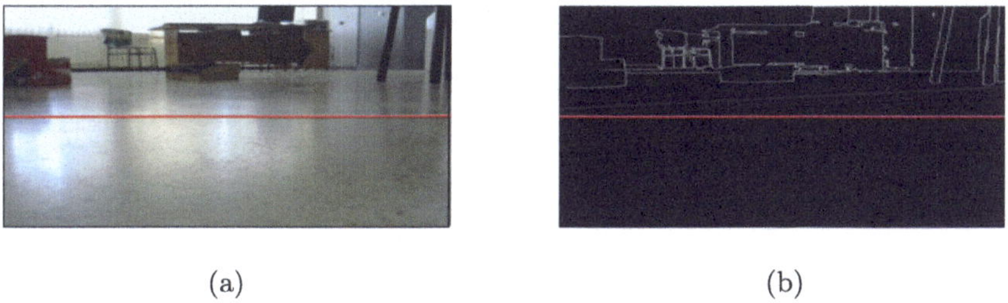

Fig. 8. Visual Representation of optimized approach: gaussian blur integration and exclusion of distant objects above the red line.

4 Target Tracking

Our project's next phase aims to enable the robot to autonomously locate, approach, and halt at its target. When the target is absent, the robot halts and communicates with wireless system. Challenges arise when relying solely on the camera. We address this by capturing, scanning, and comparing images with predefined templates [10]. A logo template stored in the Raspberry Pi's memory serves as the target indicator (Fig. 9(b)). Using an *OpenCV* matching function, the robot scans for template instances in the input image. Pixel-wise comparisons identify matches, with the search starting from the image's lower part. Upon finding a match, the robot stops, aligning with the target. If no match is found, implying obstacles, the robot engages in obstacle avoidance while searching. Another template signals navigational area end, prompting the robot to halt and signal target absence.

Our approach underwent iterative refinement to enhance effectiveness and robustness. After rigorous trials, we determined an optimal matching threshold of 70% to accommodate potential obstacles obscuring the target in images. This threshold triggers the algorithm to command the robot to halt promptly upon achieving a 70% correlation between the captured image and the stored template. This nuanced approach ensures heightened sensitivity to target matches while considering potential obstructions, as depicted in Fig. 9(a). Iterative trials allowed fine-tuning of the threshold, balancing sensitivity and reliability to minimize false positives and optimize genuine target recognition and response.

(a) The robot reaches the target.

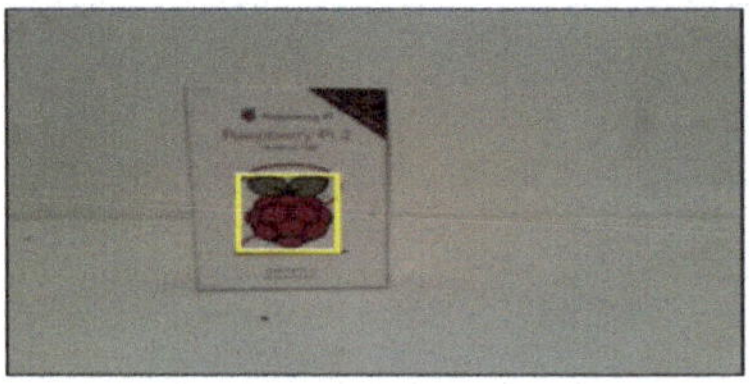

(b) The template used in the approach.

Fig. 9. Target detection.

5 Results and Discussion

The algorithm effectively identifies various obstacles but encounters image errors, mitigated by Gaussian smoothing, rectifying approximately 60% of them. Despite commendable obstacle avoidance, corner collisions persist. To enhance performance, a trapezoidal region within the camera's view is defined to identify potential obstacles, with adjustments ensuring detection around 30 cm away. Preliminary outcomes are shown in Fig. 10.

However, the trapezoidal approach faces limitations when obstacles are adjacent but not entirely within the designated region, emphasizing the need for a comprehensive solution. Refinements focus on scanning the lower image section for closer objects, incorporating image resizing and a detailed pixel-by-pixel scan for efficient target detection. Iterative enhancements optimize the matching threshold to 70%, ensuring reliable and responsive robot halting while maintaining precision. Extensive testing and experimentation have led to a proficient mobile robot. Our comprehensive approach successfully mitigates obstacles, evident in the robot's seamless navigation through complex environments. The harmonious synergy between obstacle avoidance and target tracking components underscores the efficacy of our solution.

(a) (b)

Fig. 10. Obstacle avoidance by considering just the obstacles inside the trapezoid.

6 Conclusion

This paper details the development of a mobile robot platform adept at navigating dynamic environments, avoiding obstacles, and tracking targets. By leveraging a camera as the primary sensor and a Raspberry Pi for data processing, the platform coordinates DC motors and a servomotor to achieve seamless navigation and target pursuit. Through iterative refinement, employing sophisticated algorithms and innovative techniques, the robot's capabilities have been finely tuned for obstacle avoidance and target tracking. It now navigates diverse terrains with finesse, evading obstacles and progressing towards goals with precision. This achievement underscores the synergy of hardware design, software innovation, and algorithmic precision in crafting an autonomous and intelligent mobile robot.

Looking ahead, the integration of advanced machine learning techniques promises to enhance the robot's perception and decision-making abilities, enabling adaptation to even more intricate and diverse environments. Additionally, the incorporation of multiple sensors like LiDAR or ultrasonic sensors could further augment the robot's environmental awareness and bolster its obstacle detection capabilities. Exploring swarm robotics principles offers potential for collaborative strategies among multiple robots, ultimately facilitating collective navigation and task execution within complex surroundings.

References

1. Weiss, N.: Adaptive supervision of moving objects for mobile robotics applications. Robot. Auton. Syst. **57**(10), 982–995 (2009)
2. Han, J., Han, S., Lee, J.: The tracking of a moving object by a mobile robot following the object's sound. J. Intell. Rob. Syst. **71**, 31–42 (2013)
3. Karthikeyan, M., Kudalingam, M., Natrajan, P., Palaniappan, K., Prabhu, A.: Object tracking robot by using raspberry PI with open computer vision (CV). Int. J. Trend Res. Develop. **3**, 31–35 (2016)
4. Kraußling, A.: Tracking extended moving objects with a mobile robot. In: Editor, SPRINGER Nature Switzerland AG. (eds.) Proceedings of the 3rd IEEE International Conference Intelligent Systems, in Lecture Notes in Studies in Computational Intelligence book series, vol 109, pp. 513–530. Springer, Heidelberg (2006). https://doi.org/10.1007/978-3-540-77623-9_29

5. Jung, B., Sukhatme, G.S.: Real-time motion tracking from a mobile robot. Int. J. Soc. Robot. **2**, 63–78 (2010)
6. Kawarazaki, N., Kuwae, L.T., Yoshidome, T.: Development of human following mobile robot system using laser range scanner. Proc. Comput. Sci. **76**, 455–460 (2015)
7. Nakadai, K., Lourens, T., Okuno, H.G., Kitano, H.: Active audition for humanoid, In: 17th National Conference on Artificial Intelligence (AAAI-2000), pp. 832–839, Austin, Texas, USA (2000)
8. Wang, C., Griebel, S., Brandstein, M.: Real-time automated video and audio capture with multiple cameras and microphones. J. VLSI Signal Process. Syst. Signal, Image Video Technol. **29**, 81–99 (2001)
9. Demim, F., Rouigueb, A., Belaidi, H., et al.: Smooth sliding mode control based technique of an autonomous underwater vehicle based localization using obstacle avoidance strategy, In: Proceedings of the the 20th International Conference on Informatics in Control, Automation and Robotics, vol. 1, pp. 529–537 (2023)
10. Demim, F., Rouigueb, A., Nemra, A., Bouguessa, R., Chahmi, A.: A new filtering strategy for target tracking application using the second form of smooth variable structure filter. J. Syst. Contr. Eng. **236**(6), 1224–1249 (2022)
11. Demim, F., Louadj, K., Nemra, A.: Path planning for unmanned ground vehicle. In: Proceedings of 5th IEEE International Conference on Control, Decision and Information Technologies, Thessaloniki, Greece, pp. 748–750 (2018)
12. Belaidi, H., Belkalem, J., Abed, M.A., Bentarzi, H.: IoT Path Planning Approach for Mobile Robots. In: Proceedings of International Conference on Applied Smart Systems (ICASS), pp. 1–8, Medea, Algeria (2018)
13. Belaidi, H., Demim, F.: NURBs based multi-robots path planning with obstacle avoidance. J. Comput. Theor. Appl. **1**(4), 478–491 (2024)
14. Zaki, A.M., Arafa, O., Amer, S.I.: Microcontroller-based mobile robot positioning and obstacle avoidance. J. Electr. Syst. Inform. Technol. **1**(1), 58–71 (2014)
15. Salgado Garcia, N., Menendez, J.M., Rendon, E.: efficient image segmentation for recognition-based motion estimation and compensation. In: Proceedings of IEEE Trans Circuits and Systems for Video Technology, **10**(7), pp. 1029–1039 (2000)
16. Chang, Y.H., Chung, P.L., Lin, H.W.: Deep Learning for object identification in ROS-based mobile robots. In: Proceedings of IEEE International Conference on Applied System Invention (ICASI), pp. 66–69, Chiba, Japan (2018)
17. Anand, O., Kumawat, A.K.: Object detection and position tracking in real time using Raspberry Pi. In: Proceedings of Materials Today, **47**(11), pp. 3221–3226 (2021)
18. Mosbah, A., Demim, F., Mansoul, A., Benssalah, M., Nemra, A.: Simultaneous localization and mapping navigation of unmanned ground vehicle based on second-order smooth variable structure filter with improved technique to combat fading for advanced wireless communications. J. Syst. Contr. Eng. **235**(7), 1258–1271 (2021)
19. Demim, F., Belaidi, H. et al.: Advanced trajectory planning and 3D waypoints navigation of unmanned underwater vehicles based fuzzy logic control with los guidance technique, In: Proceedings of the 20th International Conference on Informatics in Control, Automation and Robotics, vol. 1, pp. 538–545 (2023)
20. Lopez-Molina, C., De Baets, B., Bustince, H., Sanz, J., Barrenechea, E.: Multiscale edge detection based on gaussian smoothing and edge tracking. Knowl.-Based Syst. **44**, 101–111 (2013)

IoT Based Road Surface Condition Monitoring Using Spatiotemporal Feature Learning

Abdelkader Hadj-Attou(✉), Yacine Kabir, and Farid Ykhlef

Faculty of Technology, Saad Dahlab University of Blida, Ouled Yaïch, Algeria
hadjattouaek@gmail.com

Abstract. Internet of Things devices enable the monitoring of road surface condition (RSC) through a set of motion sensors that capture vehicle vibration data. These sensors' data can be converted to a much more usable form, by utilizing data preprocessing techniques. In this context, deep learning (DL) models are suitable for road anomaly detection. This study aims to evaluate the effectiveness of a spatiotemporal feature learning approach in detecting road anomalies using a new RSC-IoT dataset. Therefore, based on a combination of DL models such as Long Short-Term Memory (LSTM), Gated Recurrent Units (GRU), and Convolution Neural Network (CNN), we propose three hybrid deep learning models: ConvLSTM, TimeDistributed-CNN-LSTM, and TimeDistributed-CNN-GRU. Moreover, in order to achieve better performance, the proposed road anomaly detection models are combined into two ensemble learning methods namely averaging ensemble and weighted average ensemble. The experimental results show that the weighted average ensemble method outperformed the averaging ensemble and individual models.

Keywords: road surface condition · RSC-IoT dataset · hybrid deep learning · ensemble learning

1 Introduction

Road surface anomalies have a negative impact on vehicle performance and ride safety, while also lead to huge economic losses and traffic congestion [1–3]. One of the most common methods utilized for road surface condition (RSC) monitoring is the vibration-based approach, which uses motion sensors from Internet of Things (IoT) devices to collect the vibrations of a moving vehicle when it crosses any road surface anomaly, like speed bumps, cracks, and potholes [4–7].

Typically RSC monitoring framework consists of three main phases namely data collection, data preprocessing, and classification [8,9]. The collected data includes the vehicle vibration, road surface labels, and a timestamp for each raw data sample. However, since not all data can be accepted as desired inputs, some operations such as resampling, normalization, and noise reduction need to

B. Djamaa et al. (Eds.): CSA 2024, LNNS 1145, pp. 245–254, 2025.
https://doi.org/10.1007/978-3-031-71848-9_23

be performed on the dataset during the preprocessing phase to provide enhanced data for the classification task [8,10].

Once the phase of data preprocessing is complete, the next phase is to detect road surface anomalies using various techniques of classification. Deep learning (DL) algorithms like Recurrent Neural Networks (RNN), Gated Recurrent Unit (GRU), Long Short-Term Memory (LSTM), Artificial Neural Networks (ANN), and Convolutional Neural Networks (CNN) are popular for motion sensor data classification [6,10,11]. Recently, hybrid deep learning models based on GRU, LSTM, and CNN have achieved good results in the field of RSC monitoring [12].

In this paper, we propose 3D hybrid deep learning models to classify road anomalies, where ensemble learning methods are employed to enhance the performance of classification. The main idea of this paper is inspired by video frame classification [13], in which the sensors' time series data are reshaped into 3D format to learn spatiotemporal features from the vehicle's vibration data for better road anomaly detection. The major contributions of the paper are the following:

- Building a new dataset (RSC-IoT) for road surface condition monitoring using motion sensor data from an IoT device.
- Employing three hybrid deep learning models namely TimeDistributed-CNN-LSTM (TD-CNN-LSTM), TimeDistributed-CNN-GRU (TD-CNN-GRU), and ConvLSTM for RSC classification.
- Comparing the performance of the TD models with that of standard CNN-LSTM and CNN-GRU.
- Applying two ensemble learning methods called averaging ensemble and weighted average ensemble, which use the average of the three models' predictions to generate improved predictions.

The remainder of the paper is organized as follows. Section 2 deals with the related works. The methodology for road anomaly detection is described in Sect. 3. Section 4 provides the results and discussions. Finally, Sect. 5 concludes the paper.

2 Related Work

This section highlights some of the latest research contributions related to road surface condition monitoring systems that use the vibration-based approach, where data collection, preprocessing techniques, and classification algorithms are presented. In [14], the authors developed a method to classify segments of motion sensor data that refer to the condition of the road by using a combination of U-Net architecture and BiLSTM networks. In the data collection phase, many smartphones in various positions were utilized to record data, while cameras were used for data labeling. A reorientation process was employed in the preprocessing phase to normalize the smartphone's position.

The work proposed in [15] utilizes smartphones to collect motion sensor and road image data. Road image data were initially processed using a DL model to detect road manhole covers in various weather conditions, then machine learning (ML) methods were applied to the motion sensors data for the classification of the identified manhole covers as poor, average, and good. Moreover, in the study [16], a smartphone attached to a moving car was utilized to detect potholes in real time. The researchers developed an Android application to collect car vibration via accelerometer and gyroscope sensors, as well as the location of the vibration using a GPS sensor. They used an RF-based ML model for classification, which performed well.

In order to improve the detection performance of road cracks, Chen et al. [17] introduced a methodology based on acceleration data acquired from smartphones. For obtaining more useful information, the STFT and WT techniques were utilized to transform the acceleration data into the frequency-time-energy domain. The aim was to use CNN to distinguish between normal roads, manholes, and cracks. Their results demonstrate that the WT-CNN outperformed the STFT-CNN. Singh et al. [18] compared five ML models for road pothole detection, which are Logistic Regression (LR), Support vector machine (SVM), K nearest neighbors (KNN), Random Forest (RF), and XGBoost. A smartphone was used to retrieve accelerometer and gyroscope readings, then the obtained data were preprocessed to eliminate the noise. The results showed that the XGBoost outperformed the other models.

In contrast to previous studies, that mostly use smartphones to collect motion sensor data and cameras for data labeling. In this study, data were collected using a low-cost MPU-6050 with ESP32 chip, while data labeling was done in real-time via Bluetooth from a smartphone. This new approach for data labeling is more efficient than image data since it is unaffected by weather or light conditions. For the classification, we combine 3D hybrid deep learning models into an ensemble method to obtain more accurate prediction results.

3 Methodology

The methodology is composed of three phases, as illustrated in Fig. 1. The first phase deals with data collection. Preprocessing methods are applied to the collected data in the second phase. The third phase consists of combining the classification results of 3D hybrid deep learning models into ensemble methods and comparing them to the individual models.

3.1 Data Collection

The RSC-IoT dataset was created to monitor the road surface condition, where the ESP32 microcontroller was utilized to collect motion sensor data from the MPU-6050. Since it includes Wi-Fi and Bluetooth, the ESP32 chip is an effective choice for designing IoT devices. During driving, an MPU6050 integrated with ESP32 was placed on the top of the vehicle dashboard to transmit data from the 3-axis Accelerometer and 3-axis Gyroscope to a local server for IoT data storage. In order to guarantee reliable analysis process, all sensor data were

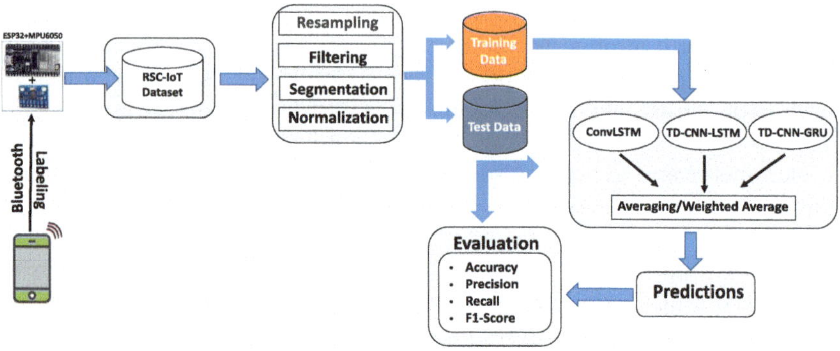

Fig. 1. The proposed framework for road surface conditions monitoring.

collected under identical circumstances. Furthermore, we performed real-time labeling of raw sensors data using an Android smartphone which communicates with the ESP32 chip via Bluetooth. While the vehicle was moving, we labeled road surfaces into three types which are "Smooth" (flat road), "Real anomalies" (potholes and cracks), and "Man-made anomalies" (speed bumps). In the end, 16 669 data samples were obtained in the form of time series, since each raw sensor data has a timestamp.

3.2 Data Preprocessing

The preprocessing is required to make the collected data clean and organized before the classification phase. In this phase, four main steps were applied. Firstly, since the sampling frequency of motion sensor data was too slow (3–5 Hz), these time series data were resampled to 20 Hz. Secondly, the median filter [19] was implemented to reduce the noise in motion sensor data. Thirdly, the data were segmented using a Static Sliding Window of 2 s (40 samples) and 66% overlap. When driving at low speeds, 2 s of road data are sufficient to represent any kind of anomaly on the road. In contrast to the Dynamic Sliding Window [20], we simply eliminated every segment that included samples with different class labels. As shown in Fig. 2, each segment consists of 3-axis accelerometer data and 3-axis gyroscope data, resulting in 240 elements for each window. Finally,

Table 1. Data distribution before and after preprocessing.

Step	Real anomalies		Man-made anomalies		Smooth	
Data collection	2747		783		13139	
Resampling	15607		6103		77473	
Segmentation	Train	Test	Train	Test	Train	Test
	918	176	313	107	4771	1006

the data segments were normalized by subtracting the mean of each value and a division by the standard deviation. Before the segmentation step, the data were manually divided into train and test to get a sufficient number of samples from minority classes because the RSC-IoT dataset is unbalanced (see Table 1).

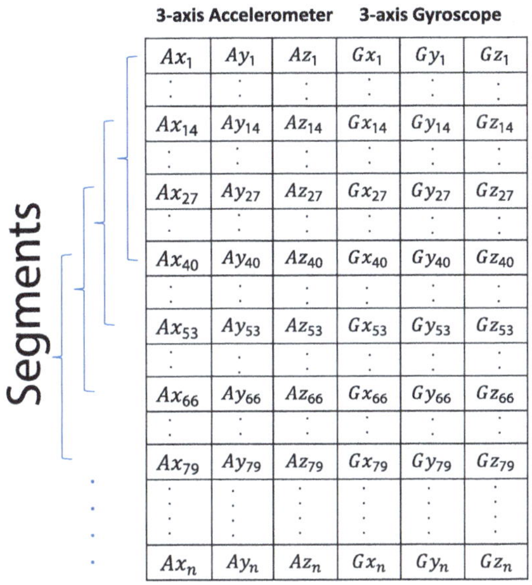

Fig. 2. Sensor data segmentation of 2 s and 66% overlap.

3.3 Classification

Hybrid Deep Learning Models. Recurrent neural network (RNN) models like LSTM and GRU were effectively applied in time series classification tasks, due to their capacity to capture long-term dependencies in data. They both include internal mechanisms known as gates that allow them to control the flow of information. Given that this study is inspired by video frame classification that usually employs 3D models to learn spatiotemporal features [21], both LSTM and GRU are inappropriate since their input data is one-dimensional (no spatial features can be captured). Therefore, we developed three hybrid deep learning models, namely TimeDistributed-CNN-LSTM, TimeDistributed-CNN-GRU, and ConvLSTM to capture spatial features from multi-dimensional data using a convolution process. The proposed hybrid deep learning models use CNN layers to extract spatial features from input data, while the LSTM/GRU layers learn temporal features from the CNN output. CNN typically includes convolutional layer, pooling layer, and fully connected layer. All CNN layers are wrapped by the TimeDistributed (TD) function, allowing any layer to be applied in each temporal frame of the input [22]. Compared with CNN-LSTM, ConvLSTM replaces the pooling layer with an LSTM cell and employs convolution operations in the fully connected layer instead of matrix multiplication in the LSTM, leading to the elimination of spatial data redundancy. However, ConvLstm model was not wrapped with a TD function in order to compare ConvLstm results fairly to standard CNN-LSTM results. Before applying the ConvLSTM and TD models, the input has to be reshaped into the 3D format that these models require. In our case, the input was reshaped to the shape (4, 10, 6, 1) which means each data frame represents 0.5 s of road data. Tables 2 and 3 illustrate the structure of the three hybrid deep learning models.

Table 2. The structure of the TD-CNN- LSTM and TD-CNN-GRU models.

Layer	Parameter Settings
Conv2D	filters: 32, filter size: [3 × 3], relu activation
Conv2D	filters: 32, filter size: [3 × 3], relu activation
MaxPooling2D	Pooling size: [2 × 2]
Dropout	0.25
Conv2D	filters: 64, filter size: [3 × 3], relu activation
Conv2D	filters: 64, filter size: [3 × 3], relu activation
Dropout	0.25
Flatten	–
LSTM/GRU	units: 64, relu activation, return sequences: True
LSTM/GRU	units: 32, relu activation, return sequences: False
Dropout	0.1
Dense	n_neurons : 1024, relu activation
Dropout	0.25
Dense	n_neurons : 1024, relu activation
Dropout	0.25
Dense	n_neurons : 3, softmax activation, adam Optimizer

Ensemble Learning. Ensemble learning is a technique that combines the outputs from multiple models to get more accurate predictions than the individual models that constitute the ensemble because each model has different strengths and weaknesses [23,24]. The most popular techniques for classification problems are bagging, boosting, stacking, weighted average, averaging, and max voting. Here, the averaging and weighted average ensembles were performed by combining TD-CNN-LSTM, TD-CNN-GRU, and ConvLSTM models. In averaging ensemble, the average of class probability predictions from the three hybrid deep learning models is taken and it is used to make the final predictions. The weighted average ensemble is similar to the averaging ensemble but here, the three models

Table 3. The structure of the ConvLSTM model.

Layer	Parameter Settings
ConvLSTM2D	filters: 32, filter size: $[3 \times 3]$, tanh activation
MaxPooling3D	Pooling size: $[1 \times 2 \times 2]$, padding: same
Dropout	0.25
ConvLSTM2D	filters: 64, filter size: $[2 \times 2]$, tanh activation
Dropout	0.25
Flatten	–
Dense	n_neurons : 1024, relu activation
Dropout	0.25
Dense	n_neurons : 1024, relu activation
Dropout	0.25
Dense	n_neurons : 3, softmax activation, adam Optimizer

are assigned with corresponding weights. Moreover, a grid search technique was applied to define the optimal weights combination for the three models. The weights in this method refer to the importance of each model.

4 Results and Discussions

The performance analysis of the experiments conducted for RSC monitoring by using our RSC-IoT dataset is presented in this section. The proposed hybrid deep learning models are evaluated using the following metrics: confusion matrix, F1-score, recall, precision, and accuracy. The comparison results in Table 4 illustrate that TD-CNN-LSTM and TD-CNN-GRU outperform the standard CNN-LSTM and CNN-GRU, while the ConvLSTM model slightly outperforms the standard models. The predictive results of TD models demonstrate the advantage of utilizing the TimeDistributed layer when dealing with time series data in the field of RSC monitoring. On the other hand, ConvLSTM's superiority over CNN-LSTM shows how replacing matrix multiplication with convolution operation gives better results. The robustness of these models demonstrates the accuracy of the Bluetooth labeling technique.

The predictive results of TD-CNN-LSTM, TD-CNN-GRU, and ConvLSTM are combined into an averaging ensemble and a weighted average ensemble, which reduce the variance in the prediction errors. Table 5 and Fig. 3 show a comparison between the averaging ensemble and the weighted average ensemble. As can be seen, the weighted average ensemble performed better than the averaging ensemble. The ConvLSTM model did not perform as well as the TD models, which caused a negative effect on the averaging ensemble's performance. In weighted average ensemble, we obtain the weights $w1 = 0.2$, $w2 = 0.1$, and $w3 = 0.1$ of TD-CNN-LSTM, TD-CNN-GRU, and ConvLSTM, respectively. Since we deal with an imbalanced dataset, we applied grid search for the best combination of

Table 4. Performance evaluation of the proposed models.

Method	Input shape	Accuracy	Precision	Recall	F1 Score
CNN-LSTM	(10,6,4)	0.9426	0.9318	0.8536	0.8887
CNN-GRU	(10,6,4)	0.9410	0.9405	0.8556	0.8933
TD-CNN-LSTM	(4,10,6,1)	**0.9573**	0.9404	**0.9201**	**0.9300**
TD-CNN-GRU	(4,10,6,1)	0.9535	**0.9427**	0.9002	0.9202
ConvLSTM	(4,10,6,1)	0.9449	0.9475	0.8611	0.8998

w1, w2, w3 that gives the maximum F1-score, which is a suitable measure for imbalanced data.

Table 5. Performance evaluation of the ensemble methods.

Method	Accuracy	Precision	Recall	F1 Score
Averaging	0.9589	**0.9567**	0.9049	0.9293
Weighted average	**0.9604**	0.9546	**0.9143**	**0.9335**

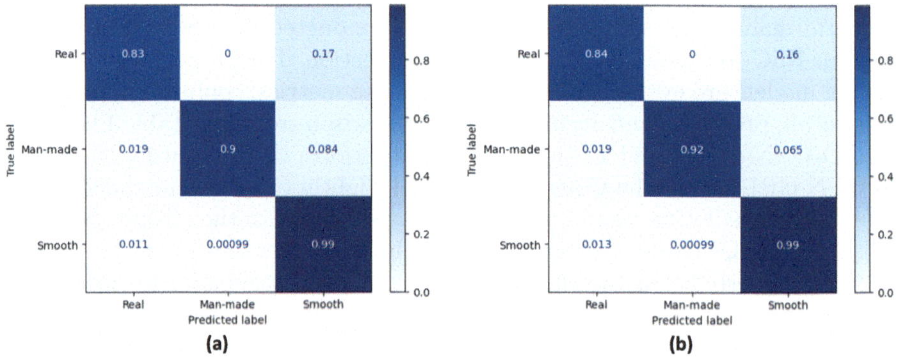

Fig. 3. Confusion matrix of the proposed ensemble methods. (a) Averaging, (b) Weighted average.

5 Conclusion and Future Work

In this study, the efficiency of a spatiotemporal feature learning approach was tested using a new RSC-IoT dataset created with an IoT device for data collection and a smartphone for data labeling. Further, we proposed ensemble learning methods that combine 3D hybrid deep learning models for road surface anomaly detection. Experiments show that the weighted average ensemble achieved the most accurate predictions compared to the averaging ensemble and all the other models. Future work will include testing the performance of the ensemble learning methods with different hybrid deep learning models to detect road surface anomalies. Also, we will apply a Dynamic Sliding Window approach to investigate the impact of vehicle velocity on road anomaly identification. Other future works will include using data augmentation techniques to balance our dataset.

References

1. Sattar, S., Li, S., Chapman, M.: Developing a near real-time road surface anomaly detection approach for road surface monitoring. Measurement **185**, 109990 (2021). https://doi.org/10.1016/j.measurement.2021.109990
2. Higashimoto, K., Fukushima, H., Kamitani, K., Chujo, N.: Identification of road surface condition on undeveloped roads :- aiming for remote car driving -. In: 2021 IEEE 10th Global Conference on Consumer Electronics (GCCE), pp. 777–781. IEEE, Kyoto, Japan (2021). https://doi.org/10.1109/GCCE53005.2021.9621967
3. Varona, B., Monteserin, A., Teyseyre, A.: A deep learning approach to automatic road surface monitoring and pothole detection. Pers. Ubiquit. Comput. **24**(4), 519–534 (2020). https://doi.org/10.1007/s00779-019-01234-z
4. Martinez-Ríos, E.A., Bustamante-Bello, M.R., Arce-Sáenz, L.A.: A review of road surface anomaly detection and classification systems based on vibration-based techniques. Appl. Sci. **12**(19), 9413 (2022). https://doi.org/10.3390/app12199413
5. Sattar, S., Li, S., Chapman, M.: Road surface monitoring using smartphone sensors: a review. Sensors **18**(11), 3845 (2018). https://doi.org/10.3390/s18113845
6. Ranyal, E., Sadhu, A., Jain, K.: Road condition monitoring using smart sensing and artificial intelligence: a review. Sensors **22**(8), 3044 (2022). https://doi.org/10.3390/s22083044
7. Gueta, L.B., Sato, A.: Classifying road surface conditions using vibration signals. In: 2017 Asia-Pacific Signal and Information Processing Association Annual Summit and Conference (APSIPA ASC), pp. 039–043. IEEE, Kuala Lumpur, Malaysia (2017). https://doi.org/10.1109/APSIPA.2017.8281999
8. Kumar, T., Acharya, D., Lohani, D.: Modeling IoT enabled classification system for road surface monitoring. In: 2022 14th International Conference on COMmunication Systems & NETworkS (COMSNETS), pp. 836–841. IEEE, Bangalore, India (2022). https://doi.org/10.1109/COMSNETS53615.2022.9668507
9. El-Kady, A., Emara, K., ElEliemy, M.H., Shaaban, E.: Road surface quality detection using smartphone sensors: Egyptian roads case study. In: 2019 Ninth International Conference on Intelligent Computing and Information Systems (ICICIS), pp. 202–207. IEEE, Cairo, Egypt (2019). https://doi.org/10.1109/ICICIS46948.2019.9014721
10. Wu, S., Hadachi, A.: Road surface recognition based on DeepSense neural network using accelerometer data. In: 2020 IEEE Intelligent Vehicles Symposium (IV), pp. 305–312. IEEE, Las Vegas, NV, USA (2020). https://doi.org/10.1109/IV47402.2020.9304737
11. Setiawan, B.D., Serdült, U.I., Kryssanov, V.: Smartphone sensor data augmentation for automatic road surface assessment using a small training dataset. In: 2021 IEEE International Conference on Big Data and Smart Computing (BigComp), pp. 239–245. IEEE, Jeju Island, Korea (2021). https://doi.org/10.1109/BigComp51126.2021.00052

12. Hadj-Attou, A., Kabir, Y., Ykhlef, F.: Hybrid deep learning models for road surface condition monitoring. Measurement **220**, 113267 (2023). https://doi.org/10.1016/j.measurement.2023.113267
13. Wang, J., Jiao, J., Bao, L., He, S., Liu, Y., Liu, W.: Self-supervised spatio-temporal representation learning for videos by predicting motion and appearance statistics. In: Proceedings of the IEEE/CVF Conference on Computer Vision and Pattern Recognition, pp. 4006–4015. IEEE, Long Beach, CA, USA (2019)
14. Setiawan, B.D., Kovacs, M., Serdült, U., Kryssanov, V.: Semantic segmentation on smartphone motion sensor data for road surface monitoring. Procedia Comput. Sci. **204**, 346–353 (2022). https://doi.org/10.1016/j.procs.2022.08.042
15. Zhou, B., et al.: Smartphone-based road manhole cover detection and classification. Autom. Constr. **140**, 104344 (2022). https://doi.org/10.1016/j.autcon.2022.104344
16. Kumar, S., Kumar, N., Barthwal, A.: Road pothole detection using smartphone sensors. J. Harbin Eng. Univ. **44**(7), 1341–1346 (2023)
17. Chen, C., Seo, H., Zhao, Y.: A novel pavement transverse cracks detection model using WT-CNN and STFT-CNN for smartphone data analysis. Int. J. Pavement Eng. **23**(12), 4372–4384 (2022). https://doi.org/10.1080/10298436.2021.1945056
18. Singh, P., Bansal, A., Kamal, A.E., Kumar, S.: Road monitoring and sensing system using mobile sensors. In: 2022 12th International Conference on Cloud Computing, Data Science & Engineering (Confluence), pp. 165–170. IEEE, Noida, India (2022). https://doi.org/10.1109/Confluence52989.2022.9734178
19. Saadeh, W., Butt, S.A., Altaf, M.A.B.: A patient-specific single sensor IoT based wearable fall prediction and detection system. IEEE Trans. Neural Syst. Rehabil. Eng. **27**(5), 995–1003 (2019). https://doi.org/10.1109/TNSRE.2019.2911602
20. Chibani, N., Sebbak, F., Cherifi, W., Belmessous, K.: Road anomaly detection using a dynamic sliding window technique. Neural Comput. Appl. **34**(21), 19015–19033 (2022). https://doi.org/10.1007/s00521-022-07436-6
21. Qiao, H., Wang, T., Wang, P., Qiao, S., Zhang, L.: A time-distributed spatiotemporal feature learning method for machine health monitoring with multi-sensor time series. Sensors **18**(9), 2932 (2018). https://doi.org/10.3390/s18092932
22. Montaha, S., Azam, S., Rafid, A.K.M.R.H., Hasan, M.Z., Karim, A., Islam, A.: TimeDistributed-CNN-LSTM: a hybrid approach combining CNN and LSTM to classify brain tumor on 3D MRI scans performing ablation study. IEEE Access **10**, 60039–60059 (2022). https://doi.org/10.1109/ACCESS.2022.3179577
23. Semwal, V.B., Gupta, A., Lalwani, P.: An optimized hybrid deep learning model using ensemble learning approach for human walking activities recognition. J. Supercomput. **77**(11), 12256–12279 (2021). https://doi.org/10.1007/s11227-021-03768-7
24. Rayhan Ahmed, M., Islam, S., Muzahidul Islam, A.K.M., Shatabda, S.: An ensemble 1d-CNN-LSTM-GRU model with data augmentation for speech emotion recognition. Expert Syst. Appl. **218**, 119633 (2023). https://doi.org/10.1016/j.eswa.2023.119633

Cyber Security and Privacy

Longitudinal Evaluation of Android Malware Detection Using Deep Learning Classifiers

Abdelhak Mesbah[1(✉)], Ibtihel Baddari[1], and Mohamed Amine Riahla[2]

[1] Faculty of Sciences, University M'Hamed Bougara of Boumerdes, Boumerdes, Algeria
{abdelhak.mesbah,i.baddari}@univ-boumerdes.dz
[2] Faculty of Technology, University M'Hamed Bougara of Boumerdes, Boumerdes, Algeria
ma.riahla@univ-boumerdes.dz

Abstract. The widespread use of Android applications has led to an increasing number of malicious variants, necessitating robust malware detection methods to enhance mobile platform security. Since Android malware and the operating system continue to evolve, designing resilient malware detection techniques that can adapt over time without frequent modifications or retraining presents a significant challenge. Nonetheless, malware authors continuously update their sample characteristics to evade detection, leading to changes in data distribution that directly impact the detection performance of machine learning models, a consideration largely absent in most academic work. In this study, we address the impact of this constant evolution by conducting a longitudinal evaluation. Our approach involves utilizing API call sequences extracted at different abstraction levels (e.g., package level, class level) by tracing the Call Graph structure to capture the order of API calls. We employ two well-established deep learning classifiers, CNN and RNN-LSTM, on a dataset comprising 80,000 Android apps. Through extensive experiments comparing API-level abstractions and classifiers, we aim to identify the most effective approach for real-world scenarios. The results of our empirical evaluation show that CNN offers better longitudinal performance than the RNN-LSTM.

Keywords: Android Malware · Deep Learning · Longitudinal Evaluation · Call Graph

1 Introduction

With billions of devices running on the Android platform occupying about 70% of the market share [1], it has become an attractive target for malicious entities aiming to exploit vulnerabilities and compromise user data. As the number of applications continues to grow, with about 2.5 million apps that can be

B. Djamaa et al. (Eds.): CSA 2024, LNNS 1145, pp. 257–269, 2025.
https://doi.org/10.1007/978-3-031-71848-9_24

downloaded [2], so does the number of their malicious counterparts, making the Android platform a continuous battleground for security.

Malware detection has traditionally relied on signature-based methods, which, while effective against known threats, often falter in the face of new, unseen malware variants. With the dynamic nature of malware evolution, there is a need for more adaptive and proactive detection mechanisms. Machine learning models have emerged as a prominent method for identifying Android malware, demonstrating notable effectiveness. However, the persistent evolution and increasing complexity of malware, which is precisely designed to avoid detection and survive, stands as an evolving obstacle for machine learning, as it reduces the overall effectiveness of the classifiers over an extended period of time.

This evolving challenge has received limited attention in academic literature, with existing studies mainly emphasizing the achievement of high accuracy levels for a temporally localized dataset. The focus has been on the baseline scenario rather than addressing the challenges posed by the continuous evolution of malware.

In this work, we present a longitudinal study that evaluates the performance over time of two deep learning classifiers, including the Convolutional Neural Network (CNN) and Recurrent Neural Network (RNN) Long-Short Term Memory (LSTM). We use Application Programming Interface (API) sequences as features to reflect malicious behavior. The API provides valuable information on the app's functionality and interaction with the underlying system. The sequence order and characteristics of these API calls can offer significant and relevant observations into the actions of applications, enabling the differentiation between benign and malware apps. On the one hand, a Call Graph, which represents the calling relationships between methods in a program structure, is used to trace the flow and order of these API calls, capturing the relations and interactions within an application's code.

On the other hand, our approach involves utilizing API call sequences extracted at different abstraction levels (e.g., package level, class level) to provide a granular and comprehensive view of an application's behavior. This granularity allows us to capture subtle nuances and patterns indicative of malicious intent. The empirical evaluation involves using a large dataset of 80000 apps collected over eight years, from 2015 to 2022. Each year is represented by 10,000 apps, ensuring a balanced dataset with 5,000 malicious and 5,000 benign apps per year. The apps collected from 2015 to 2018 are used during the training phase, while those from the subsequent years (2019 to 2022) are dedicated to the longitudinal evaluation. Hence, mirroring a zero-day scenario that offers a more authentic assessment than the conventional evaluation method. Even though the LSTM is a classic choice for sequential data, the CNN has shown unexpectedly better performance.

In this paper, our primary contributions are:

- A longitudinal comparison between two state-of-the-art deep learning classifiers, CNN and RNN-LSTM, with a feature extraction process that captures the sequence of API calls within Android applications, leveraging the Call Graph structure.

- An evaluation of the impact of API abstraction granularity at the class level and package level on malware detection performance. How granular should this data be? Would a detailed class-level view of API calls be more informative than a broader package-level perspective?
- The use of a large dataset of 80000 apps, over eight years, from 2015 to 2022, to provide empirical evidence on the efficacy of deep learning classifiers.

For the rest of the paper, the organization is as follows: Sect. 2 covers related works in Android malware detection. Section 3 presents our methodology and the data preprocessing steps. Section 4 describes our experimental setup and discusses the results. Finally, Sect. 5 concludes the paper and possibilities for future research.

2 Related Works

Machine learning-based malware detection has been extensively studied [3]. Most research aims for complete accuracy through different models, ignoring the fact that malware continually evolves over time. Only a limited number of studies consider the longitudinal assessment in this context, such as Karbab et al. [4], which treat malware detection as an API call sequences classification problem like us but propose a CNN classifier trained from consecutive API calls as they appear in the basic method blocks, in the opposite of what we do where we infer the API call sequences from the structured Call Graph to represent apps behavior as features. The authors evaluated their framework called MalDozer on datasets from four consecutive years: 2013, 2014, 2015, and 2016. The experiment involved training MalDozer on samples from one year and testing it on datasets from the other years. Specifically, when trained on the 2013 dataset, MalDozer exhibited a reasonable detection rate for the 2014 dataset, achieving an F1-score of 90% due to its proximity in time. However, the detection rate decreased for the 2015 and 2016 datasets, reaching 60–67% of F1-score. In [5], the authors conducted a longitudinal evaluation study on the structure and behaviors of Android applications on current machine learning-based malware detection techniques (e.g., Random Forest, Support Vector Machine, etc.). They employed a dataset of 26382 apps from 2010 to 2017. The results highlighted that over-time detection is more challenging than same-period detection. In most scenarios, even the best-performing technique achieved an F1-score below 80% for over-time detection. In [6], the authors introduced a LongCGDroid, an image-based method for Android malware detection, which is compared with different studies from the state-of-the-art demonstrating that deep learning models are more resilient to API changes.

Several approaches, mainly focusing on the analysis of API usage, have been proposed for Android malware detection without considering the longitudinal assessment. The authors in [7] employed dynamic malware analysis to extract API-based features, which were then used to classify malware into distinct categories. A convolutional auto-encoder is designed to learn multiple API-based malware representations with varying latent dimensionalities. These representations were then utilized to train several neural network classifiers. A decision-level fusion technique was established to consolidate individual decisions from each neural network classifier, resulting in a final classification decision. The proposed method was conventionally evaluated using 20% of the dataset. Zegzhda et al. [8] proposed a deep learning approach wherein an Android application is represented as an RGB image for analysis by a convolutional neural network. This representation is derived from a sequence of API calls and their associated protection levels. The study underscores the significance of this method by highlighting its efficiency in detecting malicious Android applications. The results from this experiment achieved an accuracy of 93.64%.

3 Methodology

In this section, we present the proposed approach for malware detection. We adopted a comprehensive approach to extract meaningful API call sequences from Android apps, aiming to facilitate the detection of potential malware behavior. Android is primarily built upon the Java programming language, forming most Android applications' foundation. The system comprises several integral components that ensure its functionality and versatility. Among these components are the META-INF directory, the res directory, and the AndroidManifest.xml. Another pivotal component is the classes.dex file, which contains the compiled bytecode of the Android application. Notably, this dex file can be decompiled to produce smali files, an intermediate representation where API-related information can be extracted and analyzed. The smali files are principally used by our approach to extract API sequences.

The overall scheme of our methodology is presented in Fig. 1, composed of four sequential steps. The first step involves gathering a large dataset of benign and malware apps. The second step consists of constructing Call Graphs and extracting the API call sequences using a static analysis. We employ CNN and RNN-LSTM to learn patterns from extracted features in the third step. Finally, we assess the models' adaptability in the evolving Android apps over multiple years.

3.1 Dataset

To conduct a comprehensive temporal assessment of the models, we constructed a dataset of Android apps that are categorized as either benign or malware and are additionally date-labeled using the first-seen date over an eight-year period from 2015 to 2022. Specifically, for each year, we collect 10,000 apps evenly distributed between benign and malware categories, resulting in a total of 40,000 benign and 40,000 malware apps. To construct this dataset, we used mainly two sources, Androzoo [9] to acquire benign apps and VirusShare [10] for malware apps. The selection of these sources was deliberate, as they provided

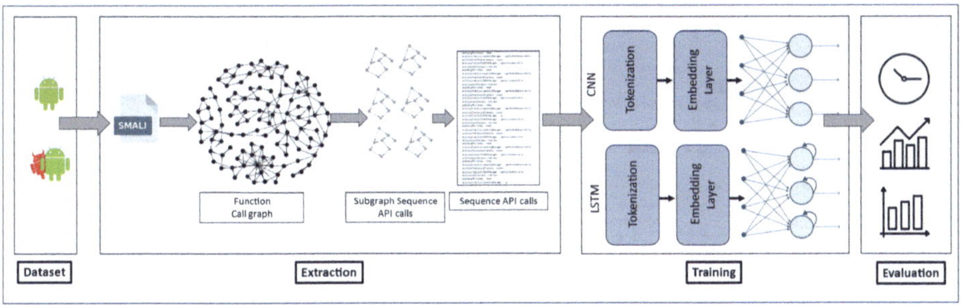

Fig. 1. The process flow of the proposed methodology

information about the date of the apps. To validate these timestamps, we cross-referenced them with the initial appearances of each sample on VirusTotal [11], a platform known for its daily analysis of numerous samples. Subsequently, we partition the dataset into two segments: one designated for feature extraction and training (from 2015 to 2018) and the other reserved exclusively for longitudinal evaluation (from 2019 to 2022).

3.2 Features Extraction

Our feature extraction process focused on capturing the order of API calls within each Android application. We used the Call Graph structure to trace the sequence of API calls, providing insights into the behavior of these apps. We extracted API call sequences at different abstraction levels, including package level and class level, to explore the impact of granularity on detection performance. Package-level abstraction simplifies analysis by grouping APIs based on the packages or namespaces they belong to, making it computationally efficient. For example, Android's "android.net" package contains networking-related functionality. On the other hand, class-level abstraction provides a high granularity and more detailed app behavior view as it goes one step deeper than package-level abstraction but requires more computational resources. For example, Android's "android.net.TelephonyNetworkSpecifier" class is used for cellular network requests. Its usage separately can provide more behavioral details.

To construct the Call Graph of each app, we use Flowdroid [12], which is a static analysis tool for Android. In this directed graph $G = (N, E)$, the set of nodes N represents methods, and the set of edges E represents method invocations. Figure 2 illustrates the Call Graph of a malware application, showcasing the code from one of its methods. Within this code, all invoked APIs are emphasized in bold.

The analysis begins at the code's entry points, defined in the AndroidManifest.xml file. Unlike Java's control flow analysis, Android code has multiple entry points. Utilizing FlowDroid to extract the Call Graph eliminates the need to handle these various entry points individually. The Algorithm 1 presents the structured representation of the process to generate API call sequences. Once

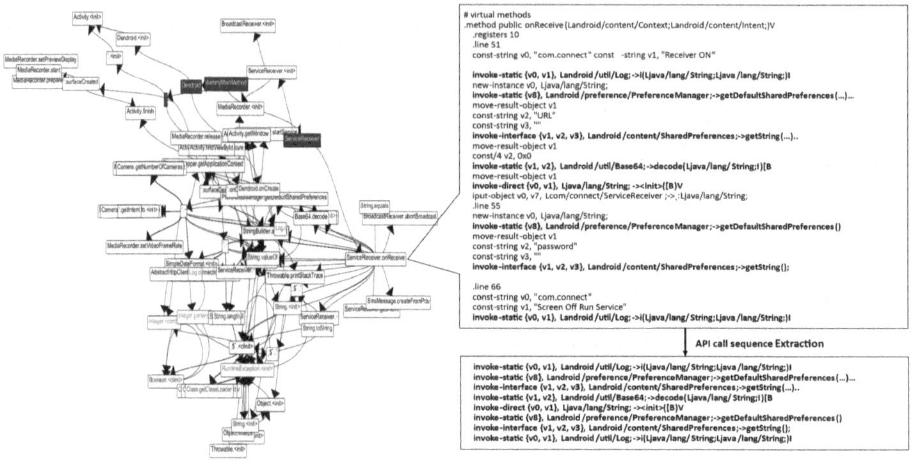

Fig. 2. The Call Graph and the subsequent method's smali code

we have the Call Graph, we traverse it to capture the order of API calls. This traversal starts from the entry points and follows method invocations in a Depth First Search (DFS). As we traverse the Call Graph, we extract information about API calls made within each method by considering both package and class levels of abstraction. Thus, organizing the extracted API calls into sequences while preserving the meaningful order.

Algorithm 1: Extract API Call Sequences from Android App's Call Graph

```
Data: Android app's call graph
Result: List of API call sequences
api_call_sequences=0 ;
visited_methods=0 ; method_stack= [call_graph.entry_point] ;
while method_stack do
    current_method=method_stack.pop();
    if current_method not in visited_methods then
        visited_methods.add(current_method);
        api_calls=extract_api_calls(current_method);
        invoked_methods=get_invoked_methods(current_method);
        method_stack.append(invoked_methods);
        api_call_sequences.append(api_calls);
    end
end
```

3.3 Classification Models

We employed two prominent deep learning classifiers, CNN and RNN-LSTM, as the core components of our malware detection models. These classifiers were chosen due to their effectiveness in handling sequential data and their potential to capture temporal patterns in API call sequences.

Before feeding the API sequences into the neural network, they must be converted from their raw format into a format the network can understand. This conversion is achieved through tokenization. Each unique API in the dataset is mapped to a unique integer. This process ensures that the model receives consistent and structured input data. An out-of-vocabulary token "unknown" is also defined to handle any APIs that might not be present in the training set but appear in the validation or test sets.

Given the variability in the length of API sequences across different apps, ensuring that all sequences fed into the model have a consistent length is essential. Figure 3 depicts a box plot of the distribution lengths in the dataset. This is achieved using the pad sequence's function. Sequences shorter than the defined length are padded, while those longer are truncated. Ideally, we want a sequence length that covers a significant portion of our data. If the sequence length is too short, we might miss essential patterns. If it is too long, it will introduce unnecessary padding for shorter sequences, which can add noise and computational overhead. On the other hand, longer sequences require more computation concerning processing time and memory. Experimenting with different sequence lengths is essential to see which performs best on a validation set. We experimented with different lengths. For instance, we tried sequence lengths at Q1, somewhere between Q1 and Q2, Q2 (Fig. 3).

The first layer of both neural networks is an embedding layer, which transforms the integer-encoded API sequences into fixed-sized dense vectors. This layer essentially captures the semantic meaning of each API, allowing the model to understand patterns based on the relationships between different APIs. The

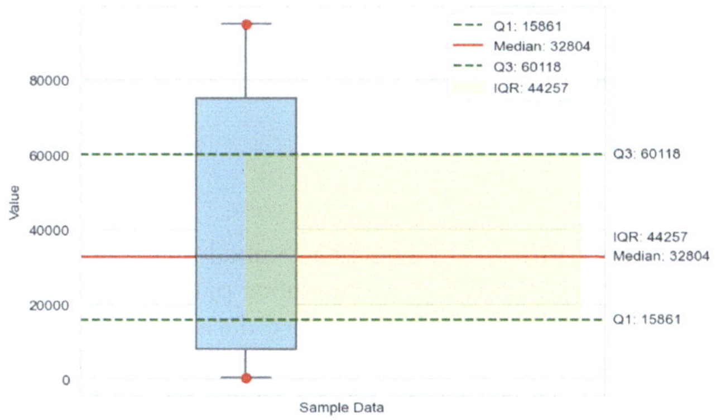

Fig. 3. Box plot highlighting lengths of the samples and outliers

embedding dimension can influence the model's capacity to capture relationships between APIs. A small dimension might not capture the complexity of the data, while a very large one can lead to overfitting and increased computational cost.

Convolutional Neural Network. The model employs 1D convolutional layers to detect local patterns within the API sequences. These layers can recognize motifs or subsequences within the API sequences that indicate benign or malicious behavior. Activation functions, like rectified linear unit (ReLU), introduce non-linearity, enabling the model to learn complex patterns. The number of filters determines the number of features or patterns the model can detect. The kernel size defines the length of the patterns. Adjusting these can impact the model's ability to recognize short or long patterns in the data. After each convolutional layer, a max-pooling layer is introduced. This layer reduces the spatial size of the representation, preserving only the most significant features and reducing computational requirements for subsequent layers. Post the convolutional operations, the model employs dense layers to make final decisions based on the features extracted by the convolutional layers. A dropout layer is also introduced to prevent overfitting by randomly setting a fraction of input units to 0 during training. The final layer uses a sigmoid activation function suitable for binary classification. Table 1 depicts the CNN architecture.

Table 1. CNN Architecture

Layers	Options	Activation
Embedding	–	–
Conv1D	Filters = 128, Kernel = 3	ReLU
MaxPooling1D	–	–
Conv1D	Filters = 256, Kernel=3	ReLU
Dropout	–	–
GlobalMaxPooling1D	–	–
Dense	units = 256	ReLU
Dense (Output)	unit = 1	Sigmoid

Long Short-Term Memory. Following the embedding layer, the dense vectors are fed into LSTM layers. LSTM can remember patterns over long sequences, making them particularly suitable for sequence data like API calls. The LSTM layers in the model are designed to identify temporal dependencies in the API call sequences, which can indicate the app's behavior. The number of LSTM units in each layer determines the capacity of the LSTM to capture information. A higher number of units can recognize more complex patterns but increases computational demands and the risk of overfitting. Stacking multiple LSTM layers can also influence the model's depth and ability to capture hierarchical patterns. However, adding more layers can also amplify the risk of overfitting and increase training times. After processing the sequences with LSTM layers, the model employs dense layers to make decisions based on the features extracted by the LSTMs. Dropout layers are interspersed between dense layers to mitigate the risk of overfitting, especially given the complexity of the model. The final layer of the model is designed for binary classification. Table 2 depicts the CNN architecture.

Table 2. RNN-LSTM Architecture

Layers	Options	Activation
Embedding	–	–
LSTM	units = 128	–
LSTM	units = 64	–
Dense	units = 128	ReLU
Dropout	–	–
Dense (Output)	unit = 1	Sigmoid

3.4 Evaluation Metrics

In our research, we employ three primary metrics to evaluate the performance of our model: Accuracy, True Positive Rate (TPR), and False Positive Rate (FPR). Accuracy provides an overall measure of model performance., calculated as the ratio of correctly predicted instances to the total instances. TPR, often referred to as Sensitivity or Recall, measures the proportion of actual positives that are correctly identified. FPR, on the other hand, quantifies the number of false alarms. These metrics are derived from four fundamental components of a confusion matrix:

- True Positives (TP): Number of malware apps correctly identified as malware.
- True Negatives (TN): Number of benign apps correctly identified as benign.
- False Positives (FP): Number of benign apps incorrectly identified as malware.
- False Negatives (FN): Number of malware apps incorrectly identified as benign.

These metrics collectively offer a comprehensive view of the model's capability to distinguish between malware and benign apps. The formulas for these metrics are presented in the Table 3.

Table 3. Evaluation Metrics Formulas

Metrics	Description
Accuracy	$\frac{TP+TN}{TP+TN+FP+FN}$
TPR(Recall)	$\frac{TP}{TP+FN}$
FPR	$\frac{TP}{TP+FP}$

4 Experimental Results

This section presents the outcomes of the conducted experiments aimed at analyzing the longitudinal performance of the deep learning classifiers at two different levels of API abstractions, focusing on three performance indicators. The training process utilized an entire four-year span, from 2015 to 2018, while the subsequent four years (2019 to 2022) were used to evaluate the trained detectors. Before delving into the detailed results, in what follows, the class abstraction is denoted as C, and the package abstraction is denoted as P.

Figure 4 presents the obtained Accuracy results over four years. The results from the very first year after training (2019) indicate that both classifiers performed exceptionally well at different abstraction levels, with accuracy above 94% for all abstractions. Table 2 shows full results, including Accuracy, TPR, and FPR. CNN with C abstraction achieved the highest accuracy at 96.8%, closely followed by RNN with C abstraction at 96.57%. The Package level abstraction slightly reduced the performance for both classifiers, but the drop was more pronounced for RNN, which had an accuracy of 94.44%. However, both classifiers demonstrated a decline in accuracy from 2019 to 2021 at different scales, with a slight rebound in 2022. For the C abstraction, CNN consistently outperformed RNN across all years, with the most significant difference observed in 2021, with a CNN C abstraction accuracy of 83.7% experiencing a decline of ≅13.5%, and RNN C abstraction accuracy of 78.28% having a decline of ≅19% in overall accuracy.

The inaccurate classification of malware is represented by the TPR in Fig. 5 (a). The TPR results for C abstraction showed that CNN and RNN were closely matched in 2019. However, while CNN's TPR declined gradually over the years, RNN experienced a more pronounced drop, especially between 2020 and 2021, achieving a TPR of 87.32%. At the package level, CNN's TPR saw a significant dip in 2020 and 2021 but remarkably rebounded in 2022 to 85.08%. RNN's TPR, at the package level, showed a consistent decline over the years, reaching 82.8% in 2022. On the other hand, both classifiers experienced a more pronounced increase in FPR (Fig. 5 (b)) from 2020 to 2021, indicating a more important inaccurate classification of benign apps compared to malware apps. For instance, the CNN with C abstraction in the year of evaluation (2021) experienced misclassification of approximately 10% for malware samples and approximately 30% for benign samples. This suggests that accuracy declines more from misclassifying benign apps than malware apps.

Based only on the findings from the obtained results of the first year of evaluation on 2019, it is evident that even the RNN utilizing the package abstraction approach has favorable outcomes, achieving an accuracy rate of 94%. However, while examining the longitudinal assessment, it becomes evident that after two years (specifically in 2021), the accuracy of the evaluation diminishes by as much as 55%, rendering it the most unfavorable outcome among all the recorded outcomes.

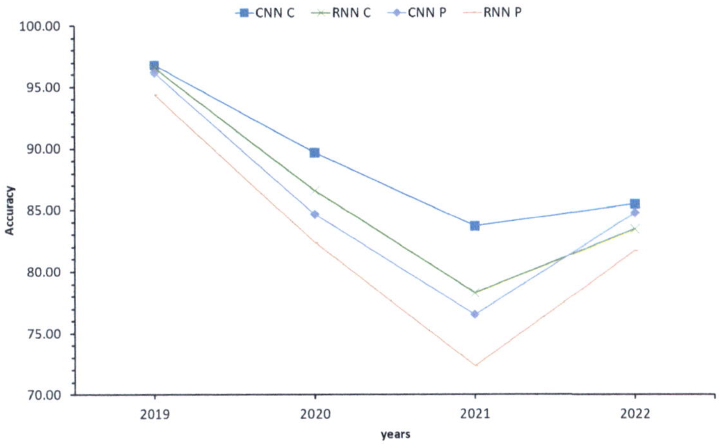

Fig. 4. The results of the accuracy over four years

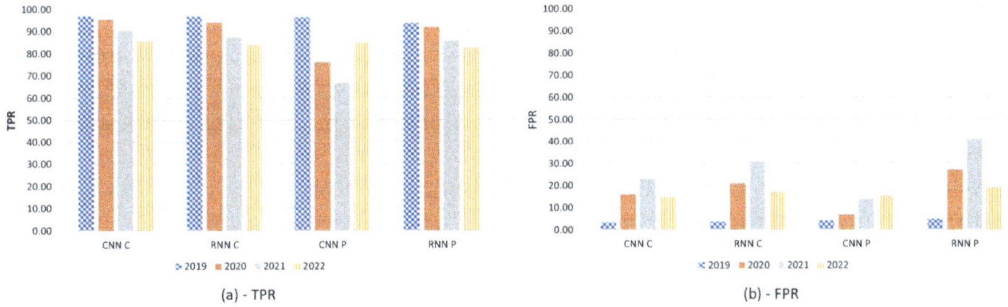

Fig. 5. The results of the TPR and FPR over four years

Overall, at the C level, where details are more pronounced, CNN consistently outperformed RNN. This suggests that CNN might be better equipped to leverage the detailed insights provided by C abstraction. At the P level, which offers a more generalized view, CNN still maintained an edge over RNN, indicating its robustness across different granularities. The observed rebound in performance metrics in 2022 for both classifiers suggests that the API sequences have been effective in adapting to newer malware characteristics.

5 Conclusion

In this research, we addressed the evolving nature of Android malware by conducting a longitudinal evaluation of deep learning classifiers, specifically CNN and RNN-LSTM, across different levels of API abstractions. Our findings revealed that while both classifiers demonstrated commendable performance in the immediate year post-training, there was a noticeable decline in accuracy in the subsequent years. Notably, CNN consistently outperformed RNN-LSTM, especially at the class abstraction, suggesting its superior capability in leveraging detailed insights. This observation was consistent even at the package level abstraction, highlighting CNN's robustness across varying granularities.

The research underscores the critical importance of continuous evaluation in malware detection. While a model might exhibit high accuracy in a temporally localized dataset, its performance can vary significantly when exposed to evolving malware samples over extended periods. This underscores the need for models that can adapt and learn from new threats, ensuring sustained protection for Android users.

In future work, we aim to explore other deep-learning architectures and hybrid models that might offer enhanced resilience against the ever-evolving nature of Android malware. Additionally, incorporating dynamic analysis techniques alongside our static analysis approach could provide a more comprehensive view of malware behavior.

References

1. Mobile Operating System Market Share Worldwide. https://gs.statcounter.com/os-market-share/mobile/worldwide/. Accessed 2023
2. App Download Data 2023. https://www.businessofapps.com/data/app-statistics/. Accessed 2023
3. Acharya, S., Rawat, U., Bhatnagar, R.: A comprehensive review of android security: threats, vulnerabilities, malware detection, and analysis. Secur. Commun. Netw. **2022**, 7775917 (2022)
4. Karbab, E.B., Debbabi, M., Derhab, A., Mouheb, D.: Android malware detection using deep learning API method sequences. Digit. Investig. **24**, S48–S59 (2017)
5. Cai, H.: Assessing and improving malware detection sustainability through app evolution studies. ACM Trans. Softw. Eng. Methodol. **29**(2), Article 8 (2020)
6. Mesbah, A., Baddari, I., Riahla, M.A.: LongCGDroid: android malware detection through longitudinal study for machine learning and deep learning. Jordanian J. Comput. Inf. Technol. (JJCIT) **09**, 328–346 (2023)
7. Belkhouche, Y.: API-based features representation fusion for malware classification. In: IEEE 47th Annual Computers, Software, and Applications Conference (COMPSAC), pp. 1658–1662 (2023)

8. Zegzhda, P., Zegzhda, D., Pavlenko, E., Ignatev, G.: Applying deep learning techniques for Android malware detection. In: Proceedings of the 11th International Conference on Security of Information and Networks, pp. Article 7. Association for Computing Machinery, Cardiff, United Kingdom (2018)
9. Allix, K., Bissyandé, T.F., Klein, J., Traon, Y.L.: AndroZoo: collecting millions of android apps for the research community. In: IEEE/ACM 13th Working Conference on Mining Software Repositories (MSR), pp. 468–471 (2016)
10. VirusShare.com - Because Sharing is Caring. https://virusshare.com/. Accessed 2019
11. Analyse suspicious files. https://www.virustotal.com/. Accessed 2019
12. Arzt, S., et al.: FlowDroid: precise context, flow, field, object-sensitive and lifecycle-aware taint analysis for Android apps. SIGPLAN Not. **49**(6), 259–269 (2014)

Addressing Poisoning Attacks Against Federated Intrusion Detection Systems in Internet of Healthcare Things

Takieddine Boumediri(✉), Islam Debicha, and Tayeb Kenaza

Ecole Militaire Polytechnique, Algiers, Algeria
boumediri.takieddine@gmail.com

Abstract. Federated learning emerges as a key technology, allowing decentralized model training in Internet of Healthcare Things (IoHT). This aims to preserve data privacy and accommodates the dynamic nature of health data, particularly in the development of sensitive systems such as Intrusion Detection Systems (IDS), known as Federated Intrusion Detection Systems (FIDS). Nevertheless, this approach is susceptible to adversarial poisoning attacks, which involve the insidious introduction of tainted data into federated models. This paper highlights the challenge of poisoning attacks against FIDS in the IoHT ecosystem. We introduce a threat scenario illustrating the potential vulnerabilities and implications of such attacks. The scenario emphasizes the need to address this nuanced cybersecurity challenge in the ongoing development and deployment of IoHT technologies. We evaluate our proposals through rigorous experiments, examining their effectiveness in simulated attack scenarios.

Keywords: Poisoning Attack · Federated Learning · Intrusion Detection Systems · Internet of Healthcare Things

1 Introduction

The integration of the Internet of Healthcare Things (IoHT) into healthcare systems offers unparalleled advantages, revolutionizing patient care through remote monitoring, treatment tracking, and virtual consultations [13].

In tandem with these advancements, the healthcare sector faces escalating cyber threats that necessitate preventive measures. Cyberattacks on healthcare systems incur substantial financial and privacy-related damages, underscoring the critical need for robust cybersecurity solutions [4].

In response to these challenges, Intrusion Detection Systems (IDS) stand out as essential cybersecurity tools, crucial for preventing and mitigating cyber threats against healthcare systems. Anomaly-based IDS [7], utilizing machine learning [11], proves particularly valuable in detecting zero-day attacks, offering a proactive defence mechanism [10].

B. Djamaa et al. (Eds.): CSA 2024, LNNS 1145, pp. 270–281, 2025.
https://doi.org/10.1007/978-3-031-71848-9_25

However, the conventional centralized machine learning approach, while effective, raises privacy concerns, especially for sensitive healthcare data. Federated learning emerges as a solution, enabling decentralized model training in the context of IoHT [14]. This approach preserves data privacy and accommodates the dynamic nature of health data, making it an ideal candidate for bolstering IDS capabilities [1].

In the context of IoHT, while federated learning addresses privacy concerns [3], it introduces a major drawback. Granting patients control over their data and sharing training parameters for federated learning models introduces a novel and consequential threat, addressed by poisoning attacks [17]. These sophisticated attacks aim to influence the learning process through the injection of manipulated data [6], posing a distinct challenge to the security framework of FIDS in healthcare environments.

To the best of our knowledge, this paper pioneers research on poisoning attacks against Federated Intrusion Detection Systems (FIDS) in the Internet of Healthcare Things (IoHT). It contributes by:

- Identifying Vulnerabilities: Meticulously pinpointing vulnerabilities associated with poisoning attacks in FIDS, laying the groundwork for targeted countermeasures to fortify IoHT networks.
- Realistic Threat Scenario: Introducing a detailed, plausible threat scenario, providing a valuable tool for practitioners and researchers to grasp the potential consequences of poisoning attacks on IoHT networks.
- Damage Evaluation: Critically analysing the potential damages inflicted by poisoning attacks on FIDS, quantifying harm, and helping prioritize the urgency of addressing these cybersecurity threats in healthcare environments.

The remainder of this paper unfolds as follows: Sect. 2 provides a comprehensive overview of the background and related work in the field. Section 3 details the implications of poisoning attacks on IoHT security, elucidating the potential risks and consequences. Section 4 presents the contextual analysis of poisoning attacks in the IoHT landscape. Section 5 concludes the paper, summarizing key findings and outlining avenues for future research in securing IoHT environments.

2 Background and Related Work

2.1 Internet of Healthcare Things

The Internet of Healthcare Things (IoHT) refers to the integration of medical devices and healthcare systems through the use of internet technologies. It leverages the capabilities of interconnected devices to collect, transmit, and exchange health-related data in real-time [12]. The widespread adoption of IoHT raises important challenges, including concerns related to data privacy, security, interoperability, and regulatory compliance [13].

2.2 Federated Learning

Federated Learning is a machine learning approach that allows a model to be trained across decentralized edge devices or servers holding local data samples, without exchanging them. Only the model updates, and not the raw data, are sent to a central server where they are aggregated [18]. Federated learning is gaining popularity in many fields, such as healthcare, finance, and telecommunications, where preserving data privacy and security are paramount concerns [14].

2.3 Intrusion Detection Systems

Intrusion Detection Systems (IDS) are security mechanisms designed to monitor network or system activities to look for malicious or unauthorized activities. In the field of cybersecurity, where the threat landscape is ever-evolving, anomaly-based detection stands out as a crucial method. By dynamically adapting to the changing patterns of normal system behaviour, anomaly-based detection enhances the capability to identify and prevent new threats that might otherwise escape traditional signature-based approaches [23].

2.4 Poisoning Attacks

Poisoning attacks, in the context of cybersecurity, refer to malicious activities aimed at manipulating or contaminating data used by machine learning models. These attacks are particularly relevant in scenarios where machine learning algorithms are employed, such as spam filters, recommendation systems, or intrusion detection systems [21]. The objective of poisoning attacks is to compromise the performance and integrity of these models by introducing maliciously crafted data during the training phase in order to influence the learning process, leading the model to make incorrect predictions or classifications during deployment [17].

2.5 Related Work

The landscape of attacks within federated learning is diverse, and past research has yielded varying classification schemes. For instance, Chen et al. [5] organized attacks according to fundamental information security principles: confidentiality, integrity, and availability (CIA). Meanwhile, Liu et al. [15] catalogued potential attacks across different Federated Learning (FL) phases, encompassing data auditing, training, and prediction. In a similar work vein, Wu et al. [22] structured their classification around the potential perpetrators within the FL system, distinguishing among local workers, the central server, and eavesdroppers. This multifaceted taxonomy underscores the breadth and complexity of security challenges inherent in FL environments.

There are different ways to classify poisoning attacks in Federated Learning. They can be categorized into data poisoning and model poisoning based on the target of the attack. Additionally, they can be classified as random attacks or backdoor attacks (targeted attacks) based on whether the attack is specifically targeted [5]. Data poisoning occurs when an attacker maliciously tampers with or adds data to the training set, leading to the disruption or takeover of the model. This type of attack is often carried out by the local worker who owns the data. Clean-label poisoning is a form of backdoor attack where the attacker adds malicious data instead of modifying existing data. In contrast, dirty data poisoning involves changing the original dataset through label-flipping or generating toxic samples [9].

Model poisoning involves the attacker directly changing the parameters of the target model, causing errors in the global model or creating a backdoor. It has been found to be more effective than data poisoning. For instance, attackers can perturb the weights of convolutional neural networks in a targeted manner to insert backdoors. Bit-flipping techniques can also be used to transform the target model into a trojan infection model [19]. Since the model needs to be repeatedly transmitted between the local worker and the central server, either of them could potentially be an attacker. These poisoning attacks violate the integrity principle of FL by making the global model unusable under certain conditions [5].

Ferrag et al. [8] conducted an anticipatory investigation on poisoning attacks in federated edge learning for digital twin 6G-enabled IoT. They explored how adversaries can impact the training and development of federated learning models in the context of digital twin 6G-enabled IoT. Tolpegin et al. [20] focuses on analysing data poisoning attacks against FL systems. The study highlights the vulnerability of FL systems to label flipping poisoning attacks and emphasizes the severe negative impact these attacks can have on the global model.

3 Poisoning Attacks Against FIDS in IoHT

3.1 Threat Scenario

We consider a scenario where an adversary with sophisticated knowledge of the IoHT ecosystem strategically infiltrates the network of a large healthcare provider. This attacker, motivated by malicious intent, gains unauthorized access to the Federated Intrusion Detection System (FIDS) governing the security of medical devices and sensitive health data.

In this threat scenario, the adversary exploits a vulnerability in the communication protocols used by the federated learning framework, allowing them to inject malicious data into the collaborative model training process. The poisoned data is carefully crafted to mimic legitimate health-related patterns but contains subtle alterations that are tailored to deceive the FIDS.

As the federated learning process continues, the poisoned data becomes an integral part of the shared model, compromising the accuracy and reliability of intrusion detection across the entire IoHT ecosystem. The adversarial injection goes undetected initially, as the poisoned data seamlessly blends with the legitimate data during the federated learning iterations.

The consequences of this poisoning attack unfold when the compromised FIDS fails to recognize and respond effectively to a real security threat within the healthcare network. False negatives increase, allowing malicious activities to go undetected, and false positives become more frequent, leading to unnecessary disruptions in medical services. The compromised FIDS may even generate misleading alerts, diverting the attention of cybersecurity personnel and resources to non-existent threats.

This threat scenario underscores the critical importance of addressing poisoning attacks against Federated Intrusion Detection Systems in the IoHT. The consequences not only impact data security but also pose a significant risk to patient safety and the overall integrity of healthcare services.

3.2 Healthcare IoT Datasets

In order to evaluate the proposed attacks in the threat scenario, we've performed our experiments on two IoHT datasets which have different features, for instance: WUSTL-EHMS-2020 and ECU-IoHT.

1. WUSTL-EHMS-2020: Hady and al. [10] created the WUSTL-EHMS-2020 dataset from a specialized testbed equipped with a multi-sensor board for patient biometrics. This dataset, containing 2046 attack and 14272 normal instances with 36 network and biometric features.
2. ECU-IoHT: Ahmed and al. [2] offered a valuable resource for analysing cyber-attacks in healthcare IoT with the ECU-IoHT dataset. This dataset features a realistic network setup. It contains 87754 normal and 23453 attack instances.

3.3 The Proposed FIDS Architecture

Our architecture, inspired by the approach presented in [16], is enhanced through integration with the TensorFlow Federated (TFF) framework. This integration allows us to leverage TFF's capabilities for anomaly detection in IoHT applications. Figure 1 illustrates the components and flow of information within our proposed system.

We outline the main components of our FIDS, including the proposed poisoning attacks, the deep neural network model, clients' data, and the federated learning environment.

- **Deep neural network model:** The proposed DNN-FL features a systematic arrangement of layers, each contributing to its learning and decision-making capabilities. The model's initial layer accepts network traffic features as input data, which serve as the foundational information for the learning process. The DNN-FL model comprises three hidden layers, the first two hidden layers consist of 64 neurons each, while the third hidden layer contains 32 neurons. The model's processing culminates at the output layer.
- **Federated learning setup:** In our experiment, 10 clients were defined to simulate the FL environment. Each client, emulating an IoT device, participates in the learning process independently. During each training round, clients utilize their local datasets to update their models. These updated models are then aggregated at a central server to create an improved global model, which is subsequently distributed back to the clients for further learning.

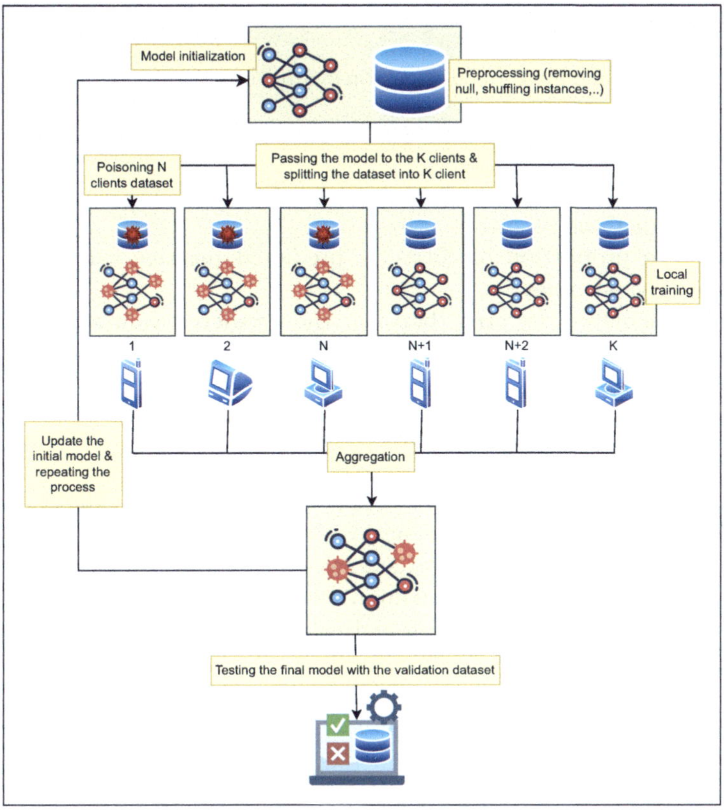

Fig. 1. Main components of the proposed FIDS architecture

3.4 The Proposed Poisoning Attacks

We explore three distinct types of poisoning attacks, each designed to manipulate the learning process and potentially compromise the performance of our FIDS.

- **Backdoor attack:** This attack involves modifying the labels of the chosen samples to be consistently labelled as "Normal". The attacker's intention is to create a backdoor in the model, biasing it towards a higher rate of negative classifications.
- **Availability attack:** Conversely, the availability attack entails altering the labels of selected samples to be consistently labelled as "Attack". The attacker aims to manipulate the model towards a higher rate of positive classifications.
- **Combined attack:** In this attack, a subset of samples' labels are maliciously flipped from "Normal" to "Attack" and from "Attack" to "Normal". This combined attack aims to challenge the learning process by introducing label mislabelling and biasing, potentially leading to incorrect decision boundaries.

3.5 Evaluation Metrics

The following metrics have been used for evaluation:

- **Accuracy:** The percentage of correct predictions out of the total predictions made by the model.
- **Precision:** The proportion of correctly predicted positive samples out of all positive predictions.
- **Recall:** The proportion of correctly predicted positive samples out of all actual positive samples.
- **Loss:** It calculates the error between the predicted values and the actual values during model training.

4 Experimental Evaluation

The primary objective of the experiments is to assess the FIDS performance under various poisoning attack scenarios. To do this, we gradually increased the number of selected clients as targets for the poisoning attacks, simulating a scenario where adversaries aim to compromise the learning process by injecting malicious data. Additionally, we systematically varied the percentage of poisoned data for each client.

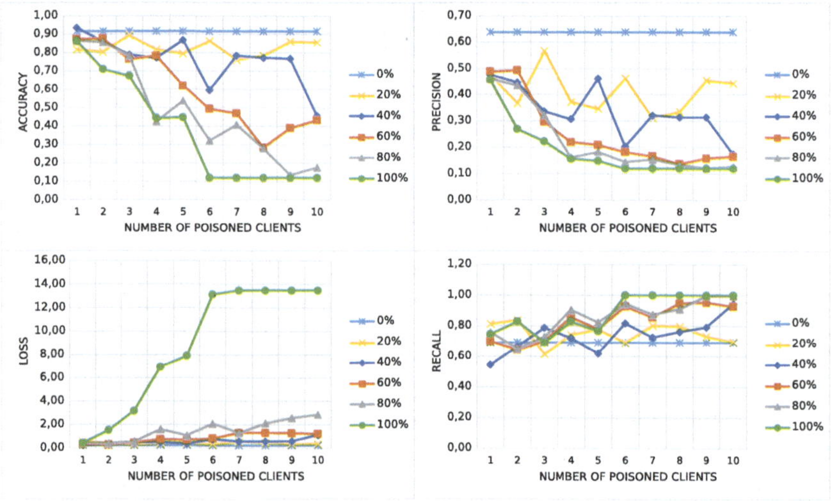

Fig. 2. Availability attacks results on the WUSTL-EHMS-2020 dataset

Figures 2 and 3 illustrate various metrics computed for availability attacks on the WUSTL-EHMS-2020 and ECU-IoHT datasets, respectively. It is noteworthy that an inverse relationship exists between the accuracy of the FIDS and the

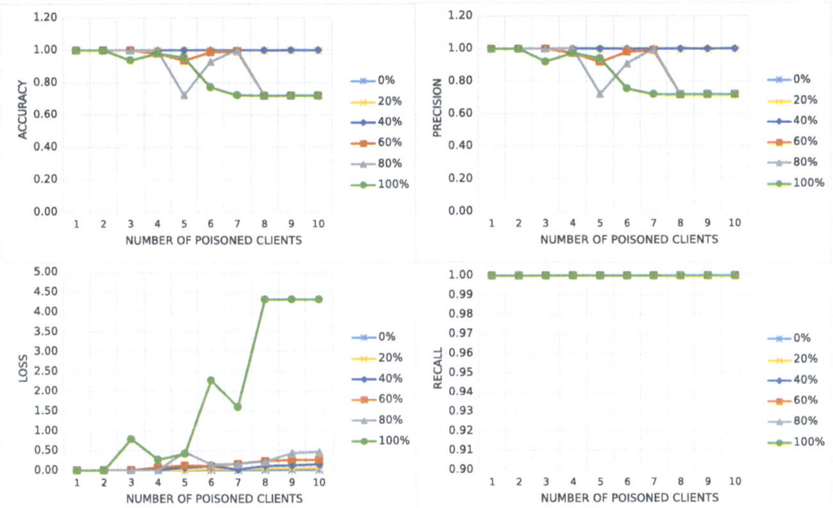

Fig. 3. Availability attacks results on the ECU-IoHT dataset

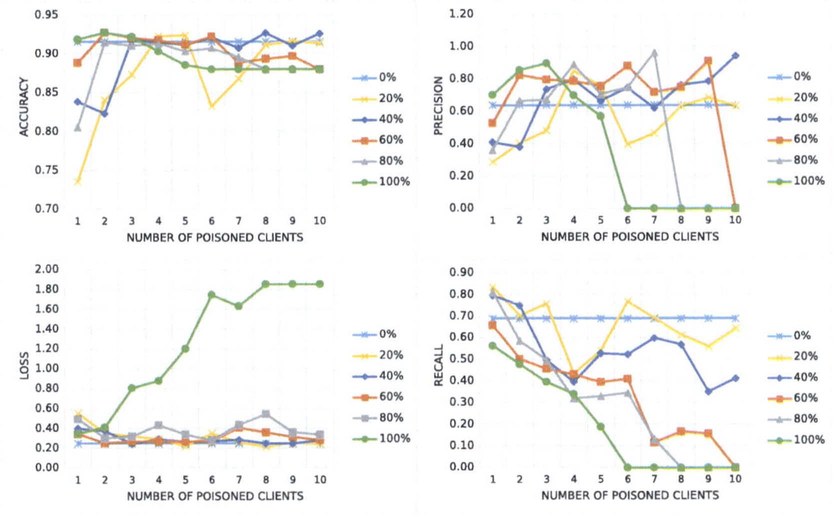

Fig. 4. Backdoor attacks results on the WUSTL-EHMS-2020 dataset

number of poisoned clients. As the number of poisoned clients increases, the model's accuracy decreases, while the recall increases. The increase in recall can be attributed to the model's heightened ability to correctly identify instances of actual attacks, even though this might come at the cost of an elevated rate of false positives due to the manipulation of labels.

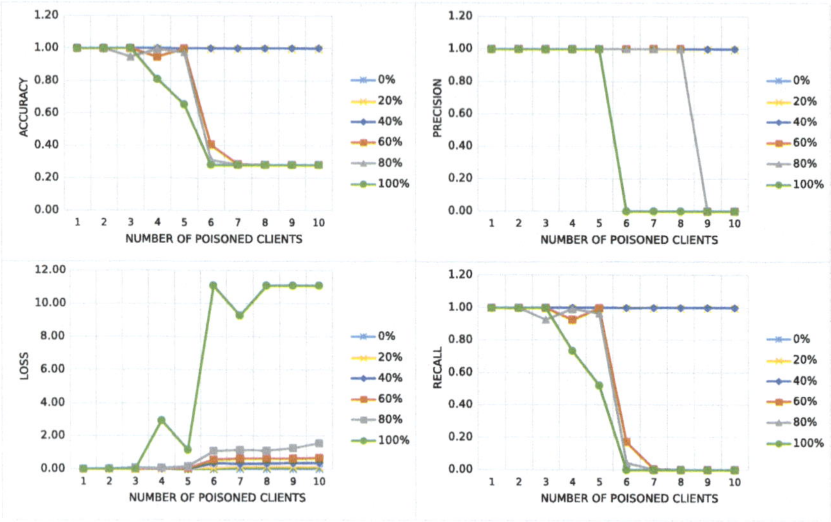

Fig. 5. Backdoor attacks results on the ECU-IoH dataset

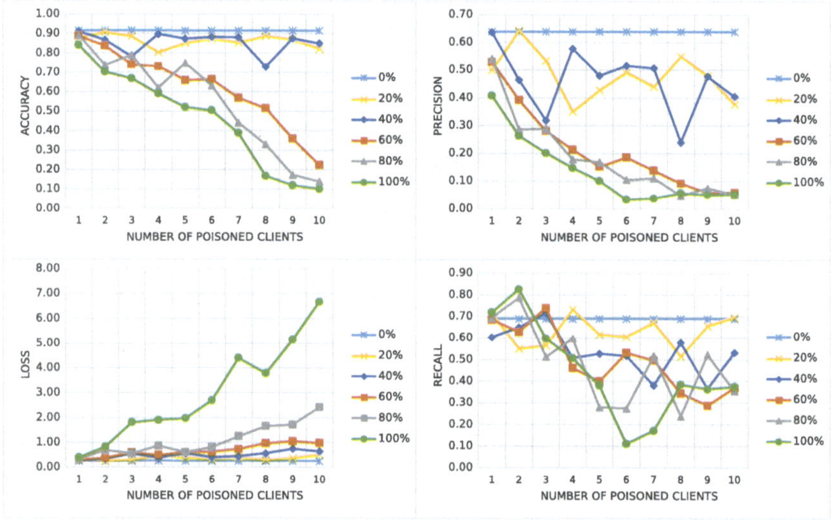

Fig. 6. Combined attacks results on the WUSTL-EHMS-2020 dataset

On the other hand, we can observe that the poisoning percentage significantly influences the system's performance. A system fully poisoned with 100% manipulation across 10 clients exhibits the poorest results in terms of accuracy and loss values. This effectively underscores the substantial impact of the poisoning attack on the classification task.

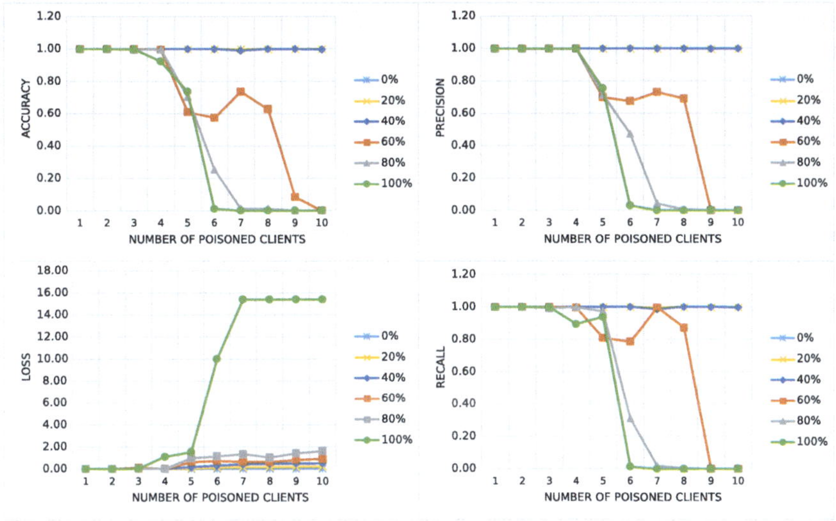

Fig. 7. Combined attacks results on the ECU-IoH dataset

In contrast to availability attacks, backdoor attacks has an opposite effect on the two system's metrics, recall and precision, as illustrated in Fig. 4 and Fig. 5. The recall decreases whenever more than 50% of the clients are poisoned, and eventually reaches zero values. The decrease in recall is driven by the model's reduced ability to identify actual attacks due to the flipped labels. The increase in precision is influenced by the correct classification of instances originally labelled as "Attack" as "Normal". These effects are typical consequences of label manipulation that shift the model's behaviour in detecting specific classes.

As depicted in Fig. 6 and Fig. 7, the system exhibits a deterioration in performance, leading to malfunctions when the number of malicious clients exceeds 5 and the poisoning percentage reaches 60%. Notably, both precision and recall decrease. These findings highlight the effectiveness of the combined attacks, where the availability attack lowers recall, and the backdoor attack diminishes precision. Additionally, it is noteworthy that the impact of the attacks is proportional to the percentage of poisoned data.

5 Conclusion and Future Work

Our research has led to some insights into the security of federated intrusion detection systems (FIDS) in the context of the Internet of Health Things (IoHT). Through our experiments, we have identified vulnerabilities in the FIDS, in particular its susceptibility to various poisoning attacks, highlighting their significant impact on system performance. This recognition revealed the challenges and complexities inherent in securing interconnected systems, underlining the continuing need for resilient and adaptable intrusion detection mechanisms.

In order to improve the security of federated learning systems, several potential lines of research can be proposed:

- Exploration of other types of attacks: We recommend that comprehensive analyses of other types of attacks, such as inference and reconstruction attacks, be carried out. This could reveal further vulnerabilities and guide the development of practical countermeasures. More studies can be conducted to address their impact on FIDS and then propose defense mechanisms to tackle these issues.
- Adoption of adversarial machine learning techniques: We propose to investigate the integration of machine learning techniques to anticipate and neutralise sophisticated attack strategies. This approach could enable FIDS to adopt a more proactive stance by adapting to the ever-changing threat landscape. For instance, we can apply continuous machine learning techniques to improve the quality of ML models in the FIDS.

Acknowledgments. Authors would like to thank students Mohammed Cherif Tifoura and Açyl Zerrouk for their hard work in the coding and system evaluation.

References

1. Agrawal, S., et al.: Federated learning for intrusion detection system: concepts, challenges and future directions. Comput. Commun. **195**, 346–361 (2022)
2. Ahmed, M., Byreddy, S., Nutakki, A., Sikos, L.F., Haskell-Dowland, P.: ECU-IoHT: a dataset for analyzing cyberattacks in Internet of Health Things. Ad Hoc Netw. **122**, 102621 (2021)
3. Ali, M., Naeem, F., Tariq, M., Kaddoum, G.: Federated learning for privacy preservation in smart healthcare systems: a comprehensive survey. IEEE J. Biomed. Health Inform. **27**(2), 778–789 (2022)
4. Cartwright, A.J.: The elephant in the room: cybersecurity in healthcare. J. Clin. Monit. Comput. **37**, 1123–1132 (2023)
5. Chen, Y., Gui, Y., Lin, H., Gan, W., Wu, Y.: Federated learning attacks and defenses: a survey. In: 2022 IEEE International Conference on Big Data (Big Data), pp. 4256–4265. IEEE (2022)
6. Debicha, I., Cochez, B., Kenaza, T., Debatty, T., Dricot, J.M., Mees, W.: Adv-Bot: realistic adversarial botnet attacks against network intrusion detection systems. Comput. Secur. **129**, 103176 (2023)
7. Debicha, I., Debatty, T., Dricot, J.M., Mees, W.: Adversarial training for deep learning-based intrusion detection systems. In: ICONS 2021: The Sixteenth International Conference on Systems (2021)
8. Ferrag, M.A., Kantarci, B., Cordeiro, L.C., Debbah, M., Choo, K.K.R.: Poisoning attacks in federated edge learning for digital twin 6G-enabled IoTs: an anticipatory study. arXiv preprint arXiv:2303.11745 (2023)
9. Goldblum, M., et al.: Dataset security for machine learning: data poisoning, backdoor attacks, and defenses. IEEE Trans. Pattern Anal. Mach. Intell. **45**(2), 1563–1580 (2022)
10. Hady, A.A., Ghubaish, A., Salman, T., Unal, D., Jain, R.: Intrusion detection system for healthcare systems using medical and network data: a comparison study. IEEE Access **8**, 106576–106584 (2020)

11. Hamache, A., et al.: Uncertainty-aware Parzen-Rosenblatt classifier for multiattribute data. In: Destercke, S., Denoeux, T., Cuzzolin, F., Martin, A. (eds.) Belief Functions: Theory and Applications: 5th International Conference, BELIEF 2018, Compiègne, France, 17–21 September 2018, Proceedings 5, pp. 103–111. Springer, Cham (2018). https://doi.org/10.1007/978-3-319-99383-6_14
12. Kaur, H., Atif, M., Chauhan, R.: An Internet of Healthcare Things (IoHT)-based healthcare monitoring system. In: Mohanty, M.N., Das, S. (eds.) Advances in Intelligent Computing and Communication. LNNS, vol. 109, pp. 475–482. Springer, Singapore (2020). https://doi.org/10.1007/978-981-15-2774-6_56
13. Ketu, S., Mishra, P.K.: Internet of Healthcare Things: a contemporary survey. J. Netw. Comput. Appl. **192**, 103179 (2021)
14. Kumar, Y., Singla, R.: Federated learning systems for healthcare: perspective and recent progress. In: Rehman, M.H., Gaber, M.M. (eds.) Federated Learning Systems. SCI, vol. 965, pp. 141–156. Springer, Cham (2021). https://doi.org/10.1007/978-3-030-70604-3_6
15. Liu, P., Xu, X., Wang, W.: Threats, attacks and defenses to federated learning: issues, taxonomy and perspectives. Cybersecurity **5**(1), 1–19 (2022)
16. Mosaiyebzadeh, F., Pouriyeh, S., Parizi, R.M., Han, M., Batista, D.M.: Intrusion detection system for IoHT devices using federated learning. In: Proceedings of IEEE INFOCOM (2023)
17. Nguyen, T.D., Rieger, P., Miettinen, M., Sadeghi, A.R.: Poisoning attacks on federated learning-based IoT intrusion detection system. In: Proceedings of Workshop Decentralized IoT Systems and Security (DISS), pp. 1–7 (2020)
18. Niknam, S., Dhillon, H.S., Reed, J.H.: Federated learning for wireless communications: motivation, opportunities, and challenges. IEEE Commun. Mag. **58**(6), 46–51 (2020)
19. Rakin, A.S., He, Z., Fan, D.: TBT: targeted neural network attack with bit trojan. In: Proceedings of the IEEE/CVF Conference on Computer Vision and Pattern Recognition, pp. 13198–13207 (2020)
20. Tolpegin, V., Truex, S., Gursoy, M.E., Liu, L.: Data poisoning attacks against federated learning systems. In: Chen, L., Li, N., Liang, K., Schneider, S. (eds.) Computer Security–ESORICS 2020: 25th European Symposium on Research in Computer Security, ESORICS 2020, Guildford, UK, 14–18 September 2020, Proceedings, Part I 25, pp. 480–501. Springer, Cham (2020). https://doi.org/10.1007/978-3-030-58951-6_24
21. Wang, Z., Ma, J., Wang, X., Hu, J., Qin, Z., Ren, K.: Threats to training: a survey of poisoning attacks and defenses on machine learning systems. ACM Comput. Surv. **55**(7), 1–36 (2022)
22. Wu, J., Si, S., Wang, J., Xiao, J.: Threats and defenses of federated learning: a survey. Big Data Res. **8**(5), 12 (2022)
23. Yang, Z., et al.: A systematic literature review of methods and datasets for anomaly-based network intrusion detection. Comput. Secur. **116**, 102675 (2022)

A Dynamic Heuristic Search for Textual Adversarial Attacks

Salim Khemis(✉), Yacine Amara, and Mohamed Akrem Benatia

Ecole Militaire Polytechnique EMP, BP 17, Bordj El Bahri, 16046 Algiers, Algeria
khemis.salim1@gmail.com

Abstract. Adversarial attacks represent a significant threat to the robustness of deep learning models by exploiting their vulnerabilities. Recent advancements have given rise to high quality attack techniques, some of which are strategically employed to enhance model robustness via adversarial training and defense strategies. This paper introduces a novel heuristic search algorithm designed for the generation of adversarial samples through iterative synonym ranking. Our approach distinguishes itself through a superior success rate, lower accuracy under attack, surpassing many existing state-of-the-art methods, while maintaining a acceptable query count. Through extensive experiments, we demonstrate the efficacy of our approach in enhancing deep learning models against adversarial attacks through adversarial training.

Keywords: Text based Adversarial Attacks · Natural Language Processing · Adversarial training · Deep learning models robustness

1 Introduction

Artificial Intelligence (AI) is recently having a significant surge, wielding a pivotal influence across diverse domains such as Health Care, Robotics, Economies, Gaming, and Social Media. Despite these advancements, the vulnerability of AI systems to adversarial attacks remains a pressing concern, potentially undermining their performance. Adversarial attacks involve the introduction of subtle and imperceptible alterations to input data. For instance, this may manifest as adding noise to images, manipulating words in Natural Language Processing (NLP) systems, or affecting other applications like autonomous cars and spam detectors. The objective is to mislead models into misclassifying these inputs.

Adversarial attacks reach into critical domains where public opinion holds substantial significance, such as online surveys, the spread of misinformation and hate speech on social media platforms, scams and fraudulent activities, especially in online communications, and votes or elections where it relies heavily on of public sentiment. Indeed, adversarial attacks on social media can significantly change public perception and influence decision-making, especially with the trust that individuals place in digital platforms for expressing their opinions. This emphasizes the importance of studying adversarial attacks to create strong

B. Djamaa et al. (Eds.): CSA 2024, LNNS 1145, pp. 282–291, 2025.
https://doi.org/10.1007/978-3-031-71848-9_26

defenses to uphold the robustness of deep learning models. Study, understand, and improve these attacks is a key to improve the robustness of deep learning models. A deeper comprehension of model vulnerabilities and the algorithms employed in attacks can pave the way for the construction of more robust AI systems.

In this paper, Our goal is to identify vulnerabilities in NLP models through generating adversarial samples more effectively achieving a higher success rate in targeted classes. Our contribution introduces a novel word-level search algorithm designed for the generation of adversarial attacks, through prioritizing the ranking of candidate synonyms in the best potential positions within the texts, rather than traditional approaches that prioritize ranking positions by importance. We conducted extensive experiments on the IMDB and Ag-News datasets, employing state-of-the-art attacks in both black-box and white-box scenarios to demonstrate the effectiveness of our algorithm across diverse conditions. Additionally, experiments were executed to illustrate the effectiveness of our enhanced attack approach in enhancing model robustness through adversarial training.

2 Background and Related Work

Word level perturbations involve generating candidate lists for each word, leveraging methods like embedding spaces, Masked Language models, synonym extraction, or character-level alterations involving swapping or deleting characters to generate visually similar words. Several algorithms proposed in literature, such as [4,9,10,13] generate attacks by perturbing each word from the input text in a specific order of importance. This order is determined by measuring the scores of each word as the difference between the original text $X = [w_1 \dots w_i \dots w_n]$ and when the specific word is deleted $X \setminus w_i = [w_1 \dots w_{i-1}, w_{i+1} \dots w_n]$ or replaced by an out-of-vocabulary word $(X \setminus w_i) \cup [unk] = [w_1 \dots [unk] \dots w_n]$ in a black-box setting. For white-box setting, the gradient method $\nabla_{ei} L(\theta, x, y)$ is used to guide the search methods [6,11,21,23,25–27]. These methods are followed by introducing perturbations by considering alternative words that share the same part of speech, and a minimum level of a similarity, following that order.

In [19], the algorithm evaluates the importance of each word w_i based on two factors. Firstly, its sensitivity when replaced by "unknown," and secondly, the maximum change $\Delta f^*(w_i^*)$ in the output scores for each of its candidate synonym words, where w_i^* is the best substitution word for w_i. Synonyms are generated through the WordNet lexical database, resulting in a method with a relatively high query count compared to simple greedy-based attacks. To enhance fluency, semantic coherence, and grammatical consistency, several research studies [5, 7,12,14,17,20] leverage BERT's masked language model for word ranking and substitutions. [12] performs perturbations at each position by executing the three actions: insertion, deletion, and merging two words. Subsequently, the positions are ranked based on the best-scoring action using a masked language model. The perturbations are then executed in that order, with each position undergoing the action that yields the highest score. Other attacks employ meta-heuristic search algorithms operating on populations, such as the Genetic Algorithm [1,8,16] Particle Swarm Optimization [28], and simulated annealing [22]. Additionally, beam-search algorithms are implemented in [3] and [15] to expand the search space during attack generation.

In [24], the recommendation is to favor greedy-based algorithms with word importance ranking over population based meta-heuristic algorithms in scenarios where minimizing run-time costs is crucial. This is particularly relevant in situations involving lengthy inputs or when generating these attacks in real-time for adversarial training and data augmentation. Building upon the recommendation in [24] to favor greedy-based algorithms with word importance ranking, our approach is carefully crafted to align with this recommendation. By doing so, we aim to strike a balance between computational efficiency and the generation of effective adversarial attacks.

3 Problem Formulation

As input, we have a text X having a sequence of words $X = [w_0, w_1, ..., w_n]$, and a pre-trained model to map each input X with a corresponding label $Y_j \in \{Y_0, Y_1, ..., Y_n\}$ as the classification output. Our objective is a text X_{adv} that is similar to X_0 yet having a different output class. Mathematically, it can be formulated by Eq. 1.

$$\begin{cases} X_{adv} = (X + \delta X) \wedge (\delta X \leq \epsilon) \\ (argmax(f(Y \mid X_{\text{adv}})) \neq Y_g) \wedge (argmax(f(Y \mid X_)) = Y_g) \end{cases} \tag{1}$$

Y_g is the ground truth label or true class of the input text, ϵ is the similarity threshold between the original and the adversarial text. It is essential for a text perturbation to adhere to lexical, grammatical, and semantic constraints to ensure stealthiness and imperceptibility to human eyes. The attack strategy outlined in Eq. 1 is categorized as an untargeted attack, where the primary objective is to misclassify the input into any class other than the true one.

4 Methodology: Heuristic Search Algorithm

In this section, we present our Heuristic Search Algorithm designed for generating textual adversarial attacks, incorporating a novel Replacement Order Strategy. The primary goal of this heuristic search is to enhance the success rate of the attack and decrease the accuracy under attack of the targeted model, all while maintaining semantic similarity and adhering to syntactic and grammatical correctness.

4.1 The Generation of Candidates Replacement Words

For each word w_i, we generate replacement words, or synonyms, through employing a) word embeddings such as Glove [10] to get the set of words S_i that share contextual similarities with w_i, or b) Using the WordNet lexical database, as seen in [19]. To ensure grammatical correctness, candidate words that do not share the same part of speech as w_i are filtered out [10]. Additionally, common stop words are excluded from consideration, as they do not serve as potential perturbations.

Algorithm 1. Heuristic Search

Input : k: Number of top words to take, X: Input text
Output : X_{adv}: Adversarial text
Begin : $m \leftarrow 0$, $X_m \leftarrow X$, $W^m \leftarrow \{w_1 \dots w_n\}$, $W_p \leftarrow \emptyset$, $Y_g \leftarrow argmax f(Y \mid X)$
do
    get the best k positions W^m_{topK} from W^m via Eq. 2
    for each position $i \in W^m_{topK}$ corresponding to the word w_i **do**
        $S_i^m \leftarrow \{syn_i^1 \dots syn_i^j \dots syn_i^u\}$ ▷ generate the candidate synonyms
    end for
    $S^m \leftarrow \bigcup_{i=1}^{k} S_i^m$ ▷ collect all the candidate synonyms of all the k positions
    $S^m \leftarrow sort(S^m)$ ▷ sort all candidate synonyms in S^m by importance via Eq. 3
    for each synonym word $syn_i^j \in S^m$ and $i \notin W_p$ **do**
        $X_{new} \leftarrow ((X_m \setminus X_m[i]) \cup syn_i^j)$ ▷ replace syn_i^j at position i
        if $f(Y_g \mid X_{new}) < f(Y_g \mid X_m)$ **then**
            $X_m \leftarrow X_{new}$ ▷ the perturbation is accepted
            $W_p \leftarrow W_p \cup \{i\}$ if i is not in W_p ▷ add the position to perturbations
        end if
        If $\arg\max f(Y \mid X_m) \neq Y_g$: return X_m
    end for
    $W^{m+1} \leftarrow W^m \setminus W^m_{topK}$, $m \leftarrow m + 1$
while $W^m \neq \emptyset$
End

4.2 New Replacement Order Method

Choosing the order of perturbations plays a crucial role in determining the overall success of the attack. The greedy algorithms presented in Sect. 2 employ a predetermined replacement order, based on the positions of the input words available for perturbations, which may exhibit a static nature. Adhering to such a fixed order for word replacement can lead to local optima and excessive word substitutions, as it neglects the context change when the text is being perturbed. Moreover, sorting positions of the input text by importance neglects the influence and impact of the candidate replacement words. While some positions w_i may have a relatively higher importance score, their candidate words w_i^j might not necessarily contribute to reducing the output confidence score of the golden label.

Our strategy seeks to enhance the attack's effectiveness, by recognizing that introducing dynamism in the selection of perturbed words significantly contributes to the overall success of the adversarial attack. Additionally, instead of solely ranking positions of words [10,12–14,17,19], our method prioritizes the ranking of synonym words, incorporating both the importance of the position within the text and the significance of each synonym word, where the ranking is performed iteratively across generations m, not solely before the initiation of the attack, taking into account the iterative context change each time perturbations are processed.

The following steps outline this process:

1. With m representing the current generation initialized at 0, X_m denoting the current text initialized by $X_0 = [w_1, w_2, \dots, w_n]$, and W^m representing the positions that are currently not visited or processed yet by the search algorithm, initialised by $W^0 \leftarrow \{w_1, w2, \dots w_n\}$;
2. For each iteration m, we choose the top k positions (words in the text) W^m_{topK} from W^m having the highest importance scores, via Eq. 2. These positions represent the current generation. We employ the 'leave one out' relevance estimation method, following the approach outlined in TextFooler [10], to calculate the importance scores of positions in case of black-box setting. In the case of white-box settings, we utilize the magnitude of the gradient of A2T-2021 [25] (Eq. 2).

$$\begin{cases} \text{black-box}: I(w_i) = f(Y_g \mid X_m) - f(Y_g \mid (X_m \setminus w_i)) \\ \text{white-box}: I(w_i) = \|\nabla_{ei} L(\theta, X_m, Y_g)\|_2 \end{cases} \tag{2}$$

3. For each word in the current generation $w_i \in W^m_{topK}$, where i denotes the position in the text X_m, we generate its candidate replacement set of synonym words $S^m_i = \{syn^1_i \dots syn^j_i \dots syn^u_i\}$. For each synonym word $syn^j_i \in S^m_i$, we calculate its importance score $I(syn^j_i)$ as the reduction of the output confidence score of the true label Y_g when the candidate word is substituted into the current text (Eq. 3).

$$I(syn^j_i) = f(Y_g \mid X_m) - f(Y_g \mid ((X_m \setminus \{w_i\}) \cup \{syn^j_i\})) \tag{3}$$

4. We optimize the search space by excluding synonyms with positive scores $I(syn^j_i) < 0$ when replaced in the text. This approach guarantees that only synonyms contributing to reducing the output score of the golden label are considered. Synonyms with negative scores may not have a significant impact on lowering the output score, potentially hindering the adversarial attack generation process and increasing query numbers;
5. We arrange the accepted candidate replacement synonyms by their importance scores, resulting in a mixed set of ordered synonyms $S^m \leftarrow Sort(\bigcup_{i=1}^{k} S^m_i)$.

4.3 The Greedy Attack

For each iteration, and after sorting perturbation words by importance following Sect. 4.2, begin the process of greedily replacing each candidate synonym $syn_i^j \in S^m$ into its position w_i. If the score of the new perturbed text improves with respect to the objective, the candidate word syn_i^j is accepted, and the current text's i'th position becomes $X_m[i] \leftarrow \{syn_i^j\}$. Subsequently, the remaining candidate synonyms of position i are discarded. At each perturbation, the newly generated text sample undergoes a similarity constraint using Google's Universal Sentence Encoder [2] to ensure semantic similarity between the original and perturbed texts by comparing their embedding.

After replacing each candidate word from the sorted set of synonyms S^m, the search algorithm stops if an adversarial sample $X_m = X_{adv}$ is successfully crafted, otherwise, a new generation is initiated.

In the new generation $m+1$, we eliminate the positions that have been visited $W^{m+1} = W^m \setminus W_{topK}^m$. Subsequently, the set of the next top k positions W_{topK}^{m+1} is calculated from the set W^{m+1}.

The next step involves computing a new generation of candidate synonym words S^{m+1} following the methodology outlined in Sect. 4.2. This process is followed by sorting the candidate synonym words by their importance scores to create S^{m+1}. The algorithm then iteratively continues this process until an adversarial sample is successfully generated or a specified termination condition is met.

The importance ranking of candidate synonyms in our approach differs from methods such as pwws [19], Clare [12], and others. Unlike these static methodologies, our ranking is dynamic and changes by each generation. This adaptability and contextual awareness contributes significantly to the effectiveness of our approach in generating adversarial samples.

5 Experiments

To validate the efficiency of our search algorithm, we performed experiments on IMDB and AG-News datasets against fine-tuned victim models: Bert-Base-Uncased and Distilbert. To ensure a fair comparison, the evaluation of the algorithms is done under identical constraints, utilizing the same datasets and victim models. These experiments provided confirmation of the effectiveness of our proposed algorithm for generating adversarial attacks. Additionally, supplementary experiments were conducted on adversarial training to illustrate the efficacy of our attack in enhancing models robustness against various types of adversarial attacks. These conducted experiments are done under [18]'s TextAttack, an open-source Python framework for generating and testing adversarial algorithms in the NLP field.

Table 1. Comparison with baseline algorithms, on AG's News and IMDB with fine-tuned BERT and DistilBert. (Acc/Atack: Accuracy under attack, Succ: Success rate of the attack.)

Victim Model and Constraints	Database	Method	Acc%	Acc/Attack%	Succ%	Query
Bert - WordNet Lexical Database No Stop-Words Modifications	IMDB	Pwws-2019	94.3%	29.0%	68.48%	1493.6
		Ours		**6.0%**	**92.39%**	**1056.07**
	Ag-News	Pwws-2019	94.2%	**52.0%**	**46.94%**	272.27
		Ours		60.0%	38.78%	**141.59**
DistilBert - WordNet Lexical Database No Stop-Words Modifications	IMDB	Pwws-2019	93.5%	29.0%	68.48%	1493.6
		Ours		**3.1%**	**96.69%**	**555.18**
Bert - Word Embedding(50) Cosine Similarity(0.5) Universal Sentence Encoder(0.84) No Stop-Words Modification	IMDB	TextFooler-2020	94.3%	13.5%	85.68%	**603.53**
		Ours		**5.4%**	**94.27%**	857.51
		Gradient-A2T-2021		6.8%	92.79%	**354.12**
		Ours (white-box)		**5.5%**	**94.09%**	370.12
	Ag-News	TextFooler-2020	94.2%	56.2%	40.34%	**202.16**
		Ours		**46.2%**	**50.96%**	304.47
		Gradient-A2T-2021		47.5%	49.58%	**160.85**
		Ours (white-box)		**47.0%**	**50.26%**	247.08
DistilBert - Word Embedding(50) Cosine Similarity(0.5) Universal Sentence Encoder(0.84) No Stop-Words Modification	IMDB	TextFooler-2020	93.7%	5.1%	94.56%	**409.18**
		Ours		**1.8%**	**98.08%**	592.15%
	Ag-News	TextFooler-2020	94.4%	58.5%	38.03%	**196.03**
		Ours		**48.2%**	**48.94%**	301.2

5.1 Experimental Setup

Our search algorithm is compared with existing algorithms under these metrics: The attack success rate as the percentage of successfully crafted attacks that mislead the model, the accuracy under attack, which is the model's classification accuracy when subjected to generated adversarial examples, and finally the query count required to successfully craft the text sample.

5.2 Attack Results and Analysis

In Table 1, we present the performance of our attack against several baseline strategies in both White-Box and Black-Box settings. Specifically, we compare our results with A2T [25] for the White-Box setting, and TextFooler [10] and PWWS [19] for the Black-Box setting. The best results are highlighted in bold. Our heuristic search consistently demonstrates positive outcomes across diverse scenarios. Notably, it exhibits a reduction in accuracy under attack and an improvement in the attack success rate in most cases. When contrasted with greedy-based attacks like TextFooler and A2T, our attack achieves a higher success rate, at the cost of a higher query count. In direct comparison to PWWS's algorithm, our approach showcases superior results in terms of success rate, lower accuracy under attack, and a significantly lower required query count in most cases. These findings highlight that our method strikes a favorable balance in terms of query count, achieving a competitive edge by simultaneously optimizing for higher attack success rates and lower accuracy under attack.

5.3 Adversarial Training Results and Analysis

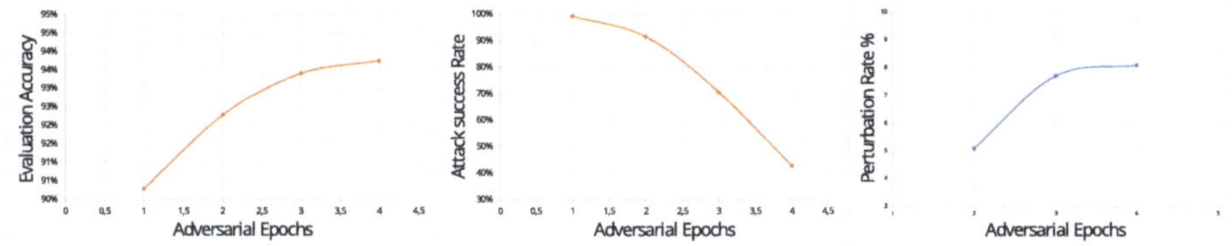

Fig. 1. The Evaluation accuracy, Attack success rate and Perturbation rate across the adversarial training epochs.

In this section, we present our experiments on adversarial training, illustrating the influence of refining adversarial attack generation in enhancing the robustness of deep learning models. The experiments conducted involve adversarial training on the BERT model, utilizing the IMDB dataset for this study. The training process consists of a combination of one clean epoch and four adversarial epochs. In each adversarial epoch, we generate 4000 adversarial examples. The adversarial training is carried out with a learning rate of 5e−5. Through adversarial training, we observed a consistent improvement in the performance of the trained model across the adversarial epochs. Starting with an initial accuracy, there was a steady increase in the evaluation accuracy as we progressed through the adversarial epochs. This substantial improvement not only underscores the efficacy of adversarial training in enhancing model robustness against adversarial attacks but also highlights its positive impact on the model's generalization to clean data. Additionally, concurrent with the notable enhancement in the evaluation accuracy, we observed a decline in attack success rates throughout the adversarial training epochs (see Fig. 1). Furthermore, our observations indicate that successful attacks necessitate more perturbations as the adversarial training progresses (see Fig. 1). Indeed, the model becomes increasingly adept at resisting attacks and consequently demands a higher degree of perturbation for an attack to succeed.

6 Conclusion

In this study, our primary aim is to expose vulnerabilities in NLP models through the more efficient generation of adversarial samples, ultimately achieving a heightened success rate in targeted classes. Our noteworthy contribution centers around the introduction of a novel word-level search algorithm specifically designed for the generation of adversarial attacks. We present an innovative replacement order strategy that prioritizes ranking candidate synonyms in the most promising positions within texts. Comprehensive experiments were conducted on the IMDB and Ag-News datasets, employing attacks in both black-box and white-box scenarios, to illustrate the efficacy of our algorithm across diverse conditions. Our findings reveal that our method attains the best success rate across all scenarios while maintaining an acceptable query count. Additionally, supplementary experiments were performed to underscore the effectiveness of our attack in fortifying model robustness through adversarial training.

References

1. Alzantot, M., Sharma, Y., Elgohary, A., Ho, B.J., Srivastava, M., Chang, K.W.: Generating natural language adversarial examples. arXiv preprint arXiv:1804.07998 (2018)
2. Cer, D., et al.: Universal sentence encoder. arXiv preprint arXiv:1803.11175 (2018)
3. Ebrahimi, J., Rao, A., Lowd, D., Dou, D.: HotFlip: white-box adversarial examples for text classification. arXiv preprint arXiv:1712.06751 (2017)
4. Fu, X., Gu, Z., Han, W., Qian, Y., Wang, B.: Exploring security vulnerabilities of deep learning models by adversarial attacks. Wirel. Commun. Mob. Comput. **2021**(1), 9969867 (2021)
5. Garg, S., Ramakrishnan, G.: BAE: BERT-based adversarial examples for text classification. arXiv preprint arXiv:2004.01970 (2020)
6. Guo, C., Sablayrolles, A., Jégou, H., Kiela, D.: Gradient-based adversarial attacks against text transformers. arXiv preprint arXiv:2104.13733 (2021)
7. Harbecke, D., Alt, C.: Considering likelihood in NLP classification explanations with occlusion and language modeling. arXiv preprint arXiv:2004.09890 (2020)
8. Jia, R., Raghunathan, A., Göksel, K., Liang, P.: Certified robustness to adversarial word substitutions. arXiv preprint arXiv:1909.00986 (2019)
9. Jin, D., Jin, Z., Zhou, J.T., Szolovits, P.: TextFool: fool your model with natural adversarial text (2019)
10. Jin, D., Jin, Z., Zhou, J.T., Szolovits, P.: Is BERT really robust? A strong baseline for natural language attack on text classification and entailment. In: Proceedings of the AAAI Conference on Artificial Intelligence, vol. 34, pp. 8018–8025 (2020)
11. Lee, D., Moon, S., Lee, J., Song, H.O.: Query-efficient and scalable black-box adversarial attacks on discrete sequential data via Bayesian optimization. In: International Conference on Machine Learning, pp. 12478–12497. PMLR (2022)
12. Li, D., et al.: Contextualized perturbation for textual adversarial attack. arXiv preprint arXiv:2009.07502 (2020)
13. Li, J., Ji, S., Du, T., Li, B., Wang, T.: TextBugger: generating adversarial text against real-world applications. arXiv preprint arXiv:1812.05271 (2018)
14. Li, L., Ma, R., Guo, Q., Xue, X., Qiu, X.: BERT-attack: adversarial attack against BERT using BERT. arXiv preprint arXiv:2004.09984 (2020)
15. Liu, H., Yu, J., Li, S., Ma, J., Ji, B.: A context-aware approach for textual adversarial attack through probability difference guided beam search. arXiv preprint arXiv:2208.08029 (2022)
16. Maheshwary, R., Maheshwary, S., Pudi, V.: Generating natural language attacks in a hard label black box setting. In: Proceedings of the AAAI Conference on Artificial Intelligence, vol. 35, pp. 13525–13533 (2021)
17. Malik, V., Bhat, A., Modi, A.: Adv-OLM: generating textual adversaries via OLM. arXiv preprint arXiv:2101.08523 (2021)
18. Morris, J.X., Lifland, E., Yoo, J.Y., Grigsby, J., Jin, D., Qi, Y.: TextAttack: a framework for adversarial attacks, data augmentation, and adversarial training in NLP. arXiv preprint arXiv:2005.05909 (2020)

19. Ren, S., Deng, Y., He, K., Che, W.: Generating natural language adversarial examples through probability weighted word saliency. In: Proceedings of the 57th Annual Meeting of the Association for Computational Linguistics, pp. 1085–1097 (2019)
20. Shi, Z., Huang, M.: Robustness to modification with shared words in paraphrase identification. arXiv preprint arXiv:1909.02560 (2019)
21. Wang, B., Xu, C., Liu, X., Cheng, Y., Li, B.: SemAttack: natural textual attacks via different semantic spaces. arXiv preprint arXiv:2205.01287 (2022)
22. Yang, X., Liu, W., Tao, D., Liu, W.: BESA: BERT-based simulated annealing for adversarial text attacks. In: IJCAI, pp. 3293–3299 (2021)
23. Ye, M., Miao, C., Wang, T., Ma, F.: TextHoaxer: budgeted hard-label adversarial attacks on text. In: Proceedings of the AAAI Conference on Artificial Intelligence, vol. 36, pp. 3877–3884 (2022)
24. Yoo, J.Y., Morris, J.X., Lifland, E., Qi, Y.: Searching for a search method: benchmarking search algorithms for generating NLP adversarial examples. arXiv preprint arXiv:2009.06368 (2020)
25. Yoo, J.Y., Qi, Y.: Towards improving adversarial training of NLP models. arXiv preprint arXiv:2109.00544 (2021)
26. Yu, Z., Wang, X., Che, W., He, K.: TextHacker: learning based hybrid local search algorithm for text hard-label adversarial attack. arXiv preprint arXiv:2201.08193 (2022)
27. Yuan, L., Zhang, Y., Chen, Y., Wei, W.: Bridge the gap between CV and NLP! A gradient-based textual adversarial attack framework. arXiv preprint arXiv:2110.15317 (2023)
28. Zang, Y., et al.: Word-level textual adversarial attacking as combinatorial optimization. arXiv preprint arXiv:1910.12196 (2019)

DDPG-PER for an IRS-Aided Secure Wireless Communication

Amina Lammari(✉), Naïma Bessah, and Anfal Bourouina

Higher School of Computer Science, LCSI Laboratory, Algeria,
BP 68M Oued Smar, Algiers, Algeria
{a_lammari,n_bessah,ka_bourouina}@esi.dz
http://www.esi.dz

Abstract. Improving the security of communications data in a wireless network is one of the most important uses of reconfigurable intelligent surfaces (IRS), especially in next generation networks like 6G. In this context, we propose an approach based on deep a reinforcement learning (DRL) algorithm to determine the optimal phase shift of IRS elements. We used the Deep Deterministic Policy Gradient (DDPG) variant, which adapts to continuous data and improves the secrecy rate by considering, in the algorithm, the best sample obtained via a Prioritized Experiment Replay (PER).

Keywords: Intelligent Reflective Surface (IRS) · 5G/6G · Data security. · DDPG-PER

1 Introduction

In recent years, efforts and research have focused on developing fifth-generation (5G) mobile wireless networks to provide ubiquitous connections, regardless of time, location, or quality of the communication channel. Meanwhile, 6G was introduced with the aim of extending 5G's capabilities to higher thresholds where millions of connected devices and applications would be able to operate seamlessly with high reliability, low latency, and high bandwidth. Nevertheless, a new paradigm, called Intelligent Reflective Surface (IRS), has been proposed as a promising technique for guaranteeing increased spectral and energy efficiency and enhance privacy rates in fifth-generation (5G) wireless communication systems and beyond. This uniform network of passive reflectors enables each element to adjust the amplitude and/or phase of an incident signal, and thus control the propagation channel [1,2]. Thus, IRS has been presented in recent studies as a promising technology also dedicated to physical layer security [5] in order to ensure maximum confidentiality of communications. In the same spirit of data security over IRS, a number of research studies were carried out on various architectures and environmental systems [4,14,15]. The essential aim was to jointly optimize active transmission and passive formation of reflected beams in order to maximize the secrecy rate.

B. Djamaa et al. (Eds.): CSA 2024, LNNS 1145, pp. 292–301, 2025.
https://doi.org/10.1007/978-3-031-71848-9_27

As the formulated problem is a non-convex one, the research work resorted the use of two sub-optimal alternating optimization (AO) algorithms to reduce complexity, and thus proved that the IRS was able to achieve a better secrecy rate. However, with the emergence of machine learning methods, and in particular deep reinforcement learning (DRL) methods dedicated to solving secrecy rate, user throughput, and power optimization problems, proposals have been put forward [6,12]. These approaches based on artificial neural networks have the particularity of learning and optimizing decision-making processes in complex and dynamic environments. In the context of data security through IRS, deep reinforcement learning (DRL) has emerged as a powerful approach for optimizing IRS phase shifts and overcoming the aforementioned problems associated with AO techniques.

In this paper, we propose a Deep Deterministic Policy Gradient (DDPG) algorithm as a deep reinforcement learning (DRL) agent that takes into account data continuity and overcomes the difficulty of the state-space dimension. Adding prioritization to DDPG algorithm in experience replay buffer has led to a considerable improvement in the secrecy rate. This is the purpose of this study.

Organization of the article: Sect. 2 presents the system model. The proposed DDPG-PER algorithm is presented in Sect. 3, as well as the parameterization of the neural networks. The simulation results and the conclusion are presented in Sects. 4 and 5, respectively.

2 System Model

The modeling of the system concerns: the environment description and (i) the modeling of the transmission channel, (ii) the formulation of the reflection coefficients of the IRS, and (iii) the signals between the base station/IRS and those between the IRS/users VS eavesdroppers, respectively.

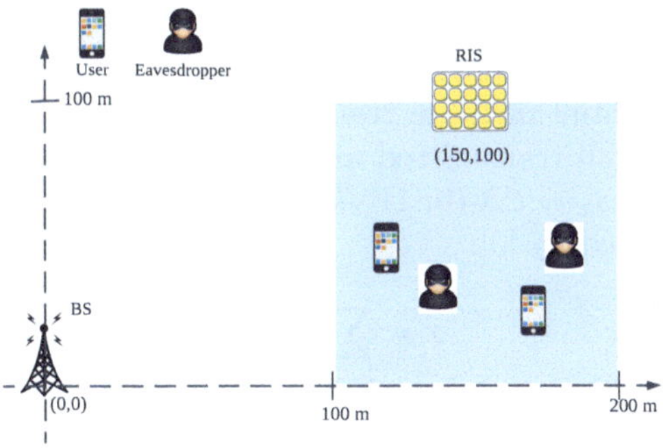

Fig. 1. IRS-assisted secure communication for multiple users and eavesdroppers.

We consider a MISO system assisted by an IRS, as illustrated in Fig. 1. The base station has M antennas, the IRS has N passive reflecting elements, and a central controller between the IRS and the base station. The controller allows joint configuration of IRS phase shifts and beamforming at the base station. The system operates in downlink, where K mobile users each have a single antenna and receive signals from the base station in the presence of P eavesdroppers. These are each equipped with a single antenna, and attempt to listen to the exchanges between the base station and mobile users. For transmission channel modeling, we'll assume: $\mathcal{K} = \{1, 2, \dots, K\}$, $\mathcal{P} = \{1, 2, \dots, P\}$ and $\mathcal{N} = \{1, 2, \dots, N\}$ the users set, the eavesdropper set, and the IRS elements set, respectively. Let's note by: $H_{BR} \in \mathbb{C}^{N\times M}$, $h_{BU,k} \in \mathbb{C}^{1\times M}$, $h_{BE,p} \in \mathbb{C}^{1\times M}$, $h_{RU,k} \in \mathbb{C}^{1\times N}$, $h_{RE,p} \in \mathbb{C}^{1\times N}$ the channels between the base station and the IRS, the base station and the kth user, the base station and the pth eavesdropper, the IRS and kth user, and the IRS and pth eavesdropper, respectively. In this scenario, the transmission occurs in an urban environment, where the IRS is installed on a wall or building facade, with a few reflectors in the surrounding area. We also assume that all the above-mentioned channels undergo fading following a Rician distribution with pathloss [4,12].

The IRS is made of passive elements, thus, the phase shift matrix of the IRS elements, denoted by Φ, is expressed by:

$$\Phi = diag(\beta_1 e^{j\theta_1}, \beta_2 e^{j\theta_2}, \dots, \beta_N e^{j\theta_N}) \tag{1}$$

The notation $diag$ designates a diagonal matrix, and $\forall i = 1, \dots, N$, $\beta_i \in [0, 1]$ and $\theta_i \in [0, 2\pi]$ represent the amplitude and phase of the signal reflected through the IRS, respectively. It should be noted that each phase is configured to ensure maximum reflection [4], i.e., $\beta_n = 1, \forall n \in \mathcal{N}$.

The beamforming at the base station is expressed by the following matrix: $W = diag[w_1, w_2, \dots, w_K] \in \mathbb{C}^{M\times K}$, where $w_{k\in\mathcal{K}} \in \mathbb{C}^{M\times 1}$ refers to the beamforming vector of the kth user. Let P_{max} denote the base station's maximum transmission power, the station's beamforming must satisfy the following constraint [7,12]:

$$Tr(WW^H) \leq P_{max} \tag{2}$$

where W^H is the Hermitian matrix of W.

Let s_k the confidential message transmitted by the base station to the kth user. We assume that all transmitted messages follow a complex circular Gaussian distribution, i.e., $s_k \sim \mathcal{CN}(0, 1)$ $\forall k \in \mathcal{K}$. The transmitted signal from the base station is expressed by [3]:

$$x = \sum_{k\in\mathcal{K}} w_k s_k \tag{3}$$

The kth user canal h_k and the pth eavesdropper canal h_p are expressed by:

$$h_k = h_{RU,k}\Phi H_{BR} + h_{BU,k} \quad \text{and} \quad h_p = h_{RE,p}\Phi H_{BR} + h_{BE,p} \tag{4}$$

Thus, the signal received at the kth user and the signal received at the pth eavesdropper (who tries to eavesdrop on the exchanges of all K user channels) are given respectively by:

$$y_k = h_k w_k s_k + \underbrace{\sum_{i \in \mathcal{K}, i \neq k} h_k w_i s_i}_{\text{interference between-users}} + n_k \tag{5}$$

$$y_p = h_p \sum_{k \in \mathcal{K}} w_k s_k + n_p \tag{6}$$

where n_k and n_p are additive complex Gaussian noise (AWGN) with variance σ_k^2 and σ_p^2 respectively [11,15]. The secrecy rate R_k^{sec} of the kth user whose channel is subject to eavesdropping by one or more eavesdroppers is formulated according to the achievable throughput R_k^u and $R_{p,k}^e$, respectively, of the kth user and the pth eavesdropper (who spies the kth user); they are expressed as follows (Shannon-Hartley Theorem) [15]:

$$R_k^u = log_2 \left(1 + \frac{|(h_k w_k|^2}{|\sum_{i \in \mathcal{K}, i \neq k} (h_k w_i)|^2 + \sigma_k^2} \right) \tag{7}$$

$$R_{p,k}^e = log_2 \left(1 + \frac{|(h_p w_k|^2}{|\sum_{i \in \mathcal{K}, i \neq k} h_p w_i)|^2 + \sigma_p^2} \right) \tag{8}$$

$$R_k^{sec} = \max[R_k^u - \underset{\forall p}{max} R_{p,k}^e, 0] \tag{9}$$

2.1 Problem Modeling and Solution Approach

In this work, our purpose is to jointly optimize the beamforming matrix W and the phase-shift matrix Φ in order to maximize the sum of the secrecy rates of all users under the constraint of the unity modulus of the phase shifts as well as the constraint of the base station transmit power (2) and (1). The optimization problem is expressed as :

$$\begin{cases} \max\limits_{W, \Phi} \sum_{k \in \mathcal{K}} R_k^{sec} \\ \text{s.t. } Tr(WW^H) \leq P_{max}, \ |\beta_n e^{j\theta_n}| = 1, \ 0 \leq \theta_n \leq 2\pi, \ 1 \leq n \leq N. \end{cases} \tag{10}$$

This problem is considered difficult because of the non-convexity of the objective function and the great number of variables to be optimized [4,12], and [15].

Machine learning (ML) methods are an efficient alternative, in view of the simplicity of their implementation and the low cost of the proposed solutions. Furthermore, by considering the problem as a reinforcement learning one, we can avoid the difficulty of labeling the entire data set; DRLs learn directly from the environment.

3 Deep Reinforcement Learning for Data Security

The data space is multidimensional and continuous, so we opted for the Deep Deterministic Policy Gradient (DDPG) model to solve the optimization problem of Eq. (10). The elements of DRL are described below:

1. Environment and Agent are, respectively, the IRS-supported communication system and the central controller between the IRS and the base station (it configures phase shifts and beamforming).
2. States and Actions: We denote the state space by $\mathcal{S}$; its elements describe the environment, and are: $\left\{h_k^{(t)}\right\}_{k\in\mathcal{K}}, \left\{h_p^{(t)}\right\}_{p\in\mathcal{P}}$ the user's and the eavesdropper's channels at time t, the sum of the user's throughputs $\sum_{k\in\mathcal{K}} R_k^{(t-1)}$, the sum of the eavesdropper's throughputs $\sum_{p\in\mathcal{P}} R_p^{(t-1)}$, the sum of the secrecy rates $\sum_{k\in\mathcal{K}} R_k^{sec(t-1)}$, and the action at time $(t-1)$, $\left\{W^{(t-1)}\right\}, \left\{\Phi^{(t-1)}\right\}$. The action space $\mathcal{A}$ has elements: $a_t = \{\{w_k^{(t)}\}_{k\in K}, \{\theta_n^{(t)}\}_{n\in N}\}$.
3. Reward: The reward $r_t(s_t, a_t)$ used is the value of the objective function of the optimization problem (10). The aim of the RL agent is to find the optimal policy π^* that maximizes the long-term cumulative reward R_t, defined by:

$$r_t(s_t, a_t) = \sum_{k\in\mathcal{K}} R_k^{sec(t)} \tag{11}$$

$$= \sum_{i=t}^{\infty} \gamma^{(i-t)} r_i \tag{12}$$

   where $\gamma \in [0, 1]$ refers to the actualization factor. This parameter determines the importance to be assigned to both immediate and future rewards, and through which we guarantee the convergence of cumulative returns in infinite horizon configurations.
4. The state-action value function (Q-function), used in RL algorithms to measure the expected long-term reward from a pair (s_t, a_t) under a given policy π, is $Q^\pi(s_t, a_t)$ [10]. The optimum policy $\pi^\star$ is recursively obtained by the Bellman equation [9] and the corresponding optimal policy $\pi^\star$ can therefore be expressed as:

$$Q^\pi(s_t, a_t) = \mathbb{E}_\pi[R_t | s_t = s, a_t = a] \tag{13}$$

$$Q^\star(s_t, a_t) = r_t + \gamma \max_{a_{t+1}\in\mathcal{A}} Q^\star(s_{t+1}, a_{t+1}) \tag{14}$$

$$Q^\star(s, a) = \arg\max_{a\in\mathcal{A}} Q^\star(s, a) \tag{15}$$

3.1 Description of DDPG Model

In the DDPG model, policy and value functions are approximated by deep neural networks (DNN). The actor network (called policy) and the critical network (also called Q network) are described below. DDPG uses two other networks, called target actor networks and target critical networks, to periodically update the weights of the actor and critical networks, which helps stabilize and improve the learning process.

Actor Network: The actor is in charge of generating the action choices (policy) a_t for each state s_t. It takes the state s_t as input and then generates the following action to perform as output: $a_t = \mu(s_t|\theta_\mu) + \mathcal{N}$, where θ_μ is the network parameter and $\mathcal{N}$ is the Ornstein-Uhlenbeck noise. This random process is very effective for continuous environments and is added to improve the exploration of the action space. The actor aims to maximize the state-value function of the equation (14).

Critical Network: with parameter θ_Q, evaluates the actor's choices. It takes as input the state s_t and the action chosen by the actor a_t and returns the value of the Q-function of (13). The goal of the critical network is to minimize the error between the Q-value obtained and the Q-value expected (target) through the cost function:

$$L(\theta_Q) = (y_t - Q(s_t, a_t|\theta_Q))^2 \tag{16}$$

$$y_t = r(s_t, a_t) + \gamma \max_{a_{t+1} \in \mathcal{A}} Q(s_{t+1}, a_{t+1}|\theta_Q) \text{ (the expected Q-value).} \tag{17}$$

The Actor and Critical neural networks use a fully connected feed-forward architecture with a ReLU activation function, at each hidden layer. At the output of the Actor network, a tanh activation function is applied to process negative elements and delimit the action, as described in [9] and [12]. The updating of DDPG network parameters are based on the gradient, so the optimizer used is ADAM [8]. Furthermore, the cost function used to calculate the error of the Critical (16) is the mean-square error MSE [9].

A normalization step is added to the input of each of the neural networks, since the elements of the state s_t and the action (phase shifts) have different scales. Furthermore, the output of the tanh activation function of the Actor lies in the interval $[-1, 1]$. To ensure compliance with the constraints of the problem (10), especially those relating to the maximum reflection of the IRS and the maximum power at the base station, the action (phase shift matrix, and beamforming matrix) needs to be normalized. For this purpose, normalization step is added to the Actor output, so that:

$$\begin{cases} |\Phi| = 1 & \text{and} \\ W = \dfrac{\sqrt{P_{max}}}{\sqrt{Tr(WW^H)}} W \text{ if } & Tr(WW^H) > P_{max} \end{cases} \tag{18}$$

3.2 Prioritized Experience Replay (PER)

The replay buffer stores experiences (tuples: (s_t, a_t, r_t, s_{t+1})) and updates them after each interaction with the environment. The agent in DDPG adjusts the weights of the actor and critical networks by learning from mini-batches of experiments kept in the replay buffer. Experiments included in the mini-batch are usually chosen at random, which encourages exploration but ignores the importance of values. Thus, we present a prioritization based on the TD error (Temporal difference error $\delta_i = (y_i - Q(s_i, a_i|\theta_Q)))$ to obtain a better selection of experiments. Thanks to this method, the experiments are prioritized depending on how they maximize rewards [13,15]. The probability $P(i)$ of selecting experience i (depending on the value of the absolute TD error) is given by:

$$P(i) = \frac{|\delta_i + \epsilon|^\alpha}{\sum_j |\delta_j + \epsilon|^\alpha} \tag{19}$$

where the exponent α determines the degree of prioritization used and ϵ is a small positive constant preventing the zero value of $P(i)$. Furthermore, we adopt Importance-Sampling weights as follows [13]:

$$W(i) = \left(\frac{1}{D} \cdot \frac{1}{P(i)}\right)^\beta \tag{20}$$

D represents the size of the replay buffer, and the parameter β is used to adjust the amount of correction used. Thus, the cost function (16) that the critic's attempt to minimize is:

$$L(\theta_Q) = \frac{1}{D}\left(\sum_{i=1}^{D} W(i)(y_i - Q(s_i, a_i|\theta_Q))^2\right) \tag{21}$$

4 Tests and Results

In this part, we first validate the applicability of the DDPG algorithm to our system, then we show the effectiveness of prioritization in improving the secrecy rate. Finally, we analyze the behavior of our algorithm by varying the number of IRS elements and the maximum power of the base station, and compare to the results obtained by (AO).

We consider the hypothesis of Sect. 2 in the case of a single user and a single eavesdropper, whose positions are randomly generated in an environment (schematized by a grid of 100 m × 100 m) including an IRS and a base station, located respectively at (150, 100) and (0, 0), as depicted in Fig. 1.

Three relevant performance parameters are used: The secrecy rate of the kth user described in the Eq. (9), the user and eavesdropper throughput described in the Eqs. (7) and (8) and the last, the average cumulative rewards: $\bar{r} = \frac{1}{T}\sum_{t=1}^{T} r_t(s_t, a_t)$ which allows us to check the convergence and performance of the RL algorithm; T is the number of iterations completed, and $r_t(s_t, a_t)$ is the reward at time t. Table 1 sets the simulation and environment parameters.

In Fig. 2, we analyze the convergence; Fig. 2a illustrates the trends in average rewards and rewards per episode. In each episode, the channels are updated, the agent is exploring the environment and experimenting with different states. After 30 episodes, the average cumulative rewards are stable around an average value of 0.57. Figure 2b shows the user and spy throughputs for episodes 0 and 1, for each of which 40, 000 iterations were used. In episode 0, (the first half of the iterations), the RL agent generates random actions (random policy) to explore the environment through various actions that do not necessarily improve the reward. However, in the last 40, 000 iterations (episode 1) the model begins to learn and apply better policy actions that allow it to increase the user's

Table 1. Environment and learning settings.

Environment settings	Learning Settings.	
Base station transmit power $P_{max} = 15dBm$	learning rate actor network (α_a)	0.001
	critic network learning rate (α_c)	0.001
Noise power $\sigma^2 = -80dBm$	Discount factor (γ)	0.99
Exponent path loss $\alpha = 2.2$ (BS-IRS) 3.0 for the remainder (BS-User, BS-Eve, IRS-User, IRS-Eve.)	Update rate of target networks (τ)	0.001
	Buffer size (D)	100000
	Mini batch size (H)	128
Path loss at reference distance $d_0 = 1m$, $(PL_0) = -30dB$	PER priority level (α_{PER})	0.6
Number of base station antennas $M = 4$	epsilon (ϵ_{PER})	10^{-6}
Number of IRS elements $N = 10$	Noise Settings (θ, σ)	$(0.15, 0.2)$
Rician Factor $K = 1$	Ornstein-Uhlenbeck (OU)	

throughput and degrade the eavesdropper's. So the agent successfully configured both the IRS phase shifts Φ and the beamforming matrix W in such a way as to improve the secrecy rate.

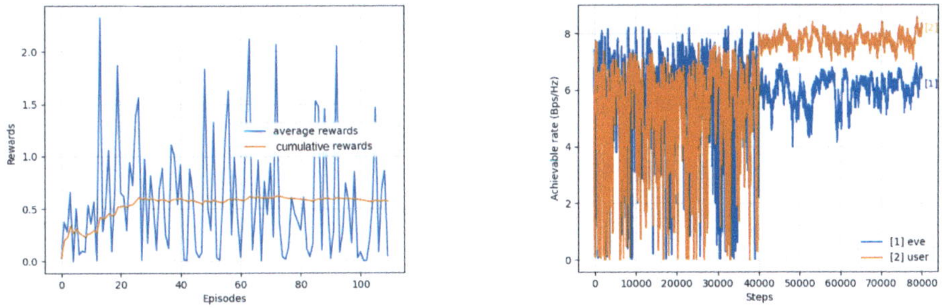

(a) Average rewards.

(b) Achievable throughput.

Fig. 2. Convergence per episode.

The weighted sampling and prioritization mechanism approach improves stability and convergence by assigning an importance weight to each selected experiment. This is illustrated in Fig. 3, where sub-Fig. 3a shows that DDPG-PER converges after only 20 episodes, while DDPG requires 40 episodes. In addition, the comparison of the secret rates of the three algorithms via the box plot of Fig. 3b, highlights some values obtained by DDPG-PER close to those of AO.

In Fig. 4 the use of DDPG maintains the contribution of RIS and its elements to increasing secrecy rate by optimizing phase shifts, increasing emission beam formation, and reflection [4]. DDPG PER outperforms DDPG by varying maximum power and number of RIS elements, as shown in Fig. 4a and Fig. 4b.

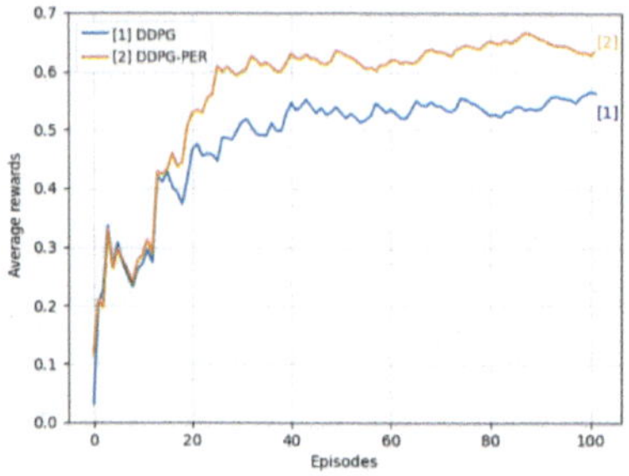

(a) Comparison between DDPG and DDPG-PER.

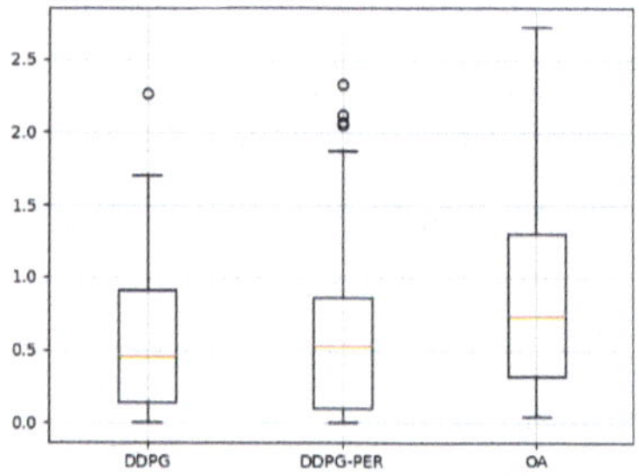

(b) Comparison between DDPG, DDPG-PER and AO.

Fig. 3. Prioritization mechanism.

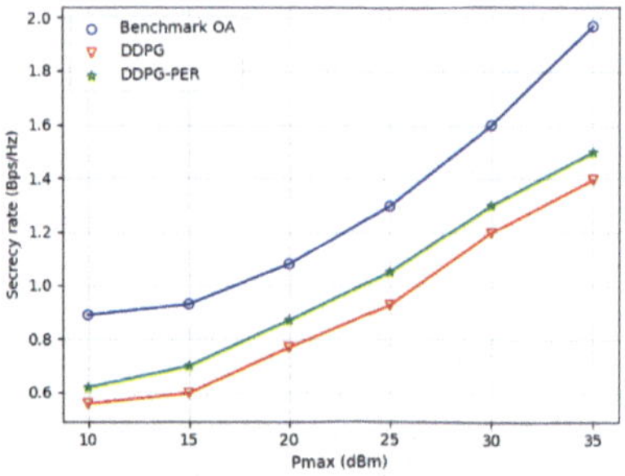

(a) Secrecy rate versus P_{max}.

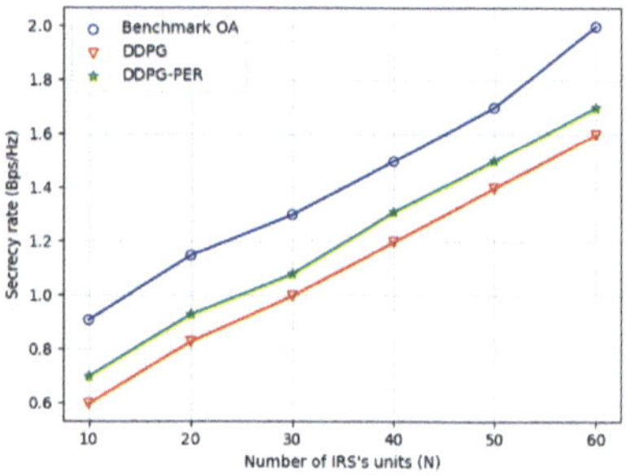

(b) Secrecy rates versus N.

Fig. 4. Variation of Pmax and N.

5 Conclusion

In this article, the problem of maximizing the secrecy rate via an IRS, in an environment comprising a user and an eavesdropper is studied. We applied a deep reinforcement learning algorithm: DDPG-PER using weighting and prioritization of buffer experiments. The results obtained are promising, even if they do not exceed those obtained by alternating optimization (AO), but are comparable to them. The proposed model is flexible and can adapt to several configurations of the environment by increasing the number of users, the number of eavesdroppers but also the transmission power and the number of IRS elements.

References

1. Almohamad, A., et al.: Smart and secure wireless communications via reflecting intelligent surfaces: a short survey. IEEE Open J. Commun. Soc. **1**, 1442–1456 (2020)
2. Basharat, S., Hassan, S.A., Pervaiz, H., Mahmood, A., Ding, Z., Gidlund, M.: Reconfigurable intelligent surfaces: Potentials, applications, and challenges for 6g wireless networks. IEEE Wirel. Commun. **28**(6), 184–191 (2021)
3. Chen, J., Liang, Y.C., Pei, Y., Guo, H.: Intelligent reflecting surface: a programmable wireless environment for physical layer security. IEEE Access **7**, 82599–82612 (2019)
4. Cui, M., Zhang, G., Zhang, R.: Secure wireless communication via intelligent reflecting surface. IEEE Wirel. Commun. Lett. **8**(5), 1410–1414 (2019)
5. Di Renzo, M., et al.: Smart radio environments empowered by reconfigurable intelligent surfaces: how it works, state of research, and the road ahead. IEEE J. Sel. Areas Commun. **38**(11), 2450–2525 (2020)
6. Faisal, K., Choi, W.: Machine learning approaches for reconfigurable intelligent surfaces: a survey. IEEE Access **10**, 27343–27367 (2022)
7. Huang, C., Zappone, A., Alexandropoulos, G.C., Debbah, M., Yuen, C.: Reconfigurable intelligent surfaces for energy efficiency in wireless communication. IEEE Trans. Wireless Commun. **18**(8), 4157–4170 (2019)
8. Kingma, D.P., Ba, J.: Adam: a method for stochastic optimization. arXiv preprint arXiv:1412.6980 (2014)
9. Lillicrap, T.P., et al.: Continuous control with deep reinforcement learning. arXiv preprint arXiv:1509.02971 (2016)
10. Mnih, V., et al.: Human-level control through deep reinforcement learning. Nature **518**(7540), 529–533 (2015)
11. Ngo, Q., Phan, T.K., Mahmood, A., Xiang, W.: Hybrid IRS-assisted secure satellite downlink communications: a fast deep reinforcement learning approach (2022). https://doi.org/10.36227/techrxiv.20478438.v2
12. Peng, Z., Zhang, Z., Kong, L., Pan, C., Li, L., Wang, J.: Deep reinforcement learning for ris-aided multiuser full-duplex secure communications with hardware impairments. IEEE Internet Things J. (2022)
13. Schaul, T., Quan, J., Antonoglou, I., Silver, D.: Prioritized experience replay. arXiv preprint arXiv:1511.05952 (2015)
14. Sheng, Z., Tuan, H.D., Nasir, A.A., Duong, T.Q., Poor, H.V.: Power allocation for energy efficiency and secrecy of wireless interference networks. IEEE Trans. Wireless Commun. **17**(6), 3737–3751 (2018)
15. Yang, H., Xiong, Z., Zhao, J., Niyato, D., Xiao, L., Wu, Q.: Deep reinforcement learning-based intelligent reflecting surface for secure wireless communications. IEEE Trans. Wireless Commun. **20**(1), 375–388 (2020)

Information Systems and Software Engineering

Model Checking-Enhanced Spectrum-Based Fault Localization

Mohammed Bekkouche(✉)

LabRI-SBA Laboratory, École Supérieure en Informatique, Sidi Bel Abbes 22000, Algeria
m.bekkouche@esi-sba.dz
https://www.esi-sba.dz/fr/index.php/personnel/bekkouche-mohammed/

Abstract. Software debugging is crucial in software development, encompassing fault detection, localization, and correction. This article proposes an innovative approach integrating model checking and testing to enhance fault localization accuracy. Our method combines model checking, model checker-based fault localization (MCFL), and spectrum-based fault localization (SBFL), prioritizing program instructions based on suspicion level. Experiments with the TCAS benchmark from Siemens demonstrate the superiority of our integrated approach over traditional spectrum-based methods, especially in challenging versions. Ochiai consistently outperforms Tarantula, and the introduction of the Enhanced Weighted Suspiciousness Ratio (EWSR) spectral metric addresses challenges in specific versions. The primary outcome is a refined fault localization approach significantly improving accuracy compared to traditional methods. Our approach achieves an average 65.47% improvement in fault localization accuracy across TCAS benchmark versions. This integrated fault localization approach represents a significant advancement in software debugging, offering developers an effective method for identifying problematic instructions in faulty programs.

Keywords: Fault Localization · Model Checking · Spectrum-based Fault Localization · TCAS Benchmark

1 Introduction

Software debugging, encompassing fault detection, localization, and correction, is a labor-intensive process. Fault localization, specifically pinpointing problematic instructions, is particularly resource-demanding. Accelerating fault location methods can profoundly impact software development and maintenance costs and quality [16].

Numerous research efforts focus on automating fault localization to alleviate debugging burdens. Prominent approaches fall into two categories: Spectrum-based and model-checking-based fault localization. Spectrum-based methods, seen in studies such as [9,16,21,24,26], use tests and spectral metrics to assess

B. Djamaa et al. (Eds.): CSA 2024, LNNS 1145, pp. 305–316, 2025.
https://doi.org/10.1007/978-3-031-71848-9_28

suspicion levels of each program instruction. Model-checking-based methods, as shown in works like [4,10,18,19], employ model checking to identify a minimal set of instructions explaining program non-conformance to its postcondition.

No existing article compares the effectiveness of methods within each class due to differing output types. This article proposes an approach combining testing and model checking for fault localization, aiming to output program instructions ordered by suspicion degree, a task unattainable with model-checking-only methods.

Failing test cases execute more non-faulty code than faulty, complicating fault localization [5]. Spectrum-based methods benefit from reduced non-faulty code execution in failing tests. Our approach uses model checking to minimize suspicious instructions from each failing test, reducing non-faulty statement coverage. The core idea is to reduce failing test cases using model checking before spectrum-based fault localization.

The remainder of the paper is structured as follows: Sect. 2 gives an overview of some related works. In Sect. 3, our approach is demonstrated using a concise example. Section 4 introduces our novel approach, which integrates test and model checking for fault localization. Our experimental results are detailed in Sect. 5. Finally, Sect. 6 encapsulates the conclusion and offers a discussion on future avenues for research.

2 Stat of the Art

Numerous studies have been undertaken to aid in fault localization within imperative programs. This section presents key methodologies in model checking and testing.

2.1 Fault Localization in Model Checking

Model checking assesses if a program adheres to a specification by traversing attainable states. This subsection delves into methodologies aimed at pinpointing errors within the context of model checking.

Bal et al. [2] demonstrated pinpointing the root cause within error traces. The works [12,13] introduce a system called Explain, operating on distance metrics applied to program executions. In [10,11], model checkers perform model-based diagnosis, facilitating fault localization.

In this subsection, three model checking-based approaches are presented, reducing fault localization to diagnosing an inconsistent system of constraints.

Firstly, Manu Jose and Rupak Majumdar [18] propose BugAssist, integrating a partial MaxSAT solver. Secondly, Si-Mohamed Lamraoui and Shin Nakajima [19,20] developed SNIPER, computing Minimal Correction Subsets. Finally, Bekkouche et al. [3,4] introduce LocFaults, amalgamating techniques from bounded model checking and constraint programming.

2.2 Spectrum-Based Fault Localization

Approaches for fault localization based on spectra employ ranking metrics to assign scores indicative of suspicion to program elements. Jones, Harrold, and Stasko [15,16] employ spectral statistics. Renieris and Reiss [22] introduced a technique for fault localization through comparable program spectra. The study in [1] explores the impact of similarity measures on diagnosis. Hao et al. [14] streamline the process of manually checking test results. The work in [26] introduces FLCN, a fault localization technique rooted in complex network theory.

In their work, Dutta et al. [9] introduce an enhanced statistical method based on Fisher's test. The ConsilientSFL [21] system synthesizes a novel ranking metric tailored to a specific program, combining a predefined set of ranking metrics.

3 Motivating Example

This section employs an illustrative example to provide an overview of our approach, which integrates both model checking and spectrum-based fault localization techniques.

Consider the "Mid" program, which calculates the median of three variables (see Listing 1.1). However, there is an error in the code at line 12. The program is provided without any preconditions, and it includes a post-condition that describes the expected outcome.

Listing 1.1. The Mid program with an error

```
1class Foo{
2  /*@ requires true;
3  @ ensures \result == (\max(\min(x, y), \min(\max(x, y), z)));
4  @*/
5  static void mid (int a, int b, int c) {
6   int m;
7   m=c;
8   if (b<c){
9     if (a<b)
10     m=b;
11    else if (a<c)
12     m=b; // error, the instruction should be m=a;
13   }
14   else{
15    if (a>b)
16     m=b;
17    else if (a>c)
18     m=a;
19   }
20   return m;
21  }}
```

Model Checking. To verify the "Mid" program's adherence to its specification, we use a model checking tool like CPBPV [7]. This tool finds counterexamples, showing where the program contradicts its post-condition. These counterexamples, like (a=2, b=1, c=3), reveal incorrect outputs due to violations of the post-condition. LocFaults [4], a constraint-based model checking tool, then examines the program's Control Flow Graph (CFG) to pinpoint suspect instructions. For example, on the path $\{m = c, (b < c), (a >= b), (a < c), m = b, \text{return } m\}$, LocFaults identifies suspicious instructions like "m=b" (line 12) and "return m". It also raises concerns about conditions at lines 8, 15, and 17. Modifying these suspect instructions or conditions could bring the output in line with the post-condition.

Suspiciousness Ranking. During model checking, five suspicious instructions were identified: $\{8, 12, 15, 17, 20\}$. For larger programs, more instructions may raise suspicion. To refine our approach, we propose scoring these instructions using spectrum analysis principles. This step uses execution spectra to rank and prioritize instructions. Alongside counterexamples, successful test cases that meet the post-condition are crucial. We selected five such cases to illustrate our method. These cases, along with the counterexample, were executed on the "Mid" program, and the resulting execution spectra are shown in Table 1. A value of 1 in a cell indicates that a statement was executed by a test case, while 0 indicates otherwise. A value of 1 in the "Result" row denotes test failure; otherwise, it's 0. Our approach focuses on instructions identified as suspicious by model checking, assuming they were executed by the counterexample. The spectral data is then used with a formula (e.g., Tarantula, Ochiai) to quantify suspicion level for each potentially faulty element (see Table 2).

Table 1. Example - Spectra, Four Counters, Scores, and Rankings in Our Approach

Statement	(3,3,5)	(3,2,1)	(5,5,5)	(5,4,3)	(2,1,3)	(4,5,3)	e_f	e_p	n_f	n_p	Ochiai	Rank	Tarantula	Rank
8	1	1	1	1	1	1	1	5	0	0	0.41	4	0.5	4
12	1	0	0	0	1	0	1	1	0	4	0.71	1	0.83	1
15	0	1	1	1	1	1	1	4	0	1	0.45	3	0.55	3
17	0	0	1	0	1	1	1	2	0	3	0.58	2	0.71	2
20	1	1	1	1	1	1	1	5	0	0	0.41	4	0.5	4
Result	0	0	0	0	1	0								

Table 2. Ochiai and Tarantula Spectral Formulas Along with Their Equations

Ochiai [1]	$\frac{e_f}{\sqrt{(e_f + n_f) \cdot (e_f + e_p)}}$
Tarantula [15,16]	$\frac{\frac{e_f}{e_f + n_f}}{\frac{e_f}{e_f + n_f} + \frac{e_p}{e_p + n_p}}$

Frequently, a formula is articulated using four counters [23], which are computed from the spectra in the following manner:

- e_f: the count of statement executions (e) in failed tests.
- e_p: the count of statement executions (e) in passed tests.
- n_f: the count of statement non-executions (n) in failed tests.
- n_p: the count of statement non-executions (n) in passed test cases.

Table 3. Illustrative Example - Four Counters, Scores, and Rankings in Pure Spectrum-Based Fault Localization

Line	**Code**	ef	ep	nf	np	**Ochiai**	**Tarantula**	**Rank**
7	`m=c;`	1	5	0	0	0.41	0.5	3
8	`if (b<c){`	1	5	0	0	0.41	0.5	3
9	`if (a<b)`	1	1	0	4	0.71	0.83	1
10	`m=b;`	0	0	1	5	0	0	7
11	`else if (a<c)`	1	1	0	4	0.71	0.83	1
12	`m=b;`	1	1	0	4	0.71	0.83	1
15	`if (a>b)`	0	0	1	1	0	0	7
16	`m=b;`	0	0	1	2	0	0	7
17	`else if (a>c)`	0	0	1	3	0	0	7
18	`m=a;`	0	0	1	4	0	0	7
20	`return m;`	1	5	0	0	0.41	0.5	3

In our approach, statements are ranked based on their suspicion scores, guiding developers to scrutinize them from the most suspicious to the least. The erroneous instruction (line 12) consistently receives the highest score, indicating its likelihood of containing bugs. Even with different spectral formulas, our methodology prioritizes this instruction. In contrast, pure spectrum-based fault localization also identifies the faulty instruction but presents challenges by ranking two additional instructions equally, leading to ambiguity in pinpointing the starting point for investigation [25]. (See Table 3 for details.)

4 Approach

Our approach requires, as input, the erroneous program along with its specification, which includes the precondition and the violated post-condition. It encompasses three principal steps (see Fig. 1): Model Checking, Model Checker-based Fault Localization (MCFL), and Spectrum-based Fault Localization (SBFL).

4.1 Model Checking Step

This step aims to produce counterexamples (failed test cases) and successful test cases from the faulty program and its specification. Tools like CPBPV [7] or

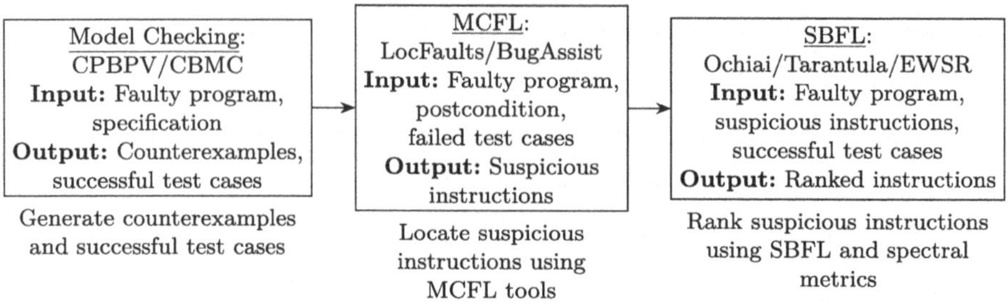

Fig. 1. Synergistic Approach - Test and Model Checking for Fault Localization

CBMC [6] are used for this purpose. CPBPV (Constraint-based Model Checker) and CBMC (Boolean-based Model Checker) utilize constraints and SAT solvers to model the program and verify compliance with its specification.

A Model Checker evaluates the Hoare triplet[1] $\{PRE, PROG, POST\}$, where PRE is the precondition, $PROG$ is the faulty program, and $POST$ is the postcondition. If the formula $\phi = PRE \wedge PROG \wedge \neg POST$ is satisfiable[2], $PROG$ fails to meet its specification. An instantiation of $PROG$'s input variables serves as a counterexample. To generate another counterexample, the negation of the previous one (denoted as CE) is included in the precondition ($PRE = PRE \wedge \neg CE$). The solver is then run with the modified ϕ. Successful test cases satisfy the precondition, program, and postcondition, defined by $\phi' = PRE \wedge PROG \wedge POST$. To avoid repeating the same successful test case, its negation is added to the precondition.

4.2 MCFL Step

In this step, a Model Checker-Based Fault Localization tool like LocFaults[3] [4] or BugAssist[4] [18] is used for each counterexample generated in the previous phase to identify suspicious instructions, shedding light on the causes of the specific counterexample.

These tools expand the model checking process to pinpoint the program section potentially responsible for the error detected during the Model Checking phase. Given a counterexample CE, which satisfies the precondition, program, and negation of the postcondition, a constraint system C is formed as $C = CE \wedge PROG \wedge POST$. Since CE satisfies $\neg POST$, C is unsatisfiable. An MCS (Minimal Correction Subset) of C constitutes a minimal set of constraints to be eliminated to make C satisfiable, serving as a potential error localization. Both LocFaults and BugAssist compute MCSs to explain the counterexample's existence but use different strategies to obtain them. LocFaults adopts

[1] A formal assertion with three parts specifying preconditions, program statements, and postconditions, used for reasoning about program behavior.

[2] Describing a formula that can be made true by assigning values to its variables.

[3] https://github.com/mdbekkouche/LocFaultsTool.

[4] https://bugassist.mpi-sws.org/.

a constraint-based and flow-driven approach, transforming the program into a Control Flow Graph (CFG) and systematically exploring this graph to detect MCSs solely from faulty paths. In contrast, BugAssist relies on SAT, transforming the entire program into a Boolean formula.

4.3 SBFL Step

This step involves ranking the suspicious instructions localized in the previous phase. It utilizes the faulty program, the suspicious instructions for each counterexample used, and the successful test cases to assign rankings.

SBFL relies on the concept of "program spectrum" or "code coverage". A program spectrum essentially serves as a record of which lines of code within a program were executed (covered) and which lines of code were not executed during the testing process (see Table 3). This information is gathered during the execution of test cases and is typically organized in the form of a matrix. In this matrix representation, each row corresponds to a specific test case, each column corresponds to individual lines of program code, and the cells within the matrix denote whether a particular program line of code was covered (executed) or left untouched during the execution of that specific test case.

In our approach, the program spectrum for the erroneous program is calculated by utilizing both failed and successful test cases obtained during the Model Checking step. For successful test cases, the coverage of various program elements during their execution is recorded, following the conventional approach. However, for failed test cases (counterexamples), a slightly different strategy is adopted. In this case, a line of code is considered as covered by a test case if it appears in the set of suspicious instructions provided by the MCFL step for that specific counterexample. Conversely, a line of code is marked as not covered if it is absent from the set of suspicious instructions. This distinction helps us identify which parts of the code contribute to the failure during error localization more precisely.

Spectral metrics are used to prioritize which program lines of code are more likely to be faulty, assisting developers in narrowing down their focus during the debugging process. Spectral metrics refer to various mathematical formulas and techniques used to quantify the degree of suspicion associated with code lines based on the program's execution spectrum. After computing the spectral metrics for each program line of code, these metrics are employed to rank these elements in descending order of suspicion. Elements with the highest suspiciousness scores are deemed the most probable sources of faults. These are the components that developers should scrutinize more closely to pinpoint and rectify errors. Among the most well-known spectral metrics are Tarantula and Ochiai (refer to Table 2).

Weighted Suspiciousness Ratio (WSR). A novel spectral metric called the Weighted Suspiciousness Ratio (WSR) is introduced (refer to Eq. 1). To understand the meaning of the operands e_f, e_p, n_f, and n_p, please refer to Sect. 3. The underlying idea behind this metric is that a suspicious statement is expected to have a higher execution frequency in failed tests as opposed to passed tests, and it is less likely to remain unexecuted (missed) in failed tests compared to passed tests.

$$WSR = \frac{e_f \times n_p + n_f \times e_p}{e_f \times n_p + n_f \times e_p + e_f \times e_p + n_f \times n_p} \tag{1}$$

It's worth noting that in our approach, a higher execution frequency of a statement in failed tests means that this statement frequently appears in the sets of suspicious instructions obtained from these failed test cases in the MCFL step. Conversely, if a statement is less likely to remain unexecuted (missed) in failed tests, it suggests that it is less frequently featured in the sets of suspicious instructions obtained from these failed test cases.

Let's further refine the WSR metric to make it even more precise in fault localization.

Enhanced Weighted Suspiciousness Ratio (EWSR). The Enhanced Weighted Suspiciousness Ratio (EWSR) represents an advancement over the fundamental Weighted Suspiciousness Ratio (WSR) metric by incorporating additional nuances from execution data and introducing normalization techniques (as depicted in Eq. 2).

$$WSR = \frac{e_f \times n_p + n_f \times e_p}{e_f \times n_p + n_f \times e_p + e_f \times e_p + n_f \times n_p} \times \frac{1}{\sqrt{(e_f + e_p) \times (n_f + n_p)}} \tag{2}$$

The second term within this formula, $\frac{1}{\sqrt{(e_f + e_p) \times (n_f + n_p)}}$, normalizes the metric based on the aggregate counts of executions and non-executions. This normalization aims to mitigate the impact of variations in non-execution counts across distinct statements, resulting in a more refined and reliable assessment of code suspiciousness.

5 Results

To evaluate the effectiveness of our approach, experiments were conducted using the well-established TCAS (Traffic Collision Avoidance System) benchmark suite provided by Siemens [8]. This Siemens test suite is widely recognized in the literature for bug localization studies [5,9,10,15,18,20–22,27]. The TCAS task within the Siemens test suite simulates an aircraft collision avoidance system and consists of 173 lines of code. The authors intentionally introduced 41 distinct versions of the program, each incorporating one or more realistic faults, drawing from their extensive experience with real-world software systems. These versions are labeled as "v1" through "v41". Certain versions have been excluded (specifically, v33 and v38) due to the fact that these versions exhibit errors related to array index out-of-bounds issues, which currently do not have the capability to address. The error in these two programs is not due to an unexpected output (violation of the postcondition), but to an array overflow. Furthermore, the TCAS benchmark encompasses a total of 1608 test cases.

In the implementation of our approach, the BugAssist model checking tool [17] was utilized to pinpoint suspicious instructions in the faulty program based on a single counterexample (faulty test case). Subsequently, the obtained instructions are classified using successful test cases and the EWSR spectral metric. The objective is to calculate a degree of suspicion for each of these instructions: the fewer they are covered by successful test cases, the more suspicious they become, and the more they are executed by successful test cases, the less suspicious they are.

The results obtained from our implementation were compared with those of an implementation based solely on testing, using the Ochiai, Tarantula, and EWSR spectral formulas[5]. To achieve this, the position of the erroneous instruction of the faulty program to be debugged is presented, which triggered the counterexample used by our implementation. The ranking of instructions is based on the degree of suspicion calculated by the spectral formulas, ranging from the most suspicious to the least suspicious instruction.

To address cases where instructions share the same degree of suspicion, known as tie elements, the average position of these instructions is considered. This position is calculated using Eq. 3 [23]:

$$\text{MID} = S + \left(\frac{E-1}{2}\right) \tag{3}$$

Here, S represents the starting position of the tie, and E denotes the tie size. This method is commonly employed to manage tied rankings.

The outcomes of our TCAS benchmark experiments are illustrated in the histogram presented in Fig. 2. Each cluster of bars corresponds to the results for individual versions within this benchmark. Within each cluster, the first, second, and third bars signify the average position of the considered faulty instruction when employing pure spectrum-based fault localization methods, specifically utilizing the Ochiai, Tarantula, and EWSR spectral formulas, respectively. The fourth bar in each cluster (MC+TEST) denotes the average position of the faulty instruction achieved through our integrated approach, which combines model checking and testing for fault localization, incorporating the EWSR spectral metric.

Observing the results, Ochiai consistently outperforms Tarantula across various versions, indicating its superior accuracy in localizing faults as a formula for pure spectrum-based fault localization methods. Our EWSR spectral metric introduces variability in its performance, demonstrating effectiveness in certain versions but facing challenges in others. The integrated approach of model checking and testing (MC+TEST) consistently excels, achieving lower average positions compared to pure spectrum-based methods, particularly in versions where other methods face challenges. This suggests the efficacy of the integrated approach in improving fault localization accuracy, especially with the incorporation of our proposed EWSR spectral metric.

[5] Experimental results are available in https://sites.google.com/esi-sba.dz/experiments-fault-localization.

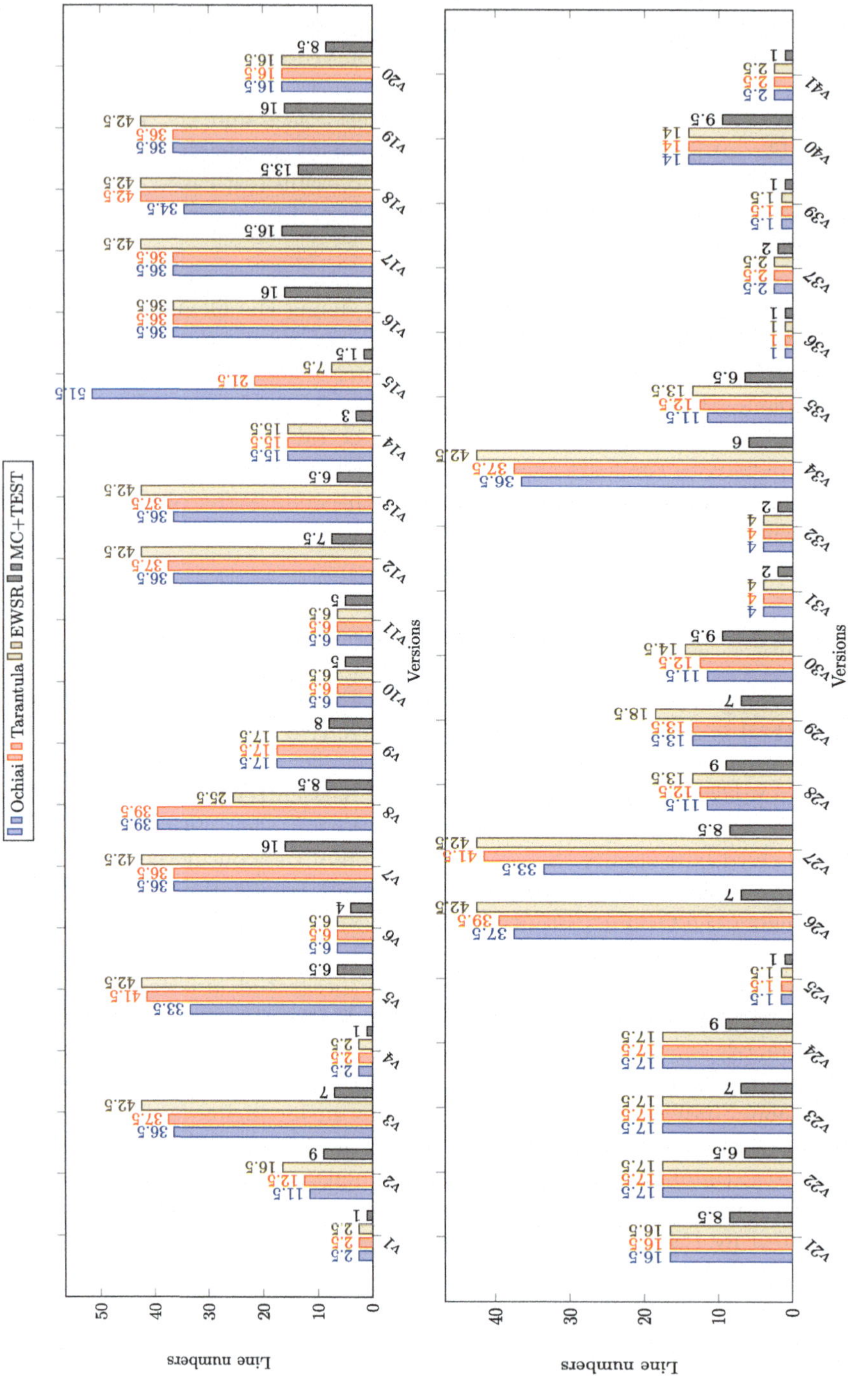

Fig. 2. The Average Position of the Faulty Instruction in Each Program Version of the TCAS Suite from Siemens

6 Conclusion

In conclusion, our integrated fault localization approach presents a significant advancement in the field of software debugging. The combination of model checking and testing, as demonstrated in the TCAS benchmark experiments, outperforms traditional spectrum-based methods, showcasing its efficacy in improving fault localization accuracy. The novel EWSR spectral metric introduces valuable variability, addressing challenges in certain versions. This research contributes to automating and improving the fault localization process, providing developers with a refined and efficient method for identifying problematic instructions in faulty programs. Overall, our integrated fault localization approach demonstrates a remarkable improvement of approximately 65.47% in fault localization accuracy compared to traditional spectrum-based methods. Future research avenues may explore the applicability of this approach to other benchmarks and further refine the integration of model checking and testing for continued advancements in software debugging.

References

1. Abreu, R., Zoeteweij, P., Van Gemund, A.J.: An evaluation of similarity coefficients for software fault localization. In: 2006 12th Pacific Rim International Symposium on Dependable Computing (PRDC'06), pp. 39–46. IEEE (2006)
2. Ball, T., Naik, M., Rajamani, S.K.: From symptom to cause: localizing errors in counterexample traces. In: Proceedings of the 30th ACM SIGPLAN-SIGACT symposium on Principles of programming languages, pp. 97–105 (2003)
3. Bekkouche, M.: Combining techniques of bounded model checking and constraint programming to aid for error localization. Ph.D. thesis, Université Nice Sophia Antipolis (2015)
4. Bekkouche, M., Collavizza, H., Rueher, M.: Locfaults: a new flow-driven and constraint-based error localization approach. In: Proceedings of the 30th Annual ACM Symposium on Applied Computing, pp. 1773–1780 (2015)
5. Christi, A., Olson, M.L., Alipour, M.A., Groce, A.: Reduce before you localize: Delta-debugging and spectrum-based fault localization. In: 2018 IEEE International Symposium on Software Reliability Engineering Workshops (ISSREW), pp. 184–191. IEEE (2018)
6. Clarke, E., Kroening, D., Lerda, F.: A tool for checking ANSI-C programs. In: Jensen, K., Podelski, A. (eds.) TACAS 2004. LNCS, vol. 2988, pp. 168–176. Springer, Heidelberg (2004). https://doi.org/10.1007/978-3-540-24730-2_15
7. Collavizza, H., Rueher, M., Van Hentenryck, P.: Cpbpv: a constraint-programming framework for bounded program verification. Constraints **15**(2), 238–264 (2010)
8. Do, H., Elbaum, S., Rothermel, G.: Supporting controlled experimentation with testing techniques: an infrastructure and its potential impact. Empir. Softw. Eng. **10**, 405–435 (2005)

9. Dutta, A., Kunal, K., Srivastava, S.S., Shankar, S., Mall, R.: FTFL: a fisher's test-based approach for fault localization. Innovations Syst. Softw. Eng. **17**(4), 381–405 (2021)
10. Griesmayer, A., Staber, S., Bloem, R.: Automated fault localization for C programs. Electron. Not. Theoret. Comput. Sci. **174**(4), 95–111 (2007)
11. Griesmayer, A., Staber, S., Bloem, R.: Fault localization using a model checker. Software Test. Verif. Reliab. **20**(2), 149–173 (2010)
12. Groce, A., Chaki, S., Kroening, D., Strichman, O.: Error explanation with distance metrics. Int. J. Softw. Tools Technol. Transfer **8**, 229–247 (2006)
13. Groce, A., Kroening, D., Lerda, F.: Understanding counterexamples with. In: Alur, R., Peled, D.A. (eds.) CAV 2004. LNCS, vol. 3114, pp. 453–456. Springer, Heidelberg (2004). https://doi.org/10.1007/978-3-540-27813-9_35
14. Hao, D., Xie, T., Zhang, L., Wang, X., Sun, J., Mei, H.: Test input reduction for result inspection to facilitate fault localization. Autom. Softw. Eng. **17**, 5–31 (2010)
15. Jones, J.A., Harrold, M.J.: Empirical evaluation of the tarantula automatic fault-localization technique. In: Proceedings of the 20th IEEE/ACM International Conference on Automated Software Engineering, pp. 273–282 (2005)
16. Jones, J.A., Harrold, M.J., Stasko, J.: Visualization of test information to assist fault localization. In: Proceedings of the 24th International Conference on Software Engineering, pp. 467–477 (2002)
17. Jose, M., Majumdar, R.: Bug-assist: assisting fault localization in ANSI-C programs. In: Gopalakrishnan, G., Qadeer, S. (eds.) CAV 2011. LNCS, vol. 6806, pp. 504–509. Springer, Heidelberg (2011). https://doi.org/10.1007/978-3-642-22110-1_40
18. Jose, M., Majumdar, R.: Cause clue clauses: error localization using maximum satisfiability. ACM SIGPLAN Not. **46**(6), 437–446 (2011)
19. Lamraoui, S.-M., Nakajima, S.: A formula-based approach for automatic fault localization of imperative programs. In: Merz, S., Pang, J. (eds.) ICFEM 2014. LNCS, vol. 8829, pp. 251–266. Springer, Cham (2014). https://doi.org/10.1007/978-3-319-11737-9_17
20. Lamraoui, S.M., Nakajima, S.: A formula-based approach for automatic fault localization of multi-fault programs. J. Inf. Process. **24**(1), 88–98 (2016)
21. Majd, A., Vahidi-Asl, M., Khalilian, A., Bagheri, B.: Consilientsfl: using preferential voting system to generate combinatorial ranking metrics for spectrum-based fault localization. Appl. Intell. **52**(10), 11068–11088 (2022)
22. Renieres, M., Reiss, S.P.: Fault localization with nearest neighbor queries. In: 18th IEEE International Conference on Automated Software Engineering, 2003. Proceedings, pp. 30–39. IEEE (2003)
23. Sarhan, Q.I., Beszédes, Á.: A survey of challenges in spectrum-based software fault localization. IEEE Access **10**, 10618–10639 (2022)
24. Wen, M., et al.: Historical spectrum based fault localization. IEEE Trans. Software Eng. **47**(11), 2348–2368 (2019)
25. Xu, X., Debroy, V., Eric Wong, W., Guo, D.: Ties within fault localization rankings: Exposing and addressing the problem. Int. J. Software Eng. Knowl. Eng. **21**(06), 803–827 (2011)
26. Zakari, A., Lee, S.P., Chong, C.Y.: Simultaneous localization of software faults based on complex network theory. IEEE Access **6**, 23990–24002 (2018)
27. Zakari, A., Lee, S.P., Hashem, I.A.T.: A single fault localization technique based on failed test input. Array **3**, 100008 (2019)

SGSN: Tool Support for GSN Development

Radouane Bouchekir(✉), Fatiha Ouazar, Mohand Cherif Boukala, and Malika Ioualalen

MOVEP, Computer Science Faculty, Université des Sciences et de la Technologie Houari Boumediene, BP 32 El-Alia, Algiers, Algeria
{rbouchekir,fouazar,mboukala,mioualalen}@usthb.dz

Abstract. Graphical argument structures such as Goal Structuring Notation (GSN) are increasingly being used in practice as a means to provide safety assurance to stakeholders, such as regulatory authorities, regarding the dependability and safety of a system. These argument-based assurance cases are utilized in practice to demonstrate that a system is suitable for its intended use. In this paper, we present Structured GSN (SGSN), an open-source tool to create and edit assurance cases. In addition to basic functionalities related to the creation of GSN, we added two types of analysis, syntactic analysis which checks if the developed GSN satisfies structured criteria, and semantic analysis which checks compartmental properties. For the semantic analysis, we mapped the GSN to a formal model i.e., Markov Decision Process (MDP), and checked properties expressed in PCTL using a probabilistic model-checker.

Keywords: Safety assurance · Goal Structuring Notation (GSN) · Markov Decision Process

1 Introduction

1.1 Context

Certification of safety-critical systems is a rigorous evaluation process to certify that these systems meet specific safety and reliability standards [15,16]. It involves in-depth analyses, tests, and assessments carried out by independent bodies or regulatory authorities. The aim is to guarantee compliance, minimize risk, and promote stakeholder confidence in the operation of complex and safety-critical systems.

Safety assurance [5,12] is a process used to assess and demonstrate that complex systems comply with safety and security requirements. This includes the identification and analysis of risks, the definition of appropriate safety measures, as well as validation and verification of safe performance. One of the motivations for using safety assurance is to explicitly capture the traceability between safety

B. Djamaa et al. (Eds.): CSA 2024, LNNS 1145, pp. 317–327, 2025.
https://doi.org/10.1007/978-3-031-71848-9_29

claims and the substantiating evidence and this simplifies the certification process. Another motivation is to make it easier to understand and critically review the safety assurance to further improve clarity.

In practice, safety assurance can be represented in different formats, such as: *text*, *table*, or graphical notations like *ACE (Argumentation, Claims, and Evidence)* [4], and *GSN (Goal Structuring Notation)* [11]. The choice of representation can depend on factors such as the audience, the complexity of the safety assurance, and the level of detail required for communication and understanding. Indeed, representation as text and table are not suitable for the safety assurance of complex systems due to their ambiguity. In such cases, graphical representation using ACE or GSN can give a clearer and more comprehensive representation. Furthermore, graphical notations have emerged over the past decade to present the elements of a safety argument, e.g., CAE and GSN. Such graphical argument structures can be thought of as a visual index into the safety reasoning and evidence comprising a safety assurance.

In this paper, we consider GSN as the graphical notation to represent safety assurance. GSN allows representing in a clear and structured way the safety objectives, the evidence needed to achieve them, along with the arguments and justifications that support this evidence. It consists of a network of interconnected elements, including *goals*, *claims*, *evidence*, and *assumptions*. Safety assurance can be represented in GSN by creating a goal node that represents the overarching safety objective. Claims and evidence nodes can be added to support the goal, with arguments and assumptions linking them together. This visual representation helps to show the logical structure of the safety assurance argument and provides a clear framework for understanding the supporting evidence.

1.2 Contributions

GSN plays an important role in the safety assurance process, it facilitates communication and understanding of safety assurance activities between stakeholders involved in the certification of safety-critical systems. However, the lack of tools supporting the creation and analysis of GSN makes the use of this notation challenging. The main challenges can be summarized as: (i) The lack of tools to create and maintain GSN structure, and (ii) The need to perform analysis (syntactic and semantic) to ensure the correctness and coherence of created GSN.

In this work, we present our tool Structured GSN (`SGSN`[1]), where we aim to address the above-mentioned challenges. Our main contributions are:

- Implementation of an open-source tool to facilitate the creation and maintenance of GSNs,
- Syntactic analysis to ensure the coherence of GSN w.r.t criteria described in the latest standard [11],
- Semantic analysis to verify compartmental properties.

[1] https://github.com/BouchekirRedouane/SGSN.

The syntactical analysis aims to check the conformity of the GSN diagrams to the notation rules described in the current standard [11], while semantic analysis aims to verify the consistency and validity of the relationships and arguments between the different elements of the GSN. To ensure the semantic analysis, we proposed to formally specify GSN models as MDP (Markov Decision Process) [20] and then check the validity of formal properties expressed in the Probabilistic Computation Tree Logic (PCTL) [2] formula. This ensures verifying the logical consistency and validity of arguments present in GSN diagrams w.r.t the specified property.

The rest of this paper is structured as follows, Sect. 2 summarizes the preliminaries used to illustrate our contributions. In Sect. 3 we illustrate the SGSN metamodel. Sections 4 and 5 illustrate the proposed syntactic and semantic analysis, respectively. Section 6 highlights related works. Finally, we provide concluding remarks and pointers to future work in Sect. 7.

2 Preliminaries

2.1 Safety Assurance and GSN

Safety Assurance is a comprehensive, defensible, and valid justification of the safety of a system for a given application in a defined operating environment [5,12]. The main aspect of this justification, which is a risk-based argument, is to link safety claims through a reasonable chain to the set of evidence that can support these claims.

The GSN standard defines the assurance case as *"a reasoned and compelling argument, supported by a body of evidence, that a system, service or organization will operate as intended for a defined application in a defined environment"* [11]. Figure 1 shows the core elements of GSN. In summary, the graphical symbols are as follows:

- Claim (known as a *Goal* in GSN-speak) represented as a rectangle,
- Strategy, represented as a parallelogram contains, that shows what strategy is used to justify the claim,
- Context, represented as a rounded rectangle contains. Usually, the claim is associated and valid in a particular context,
- Assumption represented as an oval contains,
- Solution represented as a circle, which is a description of the evidence that will be presented to justify the claim,
- Incomplete arguments are represented as a diamond represents.

2.2 Probabilistic Model

Markov Decision Processes (MDP) [20] are frequently employed to model and analyze systems characterized by non-deterministic and stochastic behavior.

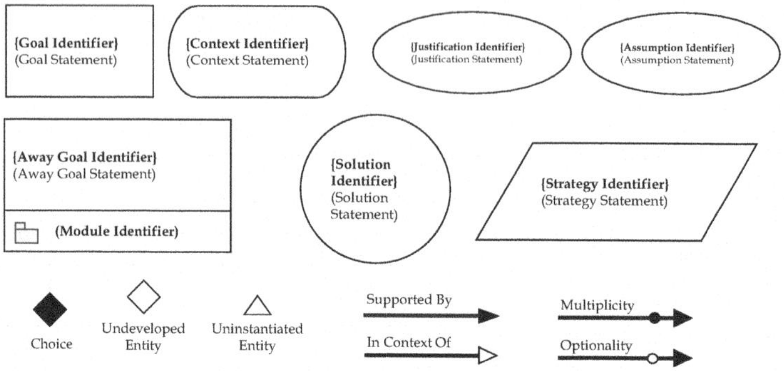

Fig. 1. Core elements of GSN.

Definition 1. *An MDP is a tuple* $M = (S_M, s_M^0, \Sigma_M, \delta_M)$*, where* S_M *is a finite set of states,* s_M^0 *is an initial state,* Σ_M *is a finite set of actions and* $\delta_M \subseteq S \times (\Sigma_M \cup \{\tau\}) \times Dist(S)$ *is a probabilistic transition relation.*

In a state s of MDP M, one or more transitions, denoted $(s, a) \rightarrow \mu$, are available, where a is an action label, μ is a probabilistic distribution over states and $(s, a, \mu) \in \delta_M$.

In this work, we focus on verifying probabilistic properties specified using Probabilistic Computation Tree Logic (PCTL) [2] in the form of $P_{\leq \rho}[\psi]$ with $\psi \in [0, 1]$ and

$$\phi ::= true|a|\phi \wedge \phi|\neg\phi$$
$$\psi ::= \phi \cup \phi$$

3 SGSN Meta-Model

A (non-modular) argument structure in GSN is illustrated in Fig. 2. It contains a top-level (root) goal stating a safety claim, which is in this example *"Safe_UAV_1: The Temporal Planning Network (TPN) solution finder should ensure that, for every generated route mission, this route should not violate geo-fences (NFZ)"*.

This goal is established under the assumption that *"all abnormal situations are listed and mitigated."*. The top-level goal is developed using two strategies, the first strategy is to verify the goal using an SMT-based verification technique, while the second is based on theorem proving. The sub-goals claim that the TPN core functions are correct (design) and implemented correctly (code). The solutions are the leaves of the GSN, and they represent the assurance evidence to support the sub-goals. The goals, strategies, and solutions are connected using the *"Is Supported by"* connectors, while context, assumption, and justification elements require an *"In Context Of"* connectors. GSN also employs a graphical annotation (◊) for goals and strategies to signify that they are to be developed (tbd), implying they are incomplete. The top level of any safety assurance

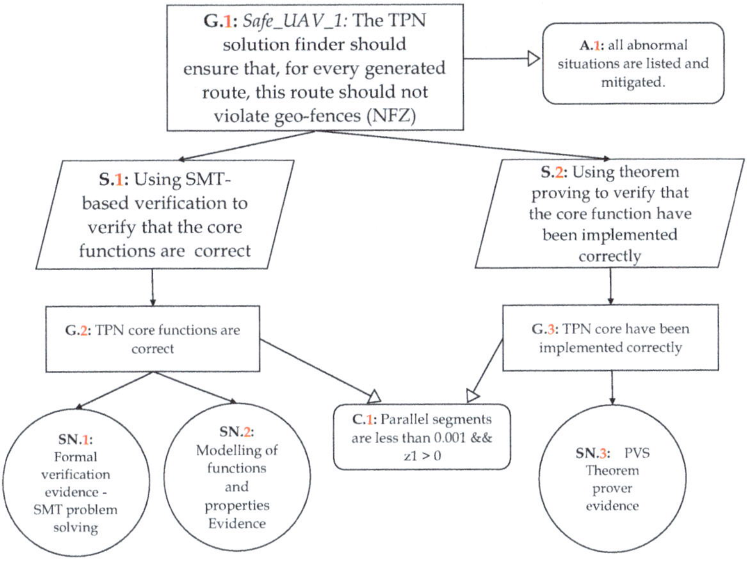

Fig. 2. A simple example of GSN used to illustrate that the AI-based planning for Unmanned Aerial Vehicles (UAV) is safe w.r.t to geo-fences constraints [6].

model based on the SGSN meta-model (Fig. 3) is a safety assurance represented as GSN. Essentially, this is the container that holds all GSN elements of the safety assurance i.e. GSN.

The GSN elements can be concrete instances of either an argument element or a relation element. An argument element generalizes the different types of GSN elements, i.e., Goal, Strategy, Assumption, Justification, Context, and Solution[2].

The attributes of an argument element are:

- *Identifier.* it contains a unique name for each element.
- *Description.* a describing associated with the argument element.
- *Attributes.* which are used to hold extra meta-data about the argument element.

In addition, we based our conceptual architecture on the MVC (Model-View-Controller) pattern. This patern helps in organizing and structuring the code-base, making it more maintainable, scalable, and easier to understand.

4 Syntactic Analysis

Syntactic analysis of GSN is the process of validating the structure of a GSN based on the current GSN standard (Version 3) [11]. The aim of syntactic analysis is to ensure that the GSN model is well-formed and respects the specific

[2] Strictly speaking, the GSN uses the term Solution. We use the term Evidence to illustrate the artifacts attached to a solution as assurance evidence. For example files or external links.

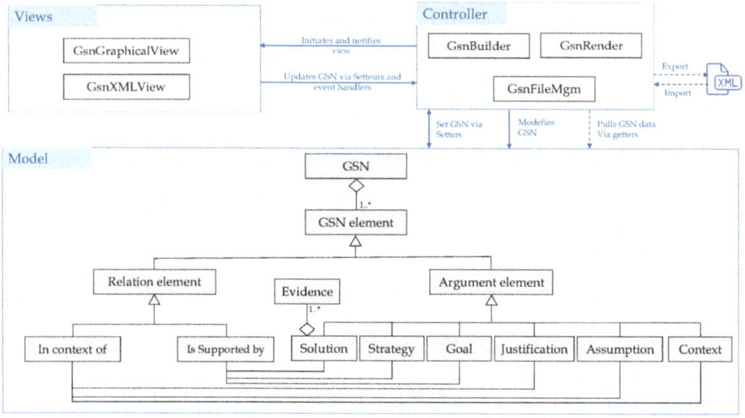

Fig. 3. Meta-Model of GSN Elements considered in our tool SGSN.

syntax rules, in order to obtain a clear and accurate representation of the safety assurance. Let first describe formally a safety assurance.

Definition 2. *We define an argument structure* A*, as a tuple* $\langle N, l, \rightarrow \rangle$*, where:*

- N *represents a set of nodes,*
- $l_X | X \in \{id, t, d, m, s\}$ *is a set of labelling functions, giving an element identifier, its type, its description, its metadata, and its status,*
- $\rightarrow$ *represents the connector.*

We denote by $\{s, g, e, a, j, c\}$ *the node types strategy, goal, evidence, assumption, justification, and context respectively. :*

- $l_{id} : N \rightarrow string$ *gives the node identifier,*
- $l_t : N \rightarrow \{\ s, g, e, a, j, c\}$ *gives the node types,*
- $l_d : N \rightarrow string$ *gives the node descriptions,*
- $l_m : N \rightarrow P(A)$ *gives the node instance attributes,*
- $l_s : N \rightarrow P(\{tbd\})$ *gives node development status.*

In addition, we consider the operation $isRoot(\rightarrow, r)$ *which check if the node* r *is a root.*

Our syntactic analysis is based on structural criteria that are checked in two phases: *on-creation criteria* and *post-creation criteria.*

4.1 On-Creation Criteria

1. Goals cannot be connected directly to other goals in the assurance argument:

$$\forall n, m \in N, (n \rightarrow m) \wedge [l_n = g] \Rightarrow l_t(m) \in \{s, e, a, j, c\}$$

2. Strategies cannot be directly linked to other strategies or evidence in the assurance argument:

$$\forall n, m \in N, (n \rightarrow m) \wedge [l_n = s] \Rightarrow l_t(m) \in \{g, a, j, c\}$$

3. Only goals and strategies are permitted to be undeveloped in the assurance argument:

$$\forall n, m \in N, tbd \in l_s(n) \Rightarrow l_t(n) \in \{g, s\}$$

4. A node cannot establish a connection with itself:

$$\forall n, m \in N, (n \rightarrow m) \Rightarrow l_{id}(n) \neq l_{id}(m)$$

5. Each element has a unique identifier

$$\forall n, m \in N, l_{id}(n) \neq l_{id}(m)$$

4.2 Post-creation Criteria

1. Root of the GSN should be a goal:

$$isRoot(\rightarrow, r) \Rightarrow l_t(n) \in \{g\}$$

2. Connectors are only permitted to originate from strategies or goals:

$$\forall n, m \in N, n \rightarrow m \Rightarrow l_n \in \{s, g\}$$

5 Semantic Analysis

The semantic analysis of a GSN may involve interpreting and understanding the meaning of elements present in the GSN model. This process includes identifying relationships between different objectives, justifications, and strategies, as well as semantic interpretation of the context in which they are placed. In this work, we use semantic analysis to verify certain behavioural properties. In that regard, we propose to model the GSN using a formal model i.e., MDP. The choice of MDP is supported by the existence of a probabilistic model-checker that can automatically verify MDP against properties expressed in PCTL. In addition, MDP will allow us to extend the current work in order to evaluate a GSN based on a *confidence score* [9]. We present the main steps of the semantic analysis in Fig. 4.

5.1 Translate GSN to MDP

Figure 5 summarises the mapping of the main elements of GSN to MDP states. Mainly, we map each GSN element to a state in MDP, and we label this state with the ID. GSN elements *Context, Assumption, Justification*, and *Solution* are

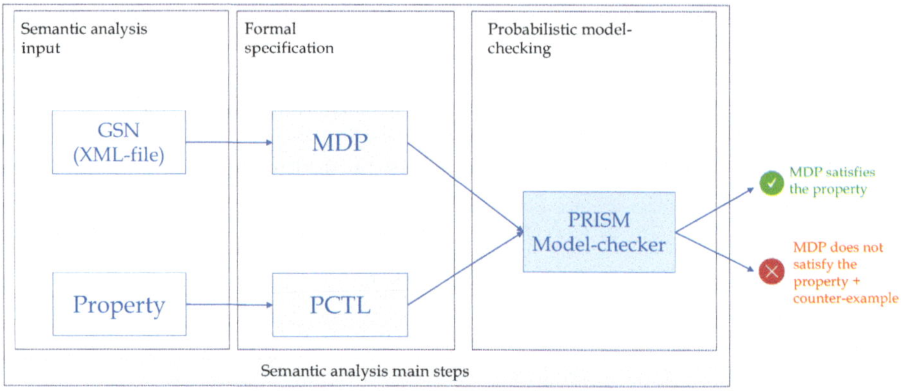

Fig. 4. Semantic analysis main steps.

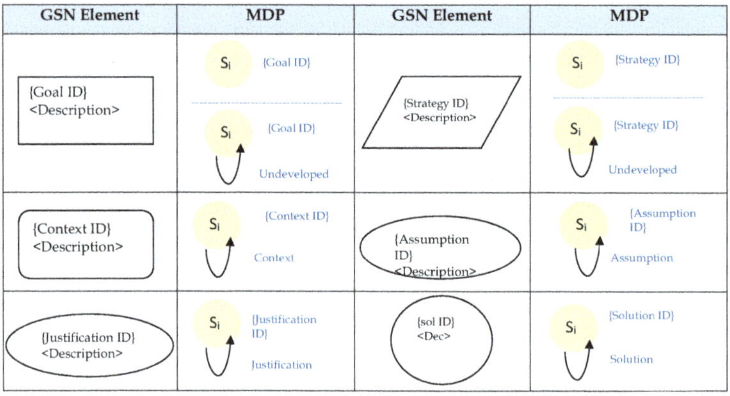

Fig. 5. Mapping of the main GSN elements to MDP states.

represented by an MDP state and loop to this state. For undeveloped *Goal* and *Strategy*, they are represented by a state with a loop labeled "Undeveloped".

After formally specifying the GSN as an MDP, we use PRISM [13] model-checker to verify this MDP against a probabilistic property. SGSN supports properties expressed in PCTL. An example of a property is to verify that all the sub-goals are supported by solutions unless specified as undeveloped.

$$P_{\geq 1}[G(true \cup \text{“}Solution \vee Undeveloped\text{”})] \quad (1)$$

5.2 Running Example

In order to illustrate the formal specification of GSN, we consider the GSN illustrated in Fig. 2, where Fig. 6 represents the MDP formally specifying this GSN. In this example, we consider the probability associated with each action as 1 since we do not consider the confidence score in this paper. However, for the confidence argument [9], the probabilities could be specified to infer the confidence score.

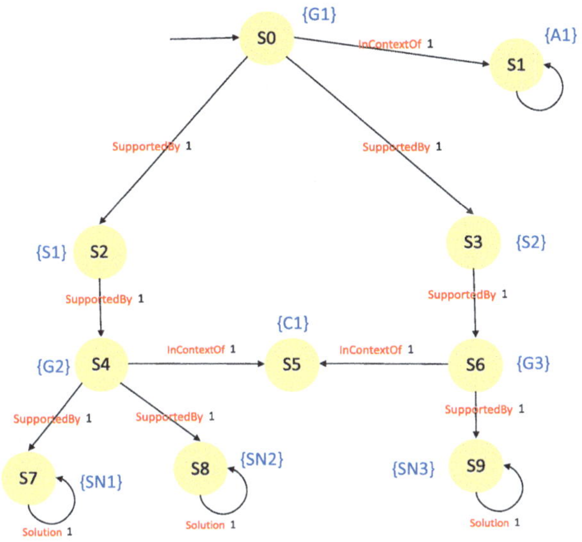

Fig. 6. MDP representing GSN in Fig. 2.

6 Related Works

Tools such as [1,3,14] facilitate manual or partly automated creation, instantiation, and management of assurance cases. Approaches built upon the AMASS platform [17–19] utilize contracts for automated assurance case creation, promoting compositionality and reusability of argumentation patterns. However, these tools do not propose syntactic and semantic analysis beyond automation.

The authors of [8] presented a set of tools supporting the automatic creation of assurance cases, where they showcase this using the tool AdvoCATE [10], however, they did not specify the formalism of the confidence argument.

The Evidential Tool Bus (ETB) [7] offers a framework for tool integration and workflow formalization to oversee the claims and evidence produced by formal tools, aiding assurance activities. The ETB is specifically designed to handle evidence generated in support of tool-specific inferences and for claims where formal tools have been employed, providing a mechanical approach to evidence management. However, this tool does not support the confidence argument, and the formal specification of context, assumption, and strategies.

7 Conclusion

In this paper, we presented Structured GSN (SGSN), an open-source tool to create and edit assurance cases. In addition to basic functionalities related to the creation of GSN, we added two types of analysis, syntactic analysis which checks if the developed GSN satisfies structured criteria, and semantic analysis which checks compartmental properties.

For the semantic analysis, we mapped the GSN to a formal model i.e., Markov Decision Process (MDP), and checked properties expressed in PCTL using PRISM probabilistic model-checker. The formal specification of GSN using MDP allows the extension of the current work to verify a larger set of properties related to the confidence argument.

We aim to enhance the existing version of SGSN to accommodate dynamic updates in the assurance argument. This extension can be particularly advantageous for systems reliant on AI, where claims and evidence may change during runtime based on operational data.

Acknowledgments. We would like to thank our students Alouani Mounir Yacine, and Ouail Mohamed Sedik for their participation in the design and implementation of the SGSN tool.

References

1. Adelard, L.: Assurance and safety case environment (ASCE) (2011)
2. Aziz, A., Sanwal, K., Singhal, V., Brayton, R.: Verifying continuous time Markov chains. In: Alur, R., Henzinger, T.A. (eds.) CAV 1996. LNCS, vol. 1102, pp. 269–276. Springer, Heidelberg (1996). https://doi.org/10.1007/3-540-61474-5_75
3. Barry, M.R.: Certware: a workbench for safety case production and analysis. In: 2011 Aerospace Conference, pp. 1–10. IEEE (2011)
4. Bloomfield, R., Bishop, P.: Safety and assurance cases: Past, present and possible future-an adelard perspective. In: Dale, C., Anderson, T. (eds.) Making Systems Safer, pp. 51–67. Springer, London (2009). https://doi.org/10.1007/978-1-84996-086-1_4
5. Bloomfield, R.E., Fletcher, G., Khlaaf, H., Hinde, L., Ryan, P.: Safety case templates for autonomous systems. arXiv:abs/2102.02625 (2021)
6. Bouchekir, R., Guzman, M., Cook, A., Haindl, J., Woolnough, R.: Formal verification for safe AI-based flight planning for UAVs. In: 2023 53rd Annual IEEE/IFIP International Conference on Dependable Systems and Networks Workshops (DSN-W), pp. 275–282. IEEE (2023)
7. Cruanes, S., Hamon, G., Owre, S., Shankar, N.: Tool integration with the evidential tool bus. In: Giacobazzi, R., Berdine, J., Mastroeni, I. (eds.) VMCAI 2013. LNCS, vol. 7737, pp. 275–294. Springer, Heidelberg (2013). https://doi.org/10.1007/978-3-642-35873-9_18
8. Denney, E., Pai, G.: Tool support for assurance case development. Autom. Softw. Eng. **25**(3), 435–499 (2018)
9. Denney, E., Pai, G., Habli, I.: Towards measurement of confidence in safety cases. In: 2011 International Symposium on Empirical Software Engineering and Measurement, pp. 380–383. IEEE (2011)

10. Denney, E., Pai, G., Pohl, J.: AdvoCATE: an assurance case automation toolset. In: Ortmeier, F., Daniel, P. (eds.) SAFECOMP 2012. LNCS, vol. 7613, pp. 8–21. Springer, Heidelberg (2012). https://doi.org/10.1007/978-3-642-33675-1_2
11. Group, A.C.W., et al.: Goal structuring notation community standard vestion 3. Technical report, Technical Report SCSC-141BA v2. 0, Safety Critical Systems Club (2018). https://scsc.uk/r141C:1?t=1 (2021)
12. Hawkins, R., Paterson, C., Picardi, C., Jia, Y., Calinescu, R., Habli, I.: Guidance on the assurance of machine learning in autonomous systems (amlas). arXiv preprint arXiv:2102.01564 (2021)
13. Kwiatkowska, M., Norman, G., Parker, D.: PRISM 4.0: verification of probabilistic real-time systems. In: Gopalakrishnan, G., Qadeer, S. (eds.) CAV 2011. LNCS, vol. 6806, pp. 585–591. Springer, Heidelberg (2011). https://doi.org/10.1007/978-3-642-22110-1_47
14. Matsuno, Y.: D-Case Editor: A Typed Assurance Case Editor. University of Tokyo (2011)
15. Palin, R., Ward, D., Habli, I., Rivett, R.: Iso 26262 safety cases: compliance and assurance (2011)
16. Robertson, A.: European organisation for the safety of air navigation (eurocontrol). Annuaire Europeen/European Yearbook: Vol. XXVII pp. 618–629 (1981)
17. Sljivo, I., Gallina, B., Carlson, J., Hansson, H., Puri, S.: Tool-supported safety-relevant component reuse: from specification to argumentation. In: Casimiro, A., Ferreira, P.M. (eds.) Ada-Europe 2018. LNCS, vol. 10873, pp. 19–33. Springer, Cham (2018). https://doi.org/10.1007/978-3-319-92432-8_2
18. Šljivo, I., Uriagereka, G.J., Puri, S., Gallina, B.: Guiding assurance of architectural design patterns for critical applications. J. Syst. Architect. **110**, 101765 (2020)
19. de la Vara, J.L., Ruiz, A., Blondelle, G.: Assurance and certification of cyber-physical systems: the amass open source ecosystem. J. Syst. Softw. **171**, 110812 (2021)
20. White, C.C., III., White, D.J.: Markov decision processes. Eur. J. Oper. Res. **39**(1), 1–16 (1989)

SysML Model-Driven Development for Digital Twins

Kanza Ouari[1(✉)], Malika Ioualalen[1], and Ahmed Hammad[2]

[1] MOVEP, USTHB University, Algiers, Algeria
{kouari,mioualalen}@usthb.dz
[2] FEMTO-ST Institute - UMR CNRS 6174, University of Bourgogne Franche-Comté, Besançon, France
ahammad@femto-st.fr

Abstract. A Digital Twin is a virtual replica of a physical system that is used to simulate its behavior, and provide services to improve its operation. However, their rigorous development is a difficult task and requires considerable efforts especially for their design. In this paper, we introduce a SysML model-driven development methodology for digital twins, a way to design digital twins at a high level of abstraction, automatically generate an essential part of their implementation, and verify their fidelity to their physical counterpart. In this methodology, the architecture followed is independent from application domains, implementation technologies, and purposes of using DTs. The physical system behavior is integrated to a SysML model in an operational language, and the fidelity verification is performed using a data lake. We explain with a case study from the literature how to implement a prototype of the architecture in the open source tool Eclipse. The case study indicates that a SysML model can facilitate the design of digital twins, improving thus their development process.

Keywords: Digital twin · Architecture · Model-driven development · SysML · Simulation · Fidelity

1 Introduction

A Digital Twin (DT) was first introduced by Michael Grieves [1] in a course for product lifecycle management in 2003 as the "virtual digital representation equivalent to physical products". Today, the advancement of digital technologies such as machine learning, cloud computing, IoT, and Big Data has led to wide adoption of DT [2]. These technologies have enabled the creation of additional services to the original system (the Physical Twin, PT) using DT such as: remote PT monitoring, user alerts, self-adapting mechanisms, etc.

Currently, the development of Digital Twins (DTs) is a difficult task due to the lack of a standard methodology for their development and their increasing complexity, which stems from the complexity of their physical counterpart.

B. Djamaa et al. (Eds.): CSA 2024, LNNS 1145, pp. 328–339, 2025.
https://doi.org/10.1007/978-3-031-71848-9_30

Adding to this, a DT must simulate the physical counterpart at a high level of fidelity. According to [3], a DT must behave exactly like its PT. It is important to ensure that the DT behaves like the PT before creating any additional service.

Model-Based Systems Engineering (MBSE) is a way to address this complexity. Its visual diagrams allow a comprehensive understanding of a system, as well as document its characteristics and enhance interdisciplinary communication among stakeholders [4]. In addition, the automated transformation of models into software implementation can improve productivity, reduce complexity, and facilitate the testing of complex system [5]. The System Modeling Language (SysML) [6] is a general-purpose modeling language used by MBSE. It enables capturing requirements, defining system structure, and describing system behavior. In the context of DT, visual diagrams created with SysML can be easily used for allowing a comprehensive understanding of DT, documenting its characteristics, and enhancing interdisciplinary communication among stakeholders. However, the automated transformation of SysML models into software implementation in general and DT implementation in particular is a challenge [4,7]. The authors in [8] refer to this utilization as "the technical utilization" of SysML models. Another challenge is to use MBSE, particularly SysML, for testing purposes in a DT context [3]. In the context of DT, the DT's behavior must be faithful to the PT's, i.e., the behavior of the DT must emulate with sufficient precision the behavior of its PT [9]. This requirement must be taken into account during the development of DTs and verified before exploitating DTs.

To the best of our knowledge, we are not aware of any comprehensive work on the development of DTs by defining a general architecture, with a focus on the automatic transformation of SysML models into DT implementations and fidelity verification before exploiting DTs. Previous works [4,7,10] have concentrated on the automatic transformation of SysML models into DT implementations in a specific context without addressing fidelity verification or proposed a standard architecture using other modeling languages with an emphasis on fidelity verification. The combination of the automatic transformation of SysML models into DT implementations with fidelity verification using a general architecture is a new approach to addressing the challenge of developing DTs. The use of a general architecture for the development of DTs has advantages in terms of making the technology of DTs, which remains promising, easy to understand, and adaptable to various contexts, regardless of application domains, implementation technologies, and purposes of using DTs. By combining the SysML language and fidelity verification, it is possible to rigorously develop a DT by modeling its characteristics, automatically generating its implementation, and ensuring its fidelity to its physical counterpart.

In this paper, we propose a new methodology to facilitate the development of DTs, structured in three steps: design, implementation, and verification. In the design step, we use a general architecture based on a data lake to represent the system components, and we use a SysML model to capture high-level information about the structure and behavior of the DT. In the implementation step, the automatic generation of usable code for the DT is possible in an operational language by directly transforming the SysML model. The behavior of the DT is integrated into the SysML model using the same operational language. In the verification step, the methodology verifies that the DT faithfully simulates the PT by exploiting the potential of the data lake. Figure 1 presents the steps of our methodology.

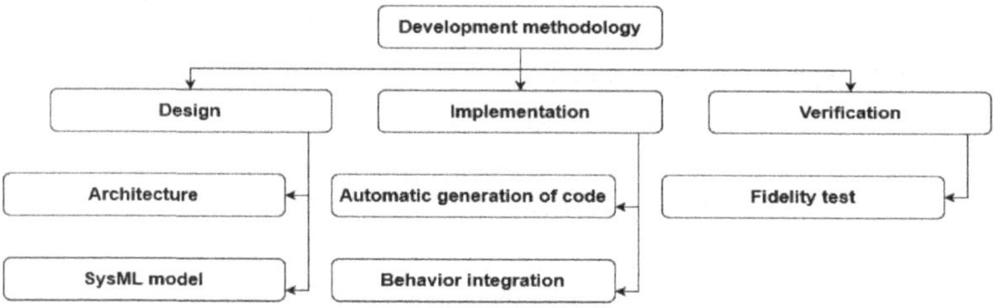

Fig. 1. Steps of the SysML model-driven development methodology.

The rest of this paper is structured as follows. After this introduction, Sect. 2 describes the background of our work. Next, Sect. 3 explicates the proposed DTs development methodology, and Sect. 4 explains the case study of a PT taken from the literature, use of the SysML model as a DT, creation of the data lake, and global implementation in Eclipse. Finally, Sect. 5 provides the related works and Sect. 6 presents the conclusion and an overview of future works.

2 Context

2.1 Digital Twin (DT)

A DT is a virtual copy that represents a physical system (its PT) and includes its characteristics. The virtual copy simulates the behavior of the PT through models and data. These models can be engineering models (e.g. CAD, Simulink) or software models (e.g. UML, SysML, MontiArc) [11]. The two twins are connected in a bidirectional way: the DT collects data from the PT and transfers it into services, and at the same time allows to interact with the PT and influence it.

Using a DT can offer many services for improving the PT operation [12]. For example, performing monitoring via real-time data collection to make better decisions and control the PT. A DT can also improve security and resilience by detecting malicious actions on a system. DT can predict future behaviors of the system and improve the process productivity using machine learning techniques. DT can also provide a testing platform for testing different scenarios to detect the most efficient one and to improve the system performances.

Additionally, the authors in [10] assert that DTs are software artifacts and, as such, require software engineering methods and practices, including rigorous processes to enhance their quality and ensure their proper functioning. An important example of a test in this context is the fidelity test, which verifies that the traces produced by the two twins are similar [9,13].

Although the meaning of DT may seem clear initially, the precise definition of a DT has been a topic of ongoing discussion and disagreement. Presently, there is no widely accepted and standardized definition for the term "DT". In our work, we consider a DT as a virtual representation that simulates the behavior of its PT. Simulation in our work represents the ability of DT to behave like its PT without providing any additional service. We define formally a concept of DT system as follows: "A system consisting of a DT connected in a bidirectional manner with its deployed PT. The purpose of this connection is to provide specific services for improving the operation of the PT. The entities involved in the system as software artifacts require specific tests to ensure that they behave as expected". In the design step of our methodology, we have defined an architecture that reflects our definition of a DT system. We consider the fidelity verification as a test to be performed on the behavior of the two twins, the objective is to have indications about the level of similarity between the behavior of the PT and the simulation of the DT.

2.2 System Modeling Language (SysML)

SysML is a general-purpose modeling language for systems engineering, based on UML. SysML reuses a subset of UML diagrams and extends it with new ones to model aspects of complex systems. It distinguishes three categories of diagrams: four structure diagrams (block definition diagram, internal block diagram, parametric diagram, package diagram), four behavior diagrams (state machine diagram, activity diagram, sequence diagram, use case diagram), and requirement diagram. Over these diagrams, in our work, we are interested in the block definition and package diagrams.

In the design phase of our methodology, we are interested in the Block Definition Diagram (BDD). The BDD is a structural diagram. It goes beyond the UML class diagram, which primarily focuses on classes and associations in an object-oriented context, by providing a more comprehensive structural representation. The BDD structures the system in blocks, sub-blocks and links. A block represents a complete system, a sub-system, or an element of a system. It can include values, properties, parts, references to other blocks, ports (properties specifying the types of interactions allowed between blocks), constraints (properties specifying constraints on other properties), and operations. Operations represent functionalities described in the behavioral part of the system. The use of the BDD has a dual objective: 1) Ensuring a comprehensive understanding of the DT structure. 2) Enabling the automatic transformation into DT implementation. In the case study section, we are interested in the Package Diagram (PD). The PD is a diagram that is used to organize and represent the various elements of a system model in a hierarchical manner. We use the PD to provide a visual representation of the possible modules that could be delivered from our proposal. It can help stakeholders to better understand the system.

3 SysML Model-Driven Development Methodology

3.1 Design

Architecture: To achieve our work, we defined a general architecture which reflects our definition of a DT system. It is composed of six components (see Fig. 2): PT component, DT component, Data Lake component (DL), Connection Components (CC), Service Components (SC), and Analysis Components (AC). The PT component represents the deployed physical device. The DT component is the virtual copy of the PT component. The DL component plays the role of a storage space to manage data collected from the other components of the DT system. The CC represent possible interactions between the other components, this component is represented in the figure as arrows. The SC represent additional services and functionalities offered by the DT component to improve the operation of the PT component. The AC represent tests on the other software components of the architecture to ensure that they behave as expected. It's worth noting that all components of the architecture can interact with the DL component through CC, which are implemented by the architecture developer.

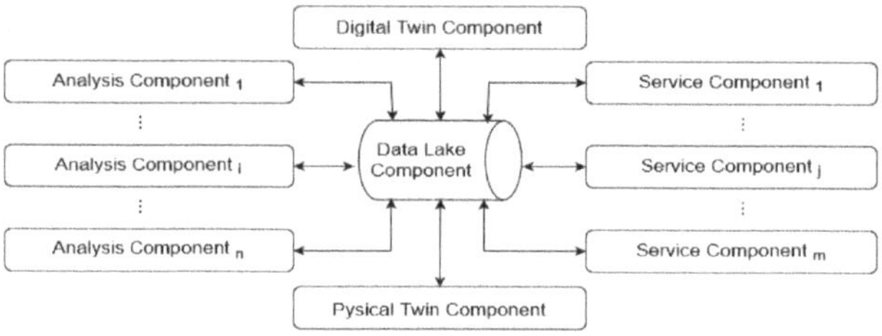

Fig. 2. Architecture for developing DT syetms.

The purpose of the methodology proposed in this work is to develop a DT that simulates the behavior of its PT and verify the fidelity, which requires the use of the PT component, DT component, DL component, a single analysis component, which we call Fidelity Test Component (FTC), CC with three interactions, between PT and DL, DL and DT, as well as between DL and FTC, without any service component. The restricted architecture that we are interested in is illustrated in Fig. 3. It is composed of five components: PT, DT, DL, CC and FTC. We consider it as an initial and essential architecture for verifying the fidelity of DT before extending it with any service component.

SysML Model: In our proposal, the developer works at a high level of abstraction by creating a SysML model as BDD to model the DT in a SysML modeling tool. Working at a high level of abstraction eliminates low-level details and takes only specific features of the PT, instead of taking the physical system as a

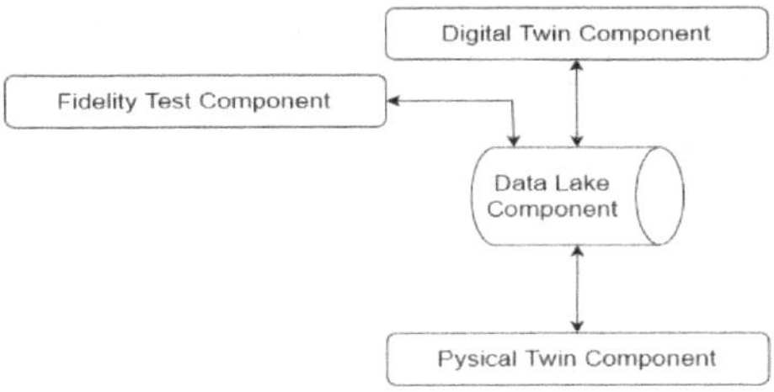

Fig. 3. Initial architecture for verifying DT fidelity.

whole, the BDD shows thus only the structural elements that interest a developer. The resulting BDD is an abstraction of the PT that can include other elements required by the simulation that the DT performs. The behavior of the DT, which reflects the behavior of the PT, can be indicated by a list of operation in each BDD block at a high level, without going into the details of how the behavior is implemented. The behavioral part of the operations is integrated in the implementation step of our methodology as we will see in the next section.

3.2 Implementation

Automatic Generation of Code: After modeling the DT using the BDD, the developer automatically generates the SysML model implementation in an operational language. This activity automatically transforms the SysML model from a graphical format to a machine-interpretable textual format, or DT implementation. This allows enhancing the productivity, saving time and minimizing coding errors and mistakes. There are tools that provide interfaces allowing code generation, thus enabling the technical utilization of SysML. In this activity, a tool that allows code generation is needed.

Behavior Integration: The behavior of the DT is completed by the developer manually in the same operational language by implementing the body of the operations modeled in the SysML model (BDD). This allows to obtain a DT that simulates the behavior of its PT. He also creates the DL in an appropriate database, and implements the CC that allow interaction between the other components. At this stage, the DL interacts with the PT that executes commands, and with the textual SysML model that simulates the behaviors of the PT.

3.3 Verification

The objective of the verification step is to ensure that the behavior of the PT and the DT are similar. To achieve this, the developer implements the FTC which uses the DL to measure the fidelity. It queries the database to retrieve the results produced following the execution of a command by the PT and the simulation of the DT. The same input data is fed into both the PT and DT, and then the traces of their behavior are compared to determine if there are any differences. At this stage, the developer also implements the CC that allows the interaction between the DL and FTC.

4 Case Study

4.1 Physical Twin (PT): Lego Mindstorms Vehicle

This case study is taken from the literature [10]. The Lego Mindstorms vehicle, shown in Fig. 4, consists of three captures: an ultrasound sensor to measure distances, a tactile sensor to detect collisions, and a light sensor to differentiate colors. The vehicle has a motor to advance and turn at a specific angle. It includes also a pose-provider to detect its position. The vehicle interacts with computers through a Bluetooth connection. The Lego software installs itself on a computer and controls it through various behaviors. This software represents the firmware used by the vehicle, and it is implemented in our case in the Java programming language. An example of a behavior to be simulated by the DT and verified by the FTC is the following: Lego follows a black line on the ground, if the line disappears, it continues to turn until it finds it.

Fig. 4. Lego Mindstorms vehicle [10].

4.2 Digital Twin (DT)

Graphical SysML Model. The BDD diagram is used as a SysML model to specify the physical vehicle structure in Papyrus tool with additional information required by the simulation. The BDD serves as the DT component in graphical format. Papyrus is a software system modeling tool developed by the Eclipse community. It allows software engineers to model software systems using UML and SysML. Papyrus is used as a modeling tool because it is open source and offers interfaces with other tools to directly transform SysML models. The BDD diagram used to model the vehicle is presented in Fig. 5. The main block presented in the BDD diagram is called Vehicle. The other main blocks are Motor, UltrasonicSensor, LightSensor and TouchSensor. The signature of each operation is indicated in the block operation compartment.

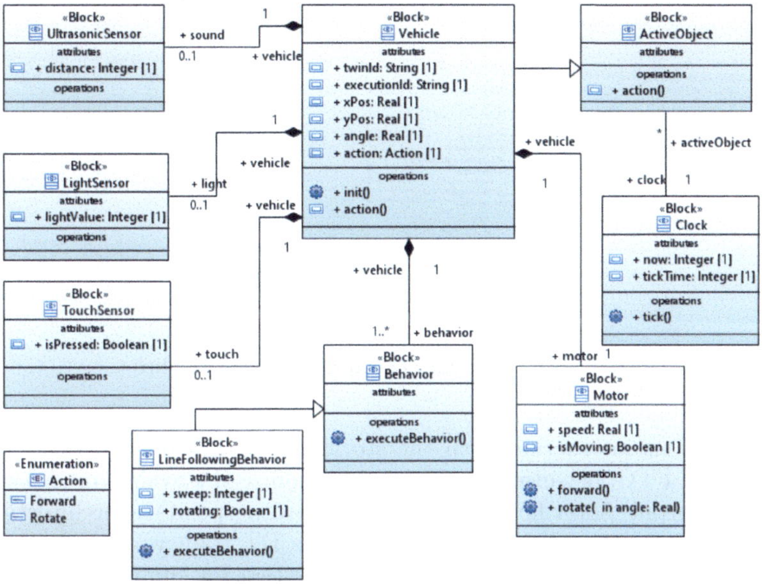

Fig. 5. Graphical SysML model of the DT of Lego.

Textual SysML Model. We have installed Papyrus in Eclipse to allow the direct exchange of SysML model, from a graphical model to a textual model, expressed as Java code. The Papyrus modeling tool features an interface that enables automatic code generation. We have chosen the Java programming language because, as mentioned before, the firmware of the vehicle uses Java. This activity presents the automatic generation of the code in the implementation step of our methodology. The code includes all the model information (block names, block properties, operation signatures and links between blocks), which ensures a lossless integration. This code is completed by the behavior of the vehicle and other behavior related to the simulation, by implementing the body of the operations. This process illustrates the behavior integration in the implementation step of the proposed methodology. To illustrate the relation between Papyrus and Eclipse, Fig. 6 shows the SysML modeling in Papyrus, automatic generation of code in Eclipse, and manual integration of behavior.

4.3 Data Lake (DL)

In order to manage all data on the DT system and ensure the interaction between all components of the architecture, we used Neo4j to create the DL. Neo4j is a graph-oriented NoSQL database. It uses nodes and arcs to store data instead of tables as in relational databases. Neo4j allows complex queries, necessary when efficient data analysis is required through its expressive Cypher language, which is one of the reasons why we chose this database, since we want to ensure an

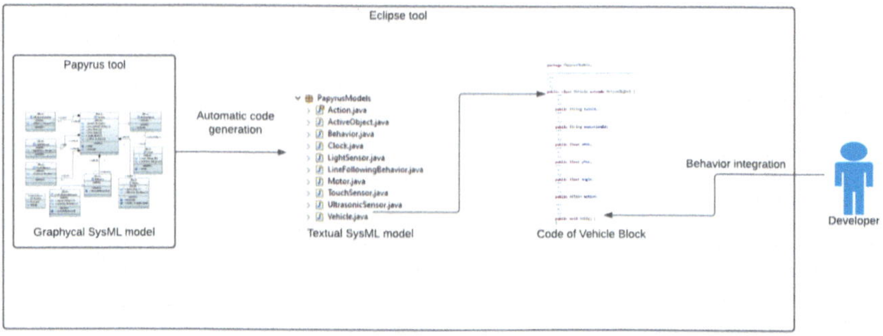

Fig. 6. The relation between Papyrus and Eclipse.

easy and efficient interaction with the storage space. In our work, we did not implement the FTC, because we did not synchronize the SysML model with the real vehicle, we only show how to implement a prototype of the architecture proposed in Eclipse to demonstrate our ideas. The DT framework that allows the SysML model to be connected to a physical system for simulation and fidelity test purposes can be implemented in Eclipse using the Java programming language by creating a project, named for example DigitalTwinFramework, with five packages. We have used the PD to illustrate these packages or modules in Fig. 7. The required knowledge to implement this project includes SysML in Papyrus, advanced concepts in Java programming language such as collections, exceptions, threads, and sockets, as well as the ability to create and interact with a Neo4j database, a real system, and the FTC using Java language.

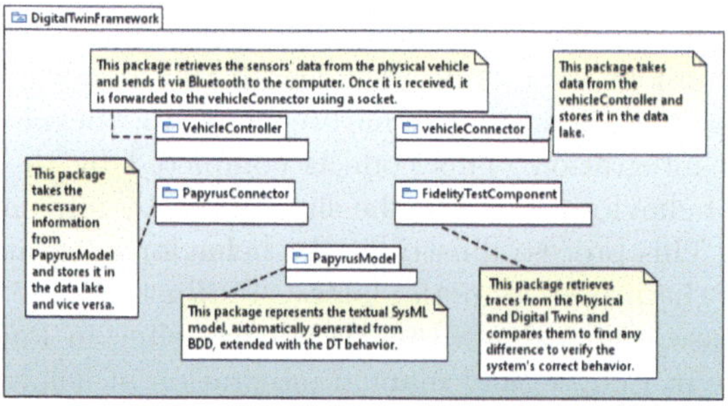

Fig. 7. PD showing how to implement a prototype of the initial architecture.

5 Related Work

We can link our work to several previous works conforming to three axes: 1) Architecture. 2) Automated transformation of models into DT implementation. 3) Fidelity verification.

Several works have described a possible architecture for developing DTs. [1] proposed an initial architecture with three dimensions: physical entity, virtual model, and connection. This same number of dimensions has been considered by other authors, such as [3]. Based on the initial architecture, works like [14,15] have proposed five dimensions: physical entity, virtual model, services, data, and connection. [10] proposed an architecture with five components: physical entity, digital twin, data lake, analysis, and services. We have taken inspiration from this work and included the AC in our architecture. Generally, there are several architectures proposed in the literature, but they do not illustrate the possibility of using the DT for fidelity verification. In our work, we introduced a formal definition for DT systems in general and associated it with an architecture consisting of six components: PT, DT, DL, CC, SC, and AC. Our architecture is independent from application domains, implementation technologies, and purposes of using DTs. We also demonstrated how to restrict this architecture to simulate the PT behavior and verify the fidelity using five components: PT, DT, DL, DL, and FTC.

The automated transformation of models into DT implementation is already used by several authors. [11,16–18] have used an architectural model of components and connectors in MontiArc with UML class diagrams specifying the data types of the objects exchanged between components as a starting point to facilitate the automatic generation of DTs, the behavior of DTs is integrated by a domain-specific language. In the context of using SysML, we have identified two works [4,7] that use a SysML diagrams to design DT and transfer them directly into machine-recognizable code in a specific context, where part of the DT behavior is generated by the SysML state machine diagram, without following any architecture or separated steps. Our work differs from these works by two main novelties: 1) Our work uses SysML as a starting point for the systematic development of DTs in general by separating the architecture from SysML usage and implementation. 2) Our methodology requires the fidelity verification before integrating services to ensure the quality of DT.

Some works have implicitly considered fidelity verification, such as [3,13,19]. [3] mentioned the possibility of performing specific tests on the DT. [13] has proposed a method for validating the behavior of the two twins. [19] has shown the importance of defining DTs at different levels of abstraction for performing tests according to the properties to be validated. Concrete proposals for measuring the fidelity between the two twins are introduced in [10] [9]. These works use the UML language to introduce a model-driven methodology for testing DTs. In [10], a verification method similar to the one proposed in our work is treated. In [9], a bioinformatics algorithm is adapted to validate that the behavior of the two twins is the same or sufficiently similar. These works use class diagrams in a tool called USE [20] to model the DT at a high level of abstraction. The behavior of the DT is integrated with an imperative language called SOIL [21]. Our work exploits the SysML language instead of the UML language to launch another research methodology for developing DTs by focusing on the fidelity verification before integrating additional services.

6 Conclusion

In this work, we have proposed a new methodology based on SysML model for systemically developing DTs. The methodology supports three main steps: design, implementation and verification. In the design step, we have adapted an independent architecture from application domains, implementation technologies, and purposes of using DTs. The DT is described using a BDD. In the implementation step, an essential part of the code was automatically generated. In the verification step, we are required to verify the fidelity of the DT before developing any other service component in order to ensure its quality in the future development steps. In the future directions of this work, we intend to synchronize the textual SysML model with the real vehicle. Also, we want to show how to model the FTC using a SysML model and automatically derive its code, following the process indicated in our methodology. In addition, we plan to analyze the traces of the two twins using the bioinformatics algorithm used in the literature. Additionally, we intend to use a requirement diagram to capture and model the requirement of our system in the design step. Also, we want to automatically generate a part of the behavior using the SysML behavior diagrams. Finally, we want to integrate other AC and SC to our initial architecture.

References

1. Grieves, M.: Digital twin: manufacturing excellence through virtual factory replication. White Paper **1**(2014), 1–7 (2014)
2. Sharma, A., Kosasih, E., Zhang, J., Brintrup, A., Calinescu, A.: Digital twins: state of the art theory and practice, challenges, and open research questions. J. Ind. Inf. Integr. 100383 (2022)
3. Grieves, M., Vickers, J.: Digital twin: mitigating unpredictable, undesirable emergent behavior in complex systems. In: Kahlen, F.-J., Flumerfelt, S., Alves, A. (eds.) Transdisciplinary Perspectives on Complex Systems, pp. 85–113. Springer, Cham (2017). https://doi.org/10.1007/978-3-319-38756-7_4
4. Wilking, F., Sauer, C., Schleich, B., Wartzack, S.: Sysml 4 digital twins-utilization of system models for the design and operation of digital twins. Proc. Des. Soc. **2**, 1815–1824 (2022)
5. France, R., Rumpe, B.: Model-driven development of complex software: a research roadmap. In: Future of Software Engineering (FOSE'07), pp. 37–54. IEEE (2007)
6. Friedenthal, S., Moore, A., Steiner, R.: Omg systems modeling language (omg sysml™) tutorial. In: INCOSE International Symposium, vol. 18, no. 1, pp. 1731–1862. Wiley Online Library (2008)
7. Wilking, F., Sauer, C., Schleich, B., Wartzack, S.: Integrating machine learning in digital twins by utilizing Sysml system models. In: 17th Annual System of Systems Engineering Conference (SOSE), pp. 297–302. IEEE (2022)
8. Handley, H.A., Khallouli, W., Huang, J., Edmonson, W., Kibret, N.: Maintaining the consistency of sysml model exports to xml metadata interchange (XMI). In: IEEE International Systems Conference (SysCon), pp. 1–8. IEEE (2021)
9. Muñoz, P., Wimmer, M., Troya, J., Vallecillo, A.: Using trace alignments for measuring the similarity between a physical and its digital twin. In: Proceedings of the 25th International Conference on Model Driven Engineering Languages and Systems: Companion Proceedings, pp. 503–510 (2022)

10. Muñoz, P., Troya, J., Vallecillo, A.: Using UML and OCL models to realize high-level digital twins. In: 2021 ACM/IEEE International Conference on Model Driven Engineering Languages and Systems Companion (MODELS-C), pp. 212–220. IEEE (2021)
11. Bibow, P., et al.: Model-driven development of a digital twin for injection molding. In: Dustdar, S., Yu, E., Salinesi, C., Rieu, D., Pant, V. (eds.) CAiSE 2020. LNCS, vol. 12127, pp. 85–100. Springer, Cham (2020). https://doi.org/10.1007/978-3-030-49435-3_6
12. Segovia, M., Garcia-Alfaro, J.: Design, modeling and implementation of digital twins. Sensors **22**(14), 5396 (2022)
13. Bordeleau, F., Combemale, B., Eramo, R., van den Brand, M., Wimmer, M.: Towards model-driven digital twin engineering: current opportunities and future challenges. In: Babur, Ö., Denil, J., Vogel-Heuser, B. (eds.) ICSMM 2020. CCIS, vol. 1262, pp. 43–54. Springer, Cham (2020). https://doi.org/10.1007/978-3-030-58167-1_4
14. Tao, F., Cheng, J., Qi, Q., Zhang, M., Zhang, H., Sui, F.: Digital twin-driven product design, manufacturing and service with big data. Int. J. Adv. Manuf. Technol. **94**, 3563–3576 (2018)
15. Qi, Q., et al.: Enabling technologies and tools for digital twin. J. Manuf. Syst. **58**, 3–21 (2021)
16. Kirchhof, J.C., Michael, J., Rumpe, B., Varga, S., Wortmann, A.: Model-driven digital twin construction: synthesizing the integration of cyber-physical systems with their information systems. In: Proceedings of the 23rd ACM/IEEE International Conference on Model Driven Engineering Languages and Systems, pp. 90–101 (2020)
17. Dalibor, M., Michael, J., Rumpe, B., Varga, S., Wortmann, A.: Towards a model-driven architecture for interactive digital twin cockpits. In: Dobbie, G., Frank, U., Kappel, G., Liddle, S.W., Mayr, H.C. (eds.) ER 2020. LNCS, vol. 12400, pp. 377–387. Springer, Cham (2020). https://doi.org/10.1007/978-3-030-62522-1_28
18. Bolender, T., Bürvenich, G., Dalibor, M., Rumpe, B., Wortmann, A.: Self-adaptive manufacturing with digital twins.' in,: International Symposium on Software Engineering for Adaptive and Self-Managing Systems (SEAMS), 156–166. IEEE (2021)
19. Arrieta, A.: Multi-fidelity digital twins: a means for better cyber-physical systems testing?" arXiv preprint arXiv:2101.05697 (2021)
20. Gogolla, M., Büttner, F., Richters, M.: Use: a UML-based specification environment for validating UML and OCL. Sci. Comput. Program. **69**(1–3), 27–34 (2007)
21. Büttner, F., Gogolla, M.: On OCL-based imperative languages. Sci. Comput. Program. **92**, 162–178 (2014)

Leveraging Large Language Models and Knowledge Graphs for Advanced Biomedical Question Answering Systems

Mohamed Chakib Amrani(✉), Abdellah Hamouda Sidhoum, M'hamed Mataoui, and Kadda Baghdad Bey

Ecole Militaire Polytechnique, BP 17, Bordj El-Bahri, Algiers, Algeria
mohamedchakib.amrani@emp.mdn.dz

Abstract. This study introduces a Knowledge Base Question Answering (KBQA) system augmented with Large Language Models (LLMs) to address the complexities of biomedical data exploration and knowledge extraction. The proposed system leverages LLMs to translate natural language questions into structured graph queries, tailored for Knowledge Graph (KG) databases. This enables medical professionals to efficiently explore stored data with no need to master any graph query language syntax. The responses retrieved are then processed through the LLMs, which rephrase and refine this information to generate human-like answers. We evaluated our system using the BioASQ benchmark dataset, employing different knowledge sources, namely, PrimeKG and Hetionet. For the LLMs, LLAMA2-70B and GPT-4 were used. Cypher, adapted for Neo4j databases, served as the graph query language for our experiments. Initial results show that our system obtains good performance in the case of List-type questions. The presented comparative experiments indicate that the success of our proposed approach depends significantly on the quality of both the KGs and the LLMs used for generating queries. This underscores the necessity for more comprehensive KGs covering a wider range of biomedical knowledge.

Keywords: Question Answering · Large Language Models · Knowledge Graphs · Cypher Queries · Biomedical Data

1 Introduction

In the rapidly development of biomedicine, researchers are increasingly confronted with the challenge of searching through an immense volume of data to extract meaningful insights. To address this issue, Question Answering systems have emerged as pivotal tools, creating a bridge between vast data and actionable knowledge. LLMs such as GPT-3.5 [1], GPT-4 [2] and LLAMA2 [3] are particularly prominent in this quest, offering exceptional linguistic capabilities. However, it appears that LLMs may sometimes produce some unexplainable,

B. Djamaa et al. (Eds.): CSA 2024, LNNS 1145, pp. 340–349, 2025.
https://doi.org/10.1007/978-3-031-71848-9_31

false or misleading results known as hallucination problem [4,5]. This issue cannot be tolerated in critical cases particularly in domain specific questions like biomedicine. Conversely, knowledge graphs are recognized as crucial, providing a structured and verifiable foundation of knowledge.

In their survey, Jin et al. [6] identify five distinctive approaches for developing Biomedical Question Answering (BQA) systems, namely: Classic, Machine Reading Comprehension based (MRC), Knowledge Base based (KB), Information Retrieval based (IR), and Question Entailment based (QE). The Classic approach is a structured, multi-stage process involving question analysis, document retrieval and content extraction, and answer ranking. The MRC-BQA approach employs contextual analysis of texts, frequently leveraging information retrieval techniques, to infer answers. The KB-BQA approach, on the other hand, retrieves answers from structured data sources such as relational databases or knowledge graphs, focusing on specific entities and their relationships. The IR-BQA based approach involves the extraction of relevant text fragments extracted from a corpus of documents assumed to be good candidates to the addressed question. In contrast, the QE-BQA approach utilizes a similarity-based technique, sourcing answers from previously resolved queries, such as those cataloged in FAQs databases. Our proposal follows the KB-BQA approach, using a combination of KGs with LLMs capabilities to improve the QA task.

In this paper, we introduce a biomedical KBQA system based on a combination of KGs and LLMs. The system employs two layers of LLMs. The first layer is for converting users' questions into structured queries. Subsequently, these queries are executed on a well-established biomedical KG database, producing a sub-graph containing relevant nodes and relations. Following this, a secondary LLM refine the returned data into coherent, natural responses. Through this pipeline, the system facilitates interaction of medical professionals with complex KG structures. Unlike traditional QA systems that source answers from LLMs, our system ensures information reliability by sourcing answers from verified KGs. This approach effectively addresses the common challenge of hallucinations in LLMs outputs. We conducted an evaluation of our system using the BioASQ biomedical benchmark dataset [7], which provides expert-level questions. For the source of biomedical knowledge, we utilized the Hetionet [8] and PrimeKG [9] KGs. GPT-4 and LLAMA2-70B were employed as the LLMs for our experiments. We selected Cypher as a structured graph queries language, due to its expressive and efficient query syntax. The results suggest that our system appears to perform effectively on List type questions, indicating a promising outlook for the methodology we have adopted. However, it seems to show less satisfactory performance in certain question categories, owing primarily to gaps in the information available within the KGs. Following an analysis of our experimental results, we conclude that the efficacy of our system is dependent on the quality of the selected LLMs and KGs. This finding underscores the necessity for more comprehensive and richly detailed biomedical KGs to enhance the system's overall performance and reliability.

2 Related Work

In recent years, the combination of KGs with LLMs has witnessed remarkable progress. Pan et al. [10] classify this integration into three types: (i) KG-Enhanced LLMs, which aim to address LLM's hallucination issues and lack of interpretability by using KGs to enhance their knowledge awareness and performance (ii) LLM-Augmented KGs, where LLMs are applied to KG-related tasks to handle incomplete KGs and process text corpus for KG construction; and (iii) Synergized LLMs + KGs, focusing on a unified framework where LLMs and KGs work together to process textual and structural data to enhance each other. While each category of approaches offers unique contributions, our focus in this research is on the KG-Enhanced LLMs category, which utilizes the natural language processing potentials of LLMs to enrich the information retrieval capabilities of KGs.

Several approaches following KG-Enhanced LLMs paradigm have been developed. Lu et al. [11] utilize LLMs to transform natural language queries from pharmacologists into SQL queries for a pharmacology KG database. This process employs Named Entity Recognition (NER) and Named-Entity Linking (NEL) models to convert entities into type-aware placeholders, facilitating the use of a GPT-4 powered SQL generator for schema linking and question decomposition, utilizing the PrimeKG knowledge graph. However, reliance on NER introduces challenges in complex reasoning scenarios. Similarly, Taffa et al. [12] demonstrate a system where LLMs are employed to convert natural language questions into SPARQL queries for a Scholarly KG based QA system. This method uses a BERT-based sentence encoder to identify similar training questions, guiding the LLM towards SPARQL query generation. Despite achieving high accuracy rates, the model's reliance on identifying similar questions can be a limiting factor for queries that diverge from the training set.

In the same context, Zhou et al. [13] propose an approach using LLMs to convert natural language queries into GraphQL queries that match the KG structure. The proposed approach involves smaller LLMs for initial processing and a larger LLM for refining the queries, showcasing the need for a balanced use of multiple models for effective KG interaction. Furthermore, Madani et al. [14] integrate LLMs with logical programming for domain-specific Question Answering over KGs. This method employs a T5-small transformer model to align natural language questions with the KG's schema. They illustrate the importance of training data coverage for the accuracy of the generated queries.

Within the varied landscape of the KG-Enhanced LLMs paradigm, our proposal aims to define a complete QA pipeline that starts with the expert's question, expressed in natural language, generates a graph query, proceeds to the selection of relevant subgraphs from the KG, and ends by returning natural language answers. Compared to existing work, our proposed system is distinguished by the following aspects: (i) it represents a unique architecture for the biomedical field that integrates KGs with double LLMs layers, where the first one is dedicated to generate Cypher queries from natural language based on the provided schema definition, and the second one is dedicated to the generation of natural language from the graph of results; (ii) can be easily configured by using multiple LLMs and KGs; (iii) based on an architecture that adapts to several domains whose knowledge is made available in the form of KG; (iv) it needs little adaptation to cover the case of other graph-oriented query languages syntax, such as Gremlin.

3 Methodology

In this study, we introduce a novel biomedical knowledge base question answering system that combines LLMs and KGs. Our system, illustrated in Fig. 1, enhances the standard KBQA framework by integrating LLMs as a bridge between expert questions and the answers. Two layers of LLMs are used: the first LLM generates structured queries from natural language questions. These resulting structured queries are then executed on the KG database to extract a sub-graph of relevant nodes and relations. The second layer of LLMs translates this data into natural language answers, utilizing both the sub-graph and the initial question. Those LLMs layers are inferred using Few-Shot Prompting technique [15–17] ensuring results reliability.

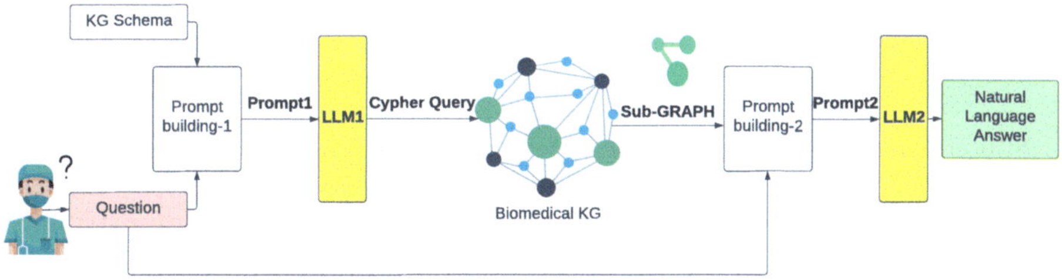

Fig. 1. Architecture of the proposed biomedical question answering system.

The foundation of our system is the Biomedical KG, denoted by $G = (V, E)$, where V represents the entities and E the semantic relations between them. This KG serves as the main source of structured biomedical data, containing detailed information about various biomedical entities, such as: Drugs, Diseases, Biological Processes, Genes, Phenotypes, Molecular Functions, and others. These entities are interconnected through biomedical relationships, for instance: Disease-Disease Interactions, Drug-Drug Interactions, and Drug-Disease Interactions.

Our pipeline process begins with the first prompt building function. This function takes a user's question q and a biomedical KG schema s, creating a tailored prompt $p1$ for generating structured queries based on the provided KG schema. This prompt includes various rules and instructions to guide the model in generating accurate queries[1]. Examples of the rules used are illustrated in Fig. 2. Rules (1, 2, 3) are employed to constrain LLM1's output by providing

[1] https://github.com/phdResearcherDz/CypherTranslator.

guidelines on the generated query format. Rule 4 is used to ensure case-insensitive matching when attempting to align nodes or attributes, meaning that the compared elements are transformed to a uniform case. Regarding wildcard search, as mentioned in Rule 4, allows for matching nodes' attributes with the terms of the question even when they do not completely alike, such as when one string is contained within the other. Rule 5 specifies the schema used, presented in textual format.

Rule 1: Ensure that Cypher query directly addresses the provided question without deviation.
Rule 2: Use only the Nodes, Relationships, and Attributes outlined in the provided schema. Avoid elements not included in the schema.
Rule 3: Provide only the Cypher query result, omitting additional explanations.
Rule 4: Employ a case-insensitive and wildcard search for property-based searches.
Rule 5: Conduct queries using the specified schema:{SCHEMA}.
Rule 6: ...

Fig. 2. Examples of rules and instructions used in the first prompt builder function.

Subsequently, as indicated in Fig. 1, the prompt building function's output $p1$ is processed by the LLM1 layer to generate a structured Cypher query c. This query is then executed on a KG database to retrieve relevant nodes, attributes, and relationships that may provide an answer to q. This result is based on the existing knowledge in the KG and represented as a relevant sub-graph of the biomedical KG. Thereafter, a second prompt building function that combines initial question q with the resulting sub-graph is employed to build a prompt $p2$. Finally, this prompt is introduced to LLM2 to construct a human-like response relevant to the initial expert question.

4 Experiments

4.1 Benchmark Dataset

BioASQ 11B Task B [7] dataset is employed to evaluate the proposed system. This dataset is a part of the BioASQ challenge supported by National Library of Medicine and Google[2]. It consists of questions in English coupled with corresponding answers and related material, tailored to the requirements of biomedical research. BioASQ contains four types of questions: factoid, list-based, summary-oriented, and yes/no, categorizing queries based on the specificity of the answer, from precise facts to broader summaries or binary responses.

[2] http://bioasq.org/.

4.2 Biomedical Knowledge Graphs Used

In order to evaluate the impact of KG coverage on the performance of our proposal, we explored the use of two graphs, namely: Hetionet and PrimeKG.

- **Hetionet** [8], is a biomedical KG that merges data from 29 publicly available sources. It is composed of 47,031 nodes with 11 distinct types, including Anatomy, Biological Process, Cellular Component, Compound, Disease, Gene, Molecular Function, Pathway, Pharmacologic Class, Side Effect, Symptom. This KG also includes 2,250,197 relationships of 24 different types, reflecting a vast network of biomedical knowledge[3].
- **PrimeKG** [9], a multimodal KG for precision medicine available through Harvard Dataverse[4], integrates 20 biomedical data resources into a heterogeneous graph with 129,375 nodes and 4,050,249 relationships, offering detailed descriptions for disease and drug nodes and featuring more recent knowledge from the latest biomedical databases compared to Hetionet.

4.3 Selection Process of the Evaluation Questions

To evaluate our system, we implemented a selection process for a set of questions based on KG content. BioASQ Task 11B Phase B test dataset [7] has been chosen. This dataset provides a collection of biomedical questions. The first step in our selection process was determining the most suitable question type for evaluation. We opted for 'List' type questions due to their inherent alignment with the nature of knowledge in KGs. Since KGs primarily consist of interconnected concepts, 'List' type questions, which typically seek a list of concepts, are well-suited for extracting this type of structured data. This makes them a natural fit for our system, unlike 'Yes/No' questions better suited for encoder models or summary questions adapted for generative models. The second step consists in identifying questions with potential responses within used knowledge graphs. This involved extracting key concept words from the questions and answers using the Spacy library, then verifying their existence in the KG database. If all the concepts exist in the KG, we select the question. Based on this process, we chose 75 questions related to Hetionet [8], and 149 questions related to PrimeKG [9]. The set of questions used in the evaluation process is available on our github repository[5].

4.4 Evaluation Metrics

To assess the efficacy of our system, we employed distinct metrics for a comprehensive performance evaluation. The analysis comprised two primary steps. Initially, We evaluated the syntax of the query translations generated by the first LLM layer that is used to convert natural language questions to structured

[3] https://github.com/hetio/hetionet.
[4] https://github.com/mims-harvard/PrimeKG.
[5] https://github.com/phdResearcherDz/CypherTranslator/.

Cypher query. This evaluation is done using the Logical Accuracy Metric [18] for assessing query syntax correctness, defined as follows:

$$\text{Logical Accuracy} = \frac{\text{number of generated graph queries with correct syntax}}{\text{total number of generated graph queries}} \quad (1)$$

Recall, Precision, and F-Score are used to assess the relevance of the system responses which are more suitable for 'List' type responses.

5 Results & Discussion

Our experiments are split into two parts, in this case, evaluation of LLM1 and LLM2 layers outputs. Initially, we assessed the LLM1 ability to generate Cypher queries by measuring logical accuracy, examining the proportion of questions syntactically converted into correct Cypher queries. Following this, we evaluated the LLM2 layer's output, representing the returned answers, by using Recall, Precision, and F1 Score metrics. This evaluation was conducted by comparing the generated answers with reference answers provided in BioASQ Task 11B.

5.1 Evaluation of Generated Queries Syntax

To define the best suitable LLM for translating natural language queries into Cypher, we conducted experiments using two leading LLMs of the moment, namely: GPT-4 [2] and LLAMA2-70B [3], focusing on logical accuracy metric. We utilized the Inference API through Hugging Face[6] and OpenAI[7] platforms for the LLAMA2-70B and GPT-4 models, respectively. We evaluated these two models on 149 selected questions from the BioASQ dataset, based on PrimeKG content. We used a Neo4j interpreter to classify each generated query as either true or false syntactically. Preliminary results suggest that GPT-4 significantly outperforms LLAMA2-70B, achieving a Logical Accuracy of 94.74% compared to 53.40%. Based on these results, we selected GPT-4 to be used as LLM1 of our architecture.

5.2 Evaluation of System Results

We conducted a comparative analysis, detailed in Table 1 and Table 2, evaluating our system based on GPT-4 for query translation and answer generation, against the direct use of GPT-4 and LLAMA2-70B as generators of answers, by presenting results according to the mentioned metrics, namely: precision, recall, and F1-score.

Table 1 presents the obtained results related to the Hetionet KG. Our system outperforms GPT-4 in many categories, such as, Compound, Gene, Biological

[6] https://huggingface.co/meta-llama/Llama-2-70b-chat-hf.
[7] https://openai.com/.

Table 1. Comparative performance analysis of our system using Hetionet [8] KG with GPT-4 query translation versus direct GPT-4 and LLAMA2-70B Responses on 75 "List" type questions from the BioASQ dataset [7], categorized by node type.

Type	Our System			GPT-4			LLAMA2-70B			Total
	Prec	Rec	F1	Prec	Rec	F1	Prec	Rec	F1	
Compound	**100**	**93.75**	**96.77**	92.59	93.52	93.05	77.78	85.19	81.31	**9**
Gene	**93.57**	**89.13**	**91.30**	58.25	52.59	55.28	55.90	49.40	52.45	**33**
SideEffect	50.00	50.00	50.00	**58.50**	**55.24**	**56.82**	55.33	51.90	53.56	**8**
BiologicalProcess	**100**	**66.67**	**80.00**	64.81	54.26	59.07	73.15	62.78	67.57	**14**
Pathway	0.00	0.00	0.00	66.67	**100**	80.00	**100**	**100**	**100**	3
Anatomy	0.00	0.00	0.00	0.00	0.00	0.00	0.00	0.00	0.00	1
CellularComponent	0.00	0.00	0.00	0.00	0.00	0.00	0.00	0.00	0.00	2
Disease	**100**	**66.67**	**80.00**	66.67	49.58	56.87	72.92	50.42	59.61	4
MolecularFunction	0.00	0.00	0.00	**88.89**	**93.75**	**91.25**	88.89	93.75	91.25	1
Global	**93.34**	**86.15**	**89.60**	63.79	58.92	61.26	62.15	56.89	59.40	**75**

Process, and Disease. Notably, in the Gene category, with 33 questions, our system achieved a precision and recall that resulted in an F1 Score of 91.30%, significantly higher than GPT-4's 55.28%. Our system also outperforms in the Compound category, achieving a very high F1 Score of 96.77%, considerably higher than both GPT-4 and LLAMA2-70B. Additionally, GPT-4 outperformed in some categories, such as, Side Effect and Molecular Function, while LLAMA2-70B excelled in Pathway category.

Similarly, Table 2 provides the same evaluation, but this time related to the PrimeKG knowledge graph as a source of knowledge. For the Drug category, with 37 questions, our system notably outperforms GPT-4 and LLAMA2-70B in precision, recall and F1-Score by more than 10% for GPT-4 and more than 16% for LLAMA2-70B. Conversely, GPT-4 excelled in the Disease category, surpassing our system's precision by nearly 50%, indicating a clear strength in this topic. For categories with a substantial number of questions, such as, Gene/Protein and Effect/Phenotype, our system showed close precision compared to GPT-4. However, in categories like Biological Process, Pathway, Anatomy, Cellular Component, and Molecular Function, where fewer questions were selected, our system failed to return any relevant items, unlike LLAMA2-70B which was able to obtain very good results.

Our comparative analysis has outlined the distinct performance characteristics of our system, integrating both PrimeKG and Hetionet KGs with GPT-4. The results reveal that while our system demonstrates improved precision and recall in many categories, GPT-4 and LLAMA2-70B excel in others. The failure of our system, in some cases, might be attributed to two main factors: the inherent limitations of the LLM generator which may generate incorrect syntax query and the depth of information in the used knowledge graphs. Addition-

Table 2. Comparative performance analysis of our system using PrimeKG [9] KG with GPT-4 query translation versus direct GPT-4 and LLAMA2-70B Responses on 149 "List" type questions from the BioASQ dataset [7], categorized by node type.

Type	Our System			GPT-4			LLAMA2-70B			Total
	Prec	Rec	F1	Prec	Rec	F1	Prec	Rec	F1	
Drug	**91.60**	**87.60**	**89.55**	80.71	76.73	78.67	74.85	67.50	70.99	**37**
Disease	13.04	30.14	18.21	**72.63**	**66.40**	**69.38**	49.77	42.24	45.70	**31**
Biological Process	0.00	0.00	0.00	**91.67**	**88.89**	**90.26**	40.00	40.00	40.00	**6**
Gene/Protein	70.69	**74.33**	**72.47**	72.12	56.98	63.66	**73.17**	51.25	60.28	**52**
Effect/Phenotype	16.95	52.38	25.61	**76.94**	**71.19**	**73.96**	65.00	54.13	59.07	**18**
Pathway	0.00	0.00	0.00	0.00	0.00	0.00	**50.00**	**50.00**	**50.00**	1
Anatomy	0.00	0.00	0.00	**50.00**	50.00	50.00	62.50	**75.00**	**68.18**	**2**
Cellular Component	0.00	0.00	0.00	66.67	**66.67**	66.67	**100**	**66.67**	**80.00**	**1**
Molecular Function	0.00	0.00	0.00	**100**	**100**	**100**	100	100	100	**1**
Global	59.89	**67.39**	63.42	**75.10**	66.73	**70.66**	66.78	54.35	59.92	**149**

ally, the richness of data within knowledge graphs like PrimeKG and Hetionet varies related to the topics, which affects the system's ability to retrieve accurate information. The possibility of fine-tuning LLMs for specialized tasks, such as, translating questions into Cypher queries, coupled with the development of more comprehensive knowledge graphs, are promising topics of research to explore.

6 Conclusion

In this study, we expanded the scope of employing LLMs like GPT-4 and LLAMA2-70B by integrating them with Knowledge Graphs, such as, PrimeKG and Hetionet, focusing specifically on biomedical KBQA. We employed LLMs to translate natural language questions into Cypher queries for exploring data within KGs and subsequently utilized LLMs to transform the structured responses from KGs into natural language answers. Utilizing the BioASQ biomedical dataset, our approach demonstrated considerable performances in processing List type questions. Despite this, we encountered challenges related to the inherent limitations of LLMs and the depth of data within KGs, leading to some inconsistencies in performance across different biomedical topics. Recognizing these limitations, our future work will focus on exploring possibilities of integrating more large KGs and further customizing LLMs for specific tasks through targeted fine-tuning with dedicated datasets.

References

1. Brown, T., et al.: Language models are few-shot learners. Adv. Neural. Inf. Process. Syst. **33**, 1877–1901 (2020)
2. OpenAI: Gpt-4 technical report. ArXiv:abs/2303.08774 (2023). https://api.semanticscholar.org/CorpusID:257532815
3. Touvron, H., et al.: Llama 2: open foundation and fine-tuned chat models. arXiv preprint arXiv:2307.09288 (2023)
4. Beutel, G., Geerits, E., Kielstein, J.T.: Artificial hallucination: GPT on LSD? Crit. Care **27**(1), 148 (2023)
5. Guerreiro, N.M., et al.: Hallucinations in large multilingual translation models. Trans. Assoc. Comput. Linguistics **11**, 1500–1517 (2023)
6. Jin, Q., Yuan, Z., Xiong, G., Yu, Q., Ying, H., Tan, C., Chen, M., Huang, S., Liu, X., Yu, S.: Biomedical question answering: a survey of approaches and challenges. ACM Comput. Surv. (CSUR) **55**(2), 1–36 (2022)
7. Krithara, A., Nentidis, A., Bougiatiotis, K., Paliouras, G.: Bioasq-QA: a manually curated corpus for biomedical question answering. Sci. Data **10**(1), 170 (2023)
8. Himmelstein, D.S., et al.: Systematic integration of biomedical knowledge prioritizes drugs for repurposing. Elife **6**, e26,726 (2017)
9. Chandak, P., Huang, K., Zitnik, M.: Building a knowledge graph to enable precision medicine. Sci. Data **10**(1), 67 (2023)
10. Pan, S., Luo, L., Wang, Y., Chen, C., Wang, J., Wu, X.: Unifying large language models and knowledge graphs: a roadmap. IEEE Trans. Knowl. Data Eng. (2024)
11. Lu, H., Li, C., Li, Y., Zhao, J.: Utilizing large language models for natural interface to pharmacology databases. arXiv preprint arXiv:2307.15717 (2023)
12. Taffa, T.A., Usbeck, R.: Leveraging LLMS in scholarly knowledge graph question answering (2023)
13. Zhou, Y., et al.: r3-nl2gql: a hybrid models approach for for accuracy enhancing and hallucinations mitigation. arXiv preprint arXiv:2311.01862 (2023)
14. Madani, N., Joseph, K.: Answering questions over knowledge graphs using logic programming along with language models. arXiv preprint arXiv:2303.02206 (2023)
15. Ye, X., Durrett, G.: The unreliability of explanations in few-shot prompting for textual reasoning. Adv. Neural. Inf. Process. Syst. **35**, 30378–30392 (2022)
16. Hasan, M.A., et al.: Zero-and few-shot prompting with LLMS: a comparative study with fine-tuned models for Bangla sentiment analysis. arXiv preprint arXiv:2308.10783 (2023)
17. Lazaridou, A., Gribovskaya, E., Stokowiec, W., Grigorev, N.: Internet-augmented language models through few-shot prompting for open-domain question answering. arXiv preprint arXiv:2203.05115 (2022)
18. Guo, A., Li, X., Xiao, G., Tan, Z., Zhao, X.: SPCQL: a semantic parsing dataset for converting natural language into cypher. In: Proceedings of the 31st ACM International Conference on Information & Knowledge Management, pp. 3973–3977 (2022)

Graph Neural Network Based Approach for Restraining Misinformation Propagation in Online Social Networks

Hichem Merini, Adil Imad Eddine Hosni(✉), Kadda Baghdad Bey, and Islam Baira

Ecole Militaire Polytechnique, Bordj El-Bahri BP 17, Algiers, Algeria
{Hichem.merini,Adilimadeddine.hosni}@emp.mdn.dz, Hosni.adil.emp@gmail.com

Abstract. This paper addresses the problem of misinformation spread within online social networks (OSNs), a critical concern impacting public opinion and societal stability. We introduce a sophisticated Graph Neural Network-based framework designed to effectively curb misinformation propagation in OSNs. This framework is composed of three integral components: the Training Dataset Generator, the Graph Convolutional Network (GCN) Model Trainer, and the Influential Node Identifier. The Training Dataset Generator is crucial for calculating the acceptance rate of misinformation across nodes, thereby setting the stage for targeted interventions. The GCN Model Trainer utilizes this data to predict acceptance rates of misinformation for each node, identifying the most influential ones likely to spread misinformation. Finally, the Influential Node Identifier leverages these predictions to apply a strategic blocking method on key nodes. Employing a trained GCN model, we introduce a greedy algorithm aimed at selecting the most susceptible nodes for effective misinformation control. Our approach, when compared against baseline methodologies, demonstrates superior performance in mitigating the spread of misinformation.

Keywords: Graph neural network · online social networks · rumor influence minimization · informtation diffusion · Misinformation containment

1 Introduction

In the current digital era, social networks have become pivotal in shaping public discourse and influencing societal aspects ranging from individual behavior to global politics. The rapid spread of misinformation within these OSNs presents a significant threat, impacting public opinion, political stability, and financial markets. The urgency of tackling this challenge is highlighted by the considerable consequences that unchecked rumors have on society [1]. Social influences play a crucial role in making individuals vulnerable to misinformation, underscoring the complex relationship between human psychology and rumor propagation [2–4].

B. Djamaa et al. (Eds.): CSA 2024, LNNS 1145, pp. 350–360, 2025.
https://doi.org/10.1007/978-3-031-71848-9_32

Addressing this pervasive issue is vital for maintaining the integrity and stability of digital discourse.

While various strategies have been proposed to mitigate the spread of misinformation, existing methods often fall short in the dynamic and complex environment of OSNs. Traditional techniques, such as centrality measures, are limited in their ability to capture the evolving nature of these networks. The shortcomings of these approaches, combined with the critical need to understand and control misinformation spread, have prompted the development of more sophisticated methods. Recent advancements in Graph Neural Networks (GNNs) provide promising avenues for addressing these challenges. GNN models [5] are particularly adept at addressing the deficiencies of traditional methods by capturing the complex, interconnected structure of social networks. Their ability to leverage node and edge information allows for a more nuanced understanding of the intricate dynamics of misinformation spread.

This paper introduces a novel approach utilizing a GNN-based framework to simulate and understand the mechanisms of misinformation propagation in OSNs. This methodology not only captures the complex dynamics of misinformation spread but also efficiently identifies and targets influential nodes driving its propagation. Our initial results demonstrate a significant improvement over traditional methods, indicating the potential of this approach to enhance our understanding and control of misinformation in online spaces. The proposed method's advantages lie in its ability to adapt and respond to the continually evolving nature of social networks.

2 Related Work

In the realm of mitigating the spread of misinformation in online social networks, research has bifurcated into two primary categories: blocking strategies and truth campaign strategies. Blocking strategies, which can be further classified into node-blocking and edge-blocking approaches [6–8], have seen a series of evolutionary developments. The Least Cost Rumor Blocking (LCRB) method introduced by Fan et al. [9] was a significant early contribution, though it struggled with scalability and adaptability. To enhance user experience, Wang et al. [10] developed the DRIMUX model, focusing on user tolerance time thresholds, yet it did not fully address the diverse user experiences. The field further advanced with Tong et al.'s [11] Reverse-Tuple Based Randomized (RBR) algorithm, which improved efficiency but faced challenges in pinpointing influential nodes in complex networks. Shi et al.'s [12] Adaptive Influence Blocking (AIB) strategy leveraged real-time data for adaptability, albeit confronting the intricacies of dynamic monitoring. Yan et al.'s [13] two-stage MIR approach and dynamic programming method offered effective solutions but were constrained in highly intricate networks. Zhu et al.'s [14] heuristic-based Activity Minimization of Misinformation Influence (AMMI) provided practical solutions with some scalability challenges. Most recently, Jiang et al. [15] introduced a deep reinforcement learning-based dynamic rumor influence minimization (DRIM) model, marking a substantial advancement towards computational efficiency and data accuracy.

In parallel, truth campaign strategies have emerged as a potent countermeasure against misinformation. Hosni et al. [16] expanded on this by integrating individual and social behaviors into the HISBmodel and introducing a truth campaign strategy (TCS) for more effective rumor influence minimization in multiplex OSNs. Subsequently, Hosni et al. [17] refined this approach by proposing the Multi-Rumors Influence Minimization (MRIM) problem and the Dynamic Blocking Period (DBP) approach, which adaptively estimates the blocking period based on node activity, offering a $(1 - 1/e)$- approximation of the optimal solution and marking a significant progression in combating rumor proliferation during breaking news events.

The existing methods in mitigating the spread of misinformation exhibit certain limitations particularly in identifying influential nodes. A deep learning based approach can potentially tackle these issues by leveraging its capacity for feature learning and structural recognition, providing a more nuanced and dynamic identification of influential nodes even in complex and evolving network environments.

3 Primarily Knowledge and Problem Formulation

3.1 Definition of Social Networks

Online social networks can be succinctly modeled as directed graphs $G = (V, E)$, where V denotes nodes representing individuals, and E denotes directed edges. Key to our discussion are the adjacency matrix A and the feature matrix X. Matrix A reflects network connections with $A_{ij} = 1$ if $e_{ij} \in E$ and 0 otherwise. The feature matrix X, tied to graph G, holds attributes pertinent to misinformation spread, with each row being a node's feature vector x_i for the i-th node.

3.2 Graph Neural Networks

Graph Neural Networks (GNNs) have been introduced as a solution to navigate the intricacies of non-Euclidean data spaces. Building on the foundations laid by traditional neural networks, researchers aim to adapt GNNs for tasks including graph classification, link prediction, and node classification [18].

Consider a graph with node features represented by vector $X_G = \{x_v | v \in V\}$. GNNs aim to generate node embeddings $Z_G = \{z_v | v \in V\}$ and a comprehensive graph embedding h_G after k layers of aggregation. For each node v at layer k, its embedding $h_v^{(k)}$ is a combination of its own embedding and that of its adjacent nodes $N(v)$ from the previous layer $k - 1$ [18]:

$$h_v^{(k)} = MERGE\left(h_v^{(k-1)}, AGGR\left(\{h_u^{(k-1)} | u \in N(v)\}\right)\right) \tag{1}$$

The final graph embedding h_G is derived by pooling the embeddings of all nodes across all layers into a singular vector:

$$h_G = POOL\left(\{h_v^{(i)} | v \in V; i = 1, .., K\}\right) \tag{2}$$

Variations among GNNs largely arise from differences in the implementation of the $MERGE$, $AGGR$, and $POOL$ functions, which dictate the specific way in which node and graph embeddings are calculated and combined.

3.3 Misinformation Propagation Model

In this work, we select the HISBmodel [19] for misinformation diffusion in OSNs due to its unique consideration of individual and social behaviors and its adaptability to various RIM strategies. The model is defined on a network $G = (V, E)$ with individual characteristics C, each comprising β (background knowledge), ω (forgetting and remembering), and δ (hesitation). Initially, individuals are infected with different opinions. Thus, the individuals attraction to the rumor or misinformation is estimated :$A_u(t) = A_{int}e^{-\beta_u t}|\sin(\omega_u t + \delta_u)|$. Each infected node u has a chance to infect a neighbor v with probability:

$$p_{u,v}(t) = p_u^{send}(t) \cdot p_{v,u}^{acc},$$

where:

$$p_u^{send}(t) = e^{-\beta_u t}|\sin(\omega_u t + \delta_u)|, p_{v,u}^{acc} = \frac{1}{1 + d_v/d_u}, \tag{3}$$

and d_u and d_v are the connection degrees of nodes u and v, respectively.

The model introduces a subjective judgment factor α, simplifying individual opinion to either support or refute rumors. The process concludes when the popularity fades, indicated by:

$$R(t) = \sum_{v \in V} A_v(t) \cdot d_v \simeq 0. \tag{4}$$

Individual opinions are modeled as a discrete-time Markov chain [17] with states representing support, questioning, denial, or neutrality. The transition probabilities are:

$$\forall v \in V, P_v = \begin{bmatrix} 1-\rho & \frac{\rho}{2} & \frac{\rho}{2} & 0 \\ \frac{1-\rho}{3} & \frac{1-\rho}{3} & \frac{1-\rho}{3} & \rho \\ \frac{1-\rho}{3} & \frac{1-\rho}{3} & \frac{1-\rho}{3} & \rho \\ 0 & \frac{1-\rho}{2} & \frac{1-\rho}{2} & \rho \end{bmatrix}. \tag{5}$$

The decision factor ρ is influenced by both negative and positive reinforcements and subjective judgment:

$$\rho = \frac{n_v^-}{n_v} \cdot J_v,$$

where n_v^- is the number of negative opinions received by node v, n_v is the total number of received misinformation, and J_v is the subjective judgment factor.

3.4 Problem Formulation

We propose in this work an approach that aims to minimize the influence of misinformation in OSNs. The goal is to find a K the number of nodes selected for the blocking strategy that minimize the most the influence of the misinformation. Formally, we formulate the problem as follows: Given an OSN, a positive constant K representing the budget, and a misinformation detection time t_{det}, the objective is to find sets K that minimize the influence of misinformation. The goal is expressed through the following objective function:

$$\min_{K} \mathbb{E}\left[\sum_{v \in V} \bar{D_v}(T_f)\right], \tag{6}$$

where $\bar{D_v}(T_f)$ represents the individual v with a supporting opinion toward a misinformation at the final stage, $\mathbb{E}[.]$ denotes the expected number of individuals at the final stage of the propagation process, and T_f indicates the final stage of the propagation process. Thus, the optimization objective is to minimize the expected number of individuals supporting misinformation.

4 Proposed Solution

4.1 Framework Overview

The proposed framework operates through a cohesive workflow presented in Fig. 1, encompassing three main components: the Training Dataset Generator, the Graph Convolutional Network (GCN) Model Trainer, and the Influential Node Identifier. Initially, the HISB Model within the Training Dataset Generator simulates misinformation propagation across the network, generating data on nodes likelihood to accept misinformation. This data forms the training set for the GCN Model Trainer, which learns to predict the acceptance rates of misinformation for each node, thereby identifying potential influential nodes. Lastly, the Influential Node Identifier utilizes the insights gained from the GCN Model to target and apply a blocking strategy to these key nodes, disrupting the misinformation's spread. This integrated approach ensures an effective response to minimize the influence of misinformation in any new social network.

4.2 Training Dataset Generation

The necessity for generating a dataset with detailed information on misinformation propagation through a network is paramount for understanding and mitigating the spread of misinformation in OSNs. A comprehensive dataset illuminates how misinformation permeate through the network, pinpointing nodes particularly susceptible to misinformation. This is where the Training Dataset Generator module comes into play, specifically designed to calculate the acceptance rate of misinformation by each node within the network. Any propagation model that effectively captures the diffusion of information could be employed, though the HISBModel is our preferred choice due to its robustness and ability to mirror the complexities of human behavior and network dynamics. The choice of a model hinges on the unique requirements and traits of the network under study.

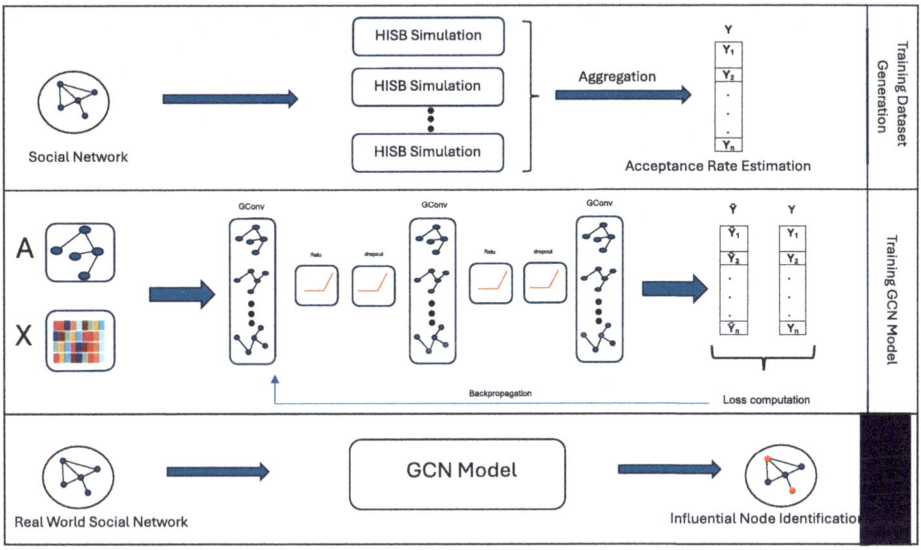

Fig. 1. Overview of the proposed framework.

The process begins with a simulation setup where the HISBModel is initialized with the network's topology and node's initial states. This includes assigning initial opinions and behavioral parameters to each node, which dictate their likelihood of engaging with and spreading misinformation. The model then simulates the spread of misinformation through the network, allowing nodes to interact and propagate the misinformation according to the model's rules. Throughout these simulations, we track how often each node accepts the misinformation. After running multiple simulations, we calculate the acceptance rate for each node reflecting its susceptibility to misinformation. Finally, the results from all simulations are aggregated, providing a comprehensive view of each node's behavior across different scenarios. The acceptance rate for each node is denoted by y_v for each node v, and collectively, these rates form the vector $Y : Y = [y_1, y_2, \ldots, y_n]$, where y_v is the acceptance rate of node v, and n is the total number of nodes in the network. The vector Y is crucial for the subsequent application of the GCN model.

4.3 Graph Neural Network Model

In our context, employing Graph Convolutional Networks (GCN), as conceptualized by Kipf et al. [20], is instrumental due to their adeptness at managing the intricacies of graph-structured data, characteristic of social networks. The GCN is particularly suited for dissecting the complex interactions and predicting the dynamics of misinformation propagation within these networks. By adapting the GCN for a regression task, we aim to predict the acceptance rate of misinformation for each node, thus identifying key influencers and mitigating misinformation spread.

The architecture and various components of the GCN are selected to capture the structure of social networks and the behavior of misinformation spread. The adjacency matrix A and the feature matrix X provide foundational structural and attribute data of the network. Coupled with the initial estimation of acceptance rates $Y = \{y_1, y_2, \ldots, y_n\}$ provided by the HISB model, the GCN is equipped to make accurate predictions. At each layer l, the GCN updates node representations by aggregating information from their neighbors. The update rule is given by: $H^{(l+1)} = \sigma(\hat{D}^{-\frac{1}{2}}\hat{A}\hat{D}^{-\frac{1}{2}}H^{(l)}W^{(l)})$. Batch normalization and dropout are applied after each convolution to stabilize training and prevent overfitting. These elements ensure the model captures both local and broader contextual information essential for accurate predictions.

The regression task is framed to predict the acceptance rates of nodes, which is crucial for identifying influential nodes. The predicted acceptance rates $\hat{Y} = \{\hat{y}_1, \hat{y}_2, \ldots, \hat{y}_n\}$ are obtained from the final layer's output. To optimize the model's parameters and improve its prediction accuracy, we employ the Mean Squared Error (MSE) as the loss function, which is defined as:

$$Loss = \frac{1}{n}\sum_{i=1}^{n}(\hat{y}_i - y_i)^2 \tag{7}$$

The regression task itself aims to minimize this loss, which is indicative of the difference between the predicted acceptance rates $\hat{Y}$ and the true rates Y from the simulations. By minimizing the MSE, the GCN refines its weights to more accurately predict each node's likelihood of accepting and propagating misinformation.

4.4 Influential Nodes Identification:

The Influential Node Identifier is an integral component of our Misinformation Influence Minimization framework. It leverages the trained Graph Convolutional Network (GCN) model to pinpoint the most influential nodes within any social network. By processing structural and feature data, it predicts each node's misinformation acceptance rate. Nodes are then ranked to identify those with the highest influence, subject to a specified budget constraint K.

Subsequently, we proposed GCN based greedy algorithm presented in Algorithm 1 that selects and blocks the most influential nodes that are the most susceptible to spread the misinformation, hence minimizing its impact.

5 Experiments

In this section, we conduct experiments to evaluate the performance and efficiency of the proposed approach. All experiments presented in this section are implemented using Pytorch and were run on a server equipped with Intel Xeon processors (3.4 GHz) and 16 GB of memory, running the Linux operating system. Table 1 provides a detailed overview of the datasets used in the experiments.

Algorithm 1: GCN based greedy algorithm

Require: Graph $G = (V, E)$, Number of nodes to block K, GCN Model M
Ensure: Blocking set B
1: Initialize an empty list for node acceptance rates A
2: **for** each node $v \in V$ **do**
3: Calculate the acceptance rate $a_v = M(v)$ using GCN model M
4: Append a_v to list A
5: **end for**
6: $B = \text{argmax}_K(A)$ Identify the top K nodes with the highest acceptance rate
7: **return** B

Table 1. Datasets Description

Dataset	Type	Nodes	Edges
Facebook [21]	Undirected	4,039	88,234
Facebook Large Page-Page Network [22]	Undirected	22,470	171,002
email-Eu-core network [23]	Directed	1,005	25,571
Social circles: Twitter [21] (used for training)	Directed	81,306	1,768,149

5.1 Performance Evaluation

In our investigation into misinformation containment strategies within social networks, we assess the efficacy of various approaches. We introduce the novel Acceptance Rate Blocking Nodes (ARBN) method, which strategically targets and blocks nodes based on their susceptibility to accept misinformation. This method is compared with other established strategies: Random Blocking Nodes (RNB), which selects nodes at random; Betweenness Centrality Blocking Nodes (BCBN) [3], defined by the number of short paths from all nodes to other nodes through node$u$; and Minimum Judgment Blocking Nodes (MJBN), a strategy that blocks nodes based on a predefined judgment parameter of each node. Additionally, we incorporate the PageRank [24] Blocking Nodes (PRBN) strategy, which employs the PageRank algorithm to target and block nodes deemed most central and influential within the network's topology. These strategies are contrasted against a baseline scenario of natural propagation (NP), where no intervention is applied. All methods were tested on HISBmodel.

The effectiveness of these strategies is evaluated based on the detection time (T_{det}), indicating the simulation timestep at which the misinformation is first identified, with values set at 2, 4, 8, and 16 timesteps in our simulations. Furthermore, we consider the percentage of nodes (K), reflecting our intervention budget and the proportion of the network that can be influenced by our blocking strategies. This percentage varies across simulations from 5% to 20%, allowing for an analysis of the impact of each strategy at different levels of network intervention. The opinion of each node is randomly assigned. To ensure robust results, individual parameters are uniformly distributed to create a homogeneous population, mitigating experimental noise. The parameters are set as follows: $\omega \in [\pi/12, \pi]$, $\beta \in [0.2, 1.2]$, $\delta \in [\pi/24, \pi/2]$, and $\alpha \in [0, 1]$. Each simulations are repeated 100 times in order to avoid randomness where the results presented in Table 2.

The data collected from the Email, Facebook, and Page-Page datasets revealed that our Acceptance Rate Blocking Nodes strategy, which employs a GCN Model to predict and block the most influential nodes based on their acceptance rates, consistently outperformed baseline methods such as centrality measures. Specifically, ARBN demonstrated lower acceptance rates of misinformation across varying detection times and intervention levels, indicating a more effective containment of misinformation spread.

Interpreting these results, it's evident that the deep learning approach embedded within ARBN provides a mechanism for identifying key propagators of misinformation. The performance of ARBN, particularly in more interconnected networks like Facebook and Page-Page, underscores the importance of understanding network dynamics and the pivotal role of influential nodes. The success of ARBN, compared to traditional methods like RNB and BCBN, highlights the advanced capability of deep learning models to capture and analyze the intricate patterns of misinformation spread in real world scenarios.

Table 2. Effectiveness Comparison Across Datasets and Detection Times.

		K											
	Networks	Email				Facebook				Page-Page			
T_det	Strategy	5%	10%	15%	20%	5%	10%	15%	20%	5%	10%	15%	20%
2	**ARBN**	**44,23%**	**40,12%**	**38,56%**	**36,19%**	**49,27%**	**45,28%**	**43,12%**	**39,46%**	**34,17%**	**33,15%**	**32,09%**	**29,43%**
	RNB	58,12%	54,93%	52,47%	50,78%	57,09%	53,86%	52,51%	50,23%	50,34%	48,27%	47,12%	45,08%
	PR	53,45%	50,67%	49,32%	47,59%	53,18%	50,23%	48,26%	46,07%	46,19%	44,32%	43,05%	41,12%
	BCBN	49,17%	48,03%	43,78%	44,19%	49,23%	47,01%	45,09%	44,82%	48,23%	44,17%	42,08%	40,34%
	MJBN	45,12%	44,89%	43,71%	43,04%	47,41%	46,19%	45,23%	44,09%	45,17%	44,83%	42,56%	38,42%
	NP	80,23%	80,23%	80,23%	80,23%	77,12%	77,12%	77,12%	77,12%	83,17%	83,17%	83,17%	83,17%
4	**ARBN**	**49,27%**	**48,75%**	**48,39%**	**47,21%**	**54,88%**	**53,89%**	**51,56%**	**47,18%**	**44,32%**	**43,18%**	**41,59%**	**38,34%**
	RNB	61,23%	61,87%	61,29%	60,24%	61,98%	61,47%	61,03%	60,91%	54,12%	52,03%	51,98%	50,56%
	PR	56,32%	56,04%	55,23%	54,08%	60,92%	60,21%	58,03%	57,09%	49,01%	47,25%	46,29%	44,18%
	BCBN	59,43%	59,28%	55,21%	54,37%	60,59%	59,23%	57,42%	55,07%	51,19%	48,56%	46,37%	45,01%
	MJBN	54,39%	54,32%	53,17%	52,24%	54,91%	54,07%	53,82%	52,09%	49,17%	47,32%	44,06%	41,53%
	NP	80,23%	80,23%	80,23%	80,23%	77,12%	77,12%	77,12%	77,12%	83,17%	83,17%	83,17%	83,17%
8	**ARBN**	**54,19%**	**50,24%**	**52,37%**	**51,32%**	**60,24%**	**60,08%**	**59,13%**	**56,19%**	**50,46%**	**65,41%**	**64,38%**	**43,59%**
	RNB	68,12%	67,19%	67,23%	66,34%	68,57%	68,32%	68,19%	67,02%	58,43%	57,26%	56,49%	54,71%
	PR	62,47%	61,54%	60,29%	59,03%	67,19%	65,42%	63,08%	62,05%	53,28%	52,37%	51,12%	49,43%
	BCBN	63,23%	62,17%	61,28%	61,09%	63,04%	62,87%	62,01%	61,53%	54,27%	52,53%	53,19%	50,58%
	MJBN	62,08%	61,34%	60,32%	60,01%	61,89%	61,12%	61,05%	60,23%	51,87%	51,46%	50,43%	49,27%
	NP	80,23%	80,23%	80,23%	80,23%	77,12%	77,12%	77,12%	77,12%	83,17%	83,17%	83,17%	83,17%
16	**ARBN**	**59,01%**	**58,72%**	**58,23%**	**58,09%**	**65,35%**	**64,55%**	**63,94%**	**60,67%**	**52,73%**	**51,52%**	**51,03%**	**49,48%**
	RNB	71,23%	70,54%	69,21%	68,42%	72,39%	71,52%	71,28%	70,71%	68,72%	67,09%	67,18%	65,89%
	PR	66,27%	65,01%	64,59%	63,48%	70,51%	69,34%	68,52%	67,43%	61,48%	60,61%	59,62%	58,37%
	BCBN	68,41%	67,92%	67,38%	67,24%	69,07%	70,23%	68,71%	68,19%	64,52%	63,17%	62,58%	59,56%
	MJBN	64,39%	63,57%	63,24%	62,17%	66,38%	65,79%	65,32%	64,89%	58,12%	56,37%	55,06%	54,61%
	NP	80,23%	80,23%	80,23%	80,23%	77,12%	77,12%	77,12%	77,12%	83,17%	83,17%	83,17%	83,17%

Lowest value Highest value

5.2 Discussion

The comparative analysis within our Misinformation Influence Minimization framework highlights the ARBN model's superior performance in reducing sus-

Table 3. The average reduction percentage in the number of individuals susceptible to being infected

	Best	Worst	Email	Facebook	Page-Page
ARBN	**53,74%**	**11,77%**	**30,55%**	**22,45%**	**37,87%**
RNB	38,09%	4,73%	12,14%	13,55%	27,22%
PR	42,83%	6,61%	22,39%	16,65%	32,69%
BCBN	42,83%	8,05%	21,97%	18,11%	31,52%
MJBN	44,75%	10,74%	19,72%	20,59%	34,40%

ceptibility to misinformation across various network scenarios. Notably excelling in Page-Page networks, the ARBN model also demonstrates robustness and adaptability in Email and Facebook networks, indicating its effectiveness in diverse social structures. Central to the ARBN approach is the utilization of the Graph Convolutional Network (GCN) model for the precise identification of key nodes within the network, enhancing our ability to target and mitigate the spread of misinformation effectively. This integration of GCN within the ARBN model positions it as a significant advancement in misinformation control strategies, providing a robust and scalable tool for real-world applications aimed at curtailing the spread and impact of misinformation in online social networks (see Table 3).

6 Conclusions

In conclusion, this paper offers a significant stride in understanding and controlling the spread of misinformation through the innovative application of GNNs in OSNs. Our exploration and the development of the Acceptance Rate Blocking Nodes method underscore the potential of deep learning techniques in identifying and mitigating influential nodes driving misinformation propagation. This research paves the way for practical tools to foster more informed and resilient digital communities. While adapting these strategies to constantly evolving network environments remains a challenge, this work establishes a solid foundation for future initiatives aimed at mitigating the influence of misinformation.

References

1. Zhang, H., Chen, Y., Rong, W., Wang, J., Tan: Effect of social media rumors on stock market volatility: a case of data mining in China. Front. Physics (2022)
2. Allport, G.W., Postman, L.: The psychology of rumor (1947)
3. Brandes, U.: On variants of shortest-path betweenness centrality and their generic computation. Soc. Networks **30**(2) (2008)
4. Bakshi, A.J., Deshmukh, J., Kumar, S.: The psychology of pandemics: preparing for the next global outbreak of infectious disease (2021)
5. Scarselli, F., Gori, M., Tsoi, A.C., Hagenbuchner, M., Monfardini, G.: The graph neural network model. IEEE Trans. Neural Networks **20**(1), 61–80 (2008)

6. Kimura, M., Saito, K., Motoda, H.: Minimizing the spread of contamination by blocking links in a network. In: AAAI, vol. 8 (2008)
7. Yan, R., Li, Y., Wu, W., Li, D., Wang, Y.: Rumor blocking through online link deletion on social networks. ACM Trans. Knowl. Disc. Data (TKDD) (2019)
8. Wang, X., Deng, K., Li, J., Yu, J.X., Jensen, C.S., Yang, X.: Efficient targeted influence minimization in big social networks. World Wide Web (2020)
9. Fan, L., Lu, Z., Wu, W., Thuraisingham, B., Ma, H., Bi, Y.: Least cost rumor blocking in social networks. In: 2013 IEEE 33rd International Conference on Distributed Computing Systems. IEEE (2013)
10. Wang, B., Chen, G., Fu, L., Song, L., Wang, X.: DRIMUX: dynamic rumor influence minimization with user experience in social networks. IEEE Trans. Knowl. Data Eng. (2017)
11. Tong, G., et al.: An efficient randomized algorithm for rumor blocking in online social networks. IEEE Trans. Network Sci. Eng. (2017)
12. Shi, Q., Wang, C., Ye, D., Chen, J., Feng, Y., Chen, C.: Adaptive influence blocking: minimizing the negative spread by observation-based policies. In: IEEE 35th International Conference on Data Engineering (ICDE). IEEE (2019)
13. Yan, R., Li, D., Wu, W., Du, D.Z., Wang, Y.: Minimizing influence of rumors by blockers on social networks: algorithms and analysis. IEEE Trans. Network Sci. Eng. (2019)
14. Zhu, J., Ni, P., Wang, G.: Activity minimization of misinformation influence in online social networks. IEEE Trans. Comput. Soc. Syst. (2020)
15. Jiang, J., Chen, X., Huang, Z., Li, X., Du, Y.: Deep reinforcement learning-based approach for rumor influence minimization in social networks. Applied Intelligence (2023)
16. Hosni, A.I.E., Li, K., Ahmad, S.: Minimizing rumor influence in multiplex online social networks based on human individual and social behaviors. Inf. Sci. **512**, 1458–1480 (2020)
17. Hosni, A.I.E., Li, K.: Minimizing the influence of rumors during breaking news events in online social networks. Knowl.-Based Syst. **193**, 105,452 (2020)
18. Do, M.T., Park, N., Shin, K.: Two-stage training of graph neural networks for graph classification. Neural Process. Lett. (2022)
19. Hosni, A.I.E., Li, K., Ahmed, S.: HISBmodel: a rumor diffusion model based on human individual and social behaviors in online social networks. In: Cheng, L., Leung, A.C.S., Ozawa, S. (eds.) ICONIP 2018. LNCS, vol. 11302, pp. 14–27. Springer, Cham (2018). https://doi.org/10.1007/978-3-030-04179-3_2
20. Kipf, T.N., Welling, M.: Semi-supervised classification with graph convolutional networks. arXiv preprint arXiv:1609.02907 (2016)
21. Leskovec, J., Mcauley, J.: Learning to discover social circles in ego networks. Advances in neural information processing systems (2012)
22. Rozemberczki, B., Allen, C.: Multi-scale attributed node embedding (2019)
23. Yin, H., Benson, A.R., Leskovec, J., Gleich, D.F.: Local higher-order graph clustering. In: Proceedings of the 23rd ACM SIGKDD International Conference on Knowledge Discovery and Data Mining, pp. 555–564 (2017)
24. Page, L.: The pagerank citation ranking: Bringing order to the web. Technical report. Stanford Digital Library Technologies Project, 1998 (1998)

Adaptive Approach for Rumors Influence Minimization in Dynamic Social Networks

Khaled Aimen Hafiani, Adil Imad Eddine Hosni(✉), Islam Baira, Hichem Merini, and Kadda Baghdad Bey

Ecole Militaire Polytechnique, Bordj El-Bahri, 16046 Algiers, Algeria
hosni.adil.emp@gmail.com

Abstract. Breaking news is a kind of rumor that can reach a large number of users and is characterized by its short-term life. Therefore, it's highly recommended to seek a solution to limit its spread through online social networks (OSNs). This work proposes an approach to mitigate the influence of breaking news rumors on dynamic social networks. Initially, we introduce a representation of a continuous node-static evolving network. Subsequently, through an analysis of existing literature gaps, we present Adaptive Graph Protection with User Experience (AGPUX). AGPUX is designed to select the optimal set of nodes in each iteration, aiming to minimize the dissemination of malicious information. Systematically, experiments have been conducted to evaluate the performance of the proposed approach on real networks. The results showed that the proposed AGPUX approach could outperform the most recent methods in the literature.

Keywords: Online social networks analysis · Data mining · Rumor influence minimization · Dynamic networks · Breaking news

1 Introduction

The rapid evolution of Internet technology (IT) has resulted in a substantial surge in the utilization of online social networks (OSNs) [1]. These platforms attract millions of users and profoundly influence daily human life. Despite their advantages, OSNs have been associated with adverse effects on mental health and user addiction. Additionally, they have been linked to malicious online activities, which threat the individual safety and national security.

In the existing literature, numerous studies have concentrated on the challenge of minimizing rumor spread in OSNs. Proposed methods predominantly employ either the blocking nodes/edges strategy or the truth campaign strategy. Among these approaches, the blocking nodes strategy [2,3] has demonstrated superior efficiency in minimizing the dissemination of malicious information, particularly when adaptive algorithms accounting for rumor flow are applied [3]. Nevertheless, the efficacy of this strategy diminishes significantly when faced with budget constraints for blocking. Conversely, the truth campaign approach [4,5] can be readily implemented to minimize rumor influence, but its effectiveness wanes with delayed rumor detection.

B. Djamaa et al. (Eds.): CSA 2024, LNNS 1145, pp. 361–372, 2025.
https://doi.org/10.1007/978-3-031-71848-9_33

Moreover, most of the existing works consider OSNs as static graphs, while real-world social networks exhibit dynamic characteristics, with vertices and edges appearing and disappearing over time. Therefore, the static representation of OSNs may result in reduced performance in rumor influence minimization. Hence, it is crucial to consider network dynamicity in the graph protection process. Some studies propose discrete representations to emulate the evolving pattern of the network's edge set [6], while others assume continuous dynamic representations for either nodes or edges [7].

This paper introduces an Adaptive Graph Protection with User Experience (AGPUX) approach for dynamic social networks, aiming to minimize the spread of breaking news and its influence. In contrast to previous approaches that considered the dynamic aspect of OSNs, our method incorporates rumor flow by applying temporary adaptive blocking of nodes across multiple turns. The objective is to minimize the set of rumor supporting nodes. To achieve this, we propose a heuristic based on survival theory that considers both the structural features of nodes, their behavior, and individual opinions. The proposed method's efficiency is validated through experiments conducted on real-world datasets, demonstrating that AGPUX outperforms existing methods in dynamic networks with a limited minimization budget.

The rest of the paper is organized as follows: Sect. 2 provides a comprehensive review of related work in the field of rumor influence minimization. Section 3 presents the problem formulation and describes the proposed AGPUX approach. Section 4 presents the experimental setup and discusses the results. Finally, Sect. 5 concludes the paper and outlines directions for future research.

2 Related Works on Rumor Influence Minimization

Existing strategies for Rumor Influence Minimization (RIM) in the literature predominantly revolve around blocking nodes or edges, involving the removal of critical nodes or links to curb the propagation of rumors and breaking news. Various methods have been proposed based on diverse criteria, including the structural features of nodes [2,8], calculations for blocking gain [3], and minimizing betweeness centrality of edges [9].

Furthermore, several RIM approaches consider the source and flow of misinformation. Some methods concentrate on minimizing the global activation probability of nodes through the strategic blocking a set of edges [10]. Additionally, there exist clarification-based algorithms aiming to protect against rumors or execute anti-rumor campaigns within the network [4,5,11,12].

While most previous works operate under the static assumption of network's graphs, a few contributions have been made to minimize rumor spread through dynamic networks [6,7]. These approaches address the evolving nature of dynamic OSNs by considering snapshots or discrete representations of the network. Adaptive algorithms, founded on vaccination strategies, have been introduced [7], alongside multi-turn RIM methods utilizing the Minimum Vertex Cover (MVC) [6]. Reinforcement learning has been leveraged to estimate the MVC of the network in each snapshot [6].

3 Primarily Knowledge and Problem Formulation

3.1 Rumor Propagation Model

HISBmodel is one of the few models that considered both individual and social influencing factors, besides, the individuals' opinions.

HSIBmodel Overview. Hosni et al. [13] created the Human Individual and Social Behavior Model (HISBmodel), which is a stochastic rumor propagation simulation model.

The simulation of an individual's attraction to the rumor adopts the analogy of damped harmonic oscillation motion, a manifestation influenced by the interplay of the aforementioned factors. The attraction expression is defined as follows:

$$A(t) = A_{int}e^{-\beta t}|\cos(\omega t + \delta)|, \tag{1}$$

where: $\mathbf{A_{int}}$: denotes the individual initial attraction to the rumor, β: or also noted by *IBK*, which is the individual background knowledge, ω: denotes the FR factor and δ: is the hesitation mechanism factor (HM).

Rumor Transmission Rule. In this model, the rule of transmission of rumors between two nodes is defined by the probability of sending, the probability of acceptance, and the probability of having a link between these two nodes at time t. And it's calculated using the following expression:

$$P_{u,v}(t) = P_{(u,v)\in E}(t) \cdot P_u^{send}(t) \cdot P_{v,u}^{acc} \tag{2}$$

$P_{(u,v)\in E}(t)$ express the probability of a link existence between u and v at time t, the sending probability $P_u^{send}(t)$ is an estimation of the likelihood of an individual u sending a rumor I at time t, which is defined as follows:

$$P_u^{send}(t) = e^{-\beta t}|\cos(\omega t + \delta)| \tag{3}$$

A balanced weighted probability is used to calculate the acceptance probability, simulating the influence of follower-following according to the *celebrity affect rule*. It's defined as:

$$P_{v,u}^{acc} = \frac{1}{1 + \dfrac{d_v}{d_u}} \tag{4}$$

Individuals' Opinion Formulation. An additive model is adopted to estimate B_v, the opinion of an individual v towards the rumor. $B_v \in]-\infty, 0]$ corresponds to denying, $B_v \in [0, 10]$ is neutral, $B_v \in [10, 20]$ corresponds to questioning, and $B_v \in [20, +\infty[$ corresponds to the rumor support [13].

Based on the work of Hosni et al. [14], which emphasized the role of individuals' subjective judgment in rumor spreading and suggested that an individual utilizes his or her knowledge in judging the rumor veracity. They introduced the subjective judgment factor $J \in [0, 1]$ which represents the ability of an individual to judge the rumor validity [14]. On this basis, they introduced $rho \in [0, 1]$ the decision factor that estimates the supporting or the denying rate. Thus, the decision factor of a node u is defined by the formula (5).

$$\rho = \frac{\eta_u^-}{\eta_u} \cdot J_u \tag{5}$$

where η_u denotes the total number of rumor accepting times by a node u, and η_u^- is the number of the rumor acceptance times with a negative opinion.

3.2 Problem Formulation

We define a dynamic graph $G = (V, E)$ where V denotes the set of the graph's nodes i.e. the set of OSN's users $E = (u, v, t_s, t_e)$ exhibits the existence of an edge linking u to v between two timestamps t_s and t_e, with $u, v \in V$. A constant k ($k \geq 0$) which is the protection strategy budget, in other word, it denotes the maximum number of nodes that can be selected to be used in the RIM process. And t_{det} is the rumor detection time. The goal is to select k_t most influencing nodes in the network at t, where $t \geq t_{det}$, whose their blocking minimize the most the rumor influence in the network at t, i.e. the global rumor influence minimization. We define the objective function as follows:

$$For\ each\ t,\ with t_{det} \leq t \leq T_f : \min \quad \mathbb{E}_t \left[\sum_{v \in V} S_v(t) \right] \tag{6}$$

$S_v(t)$ represents an individual v having a favorable opinion towards a rumor at t, where T_f represents the final stage of the propagation process. $\mathbb{E}_{\approx}[.]$ denotes the expected number of individuals at each instant t.

4 Solution Formulation

The approach we propose is denoted as AGPUX, which stands for Adaptive Graph Protection with User Experience. AGPUX is a RIM method involving the adaptive temporary blocking of nodes. The primary objective is to safeguard dynamic network graphs by mitigating the influence of rumors, thereby restricting the expansion of belief in rumors within the users' set. Figure 1 resumes the iterative process of the proposed approach.

Temporary Blocking Nodes. The AGPUX methodology adopts a temporary blocking nodes approach across multiple iterations, wherein the blocking time threshold is individualized for each user. In this framework, the node degree serves as a heuristic determining these blocking thresholds, as discussed in [15]. Specifically, the node with the lowest degree is assigned a blocking time threshold approximately equal to the rumor

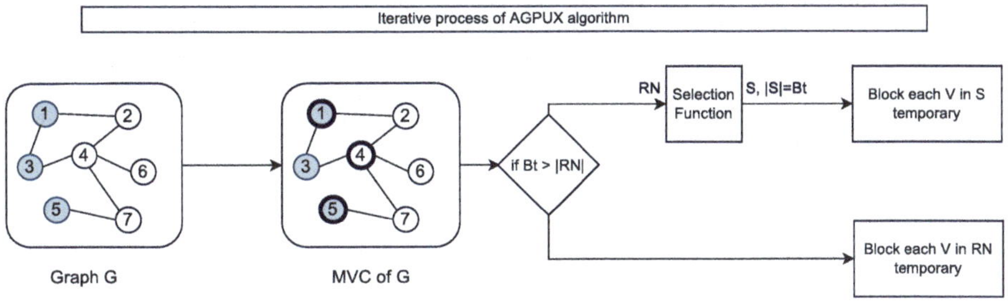

Fig. 1. Illustrative schema of AGPUX process.

lifetime, particularly targeted at breaking news rumors characterized by their short lifespan. Conversely, the node with the highest degree is assigned an approximately null blocking time threshold. These two values establish a scale against which the blocking time of other nodes in the graph is evaluated. Consequently, the Rumor Influence Minimization (RIM) problem incorporates an additional condition, i.e. restricting user experience beyond a specified threshold U_{th}:

$$U \geq U_{th}, \ \ with \ \ U = \sum_{u=1}^{N} \frac{T_{th} - T_b(u)}{T_{th}}[15] \tag{7}$$

where $T_b(u)$ denotes the blocking time attributed to the node u, and T_{th} evaluates the blocking time tolerance. Thus, the selected blocking times must respect the aforementioned constraint in Eq. 7.

Selection Function. This function aims to identify the optimal set of nodes K whose blocking minimizes the most the rumor influence. It introduces a probabilistic method by leveraging survival theory, specifically estimating the likelihood of node infection likelihood after an observation time [15, 16]. Under this premise, we articulate the hazard function $h(t)$, representing the conditional probability of an individual's infection between t and $t + dt$ if it has not been infected at time t.

$$h(t) = \lim_{dt \to 0} \frac{\Pr\{t \leqslant T < t + dt \mid T \geqslant t\}}{dt} = -\frac{S'(t)}{S(t)}. \tag{8}$$

Given that a rumor is propagating in an OSN represented by a dynamic network $\{\mathbb{G}\}_{t=0,1,2,\ldots.,T}$. The following expression defines the hazard rate of a given node v getting infected by the rumor at time t with a positive opinion toward it

$$h_v^+(t) = \sum_{u \in \mathbb{N}^v} (p_{u,v})(t) \cdot \pi_v^+. \tag{9}$$

Then, yield the cumulative distribution function

$$F_v^+(t) = 1 - \prod_{u \in \mathbb{N}^v} e^{-\pi_v^+ \int_0^t (p_{u,v})(\tau) d\tau}. \tag{10}$$

The likelihood function of a node v becoming infected at time t is then given.

$$f_v^+(t) = \sum_{u \in \mathbb{N}^v} (\pi_v^+)(p_{u,v})(t) \cdot \prod_{w \in \mathbb{N}^v} e^{-\pi_v^+ \int_0^t (p_{u,w})(\tau) d\tau}. \tag{11}$$

The Eq. 11 provide the likelihood of the infection of any node in the network at time t and supports the rumor. Additionally, a node u with $A_u(t) = 0$, i.e. it non-infected at t, we estimate its impact at time $t+1$. Thus, for a given node $u \in \{\mathbb{G}\}_{t=0,1,2,\ldots,T}$ where $h_u(t) > 0$ (at least one of the neighbors of u is infected), the hazard rate of the individuals' infection at time $t+1$ by a node u with a supporting opinion toward the rumor is defined as follows

$$r_u^+(t+1) = \sum_{v \in \mathbb{N}^u} (p_{u,v})(t+1)\pi_v^+. \tag{12}$$

Continuing with the aforementioned demonstration approach, the cumulative distribution function is presented as follows:

$$G_v+(t) = 1 - e^{-\int_t^{t+1} p_u^{send} \sum_{v \in \mathbb{N}^u} \pi_v^+(p_u)(p_{u,v}^{acc}) d\tau}. \tag{13}$$

Then, the likelihood function of nodes getting infected by u, and influence positively on their neighbors, is given as follow

$$g_u^+(t) = \sum_{v \in \mathbb{N}^u} (p_{u,v})(t+1)\pi_v^+ \prod_{w \in \mathbb{N}^u} e^{-\pi_v^+(p_u)(p_{u,w}^{acc}) \int_t^{t+1} p_u^{send}(\tau) d\tau}. \tag{14}$$

Therefore, we estimate the likelihood of a node v getting infected at time t, and its positive impact toward a rumor at time $t+1$ presented as

$$l_v^+(t) = f_v^+(t) \cdot g_v^+(t+1), \tag{15}$$

For a number of infected nodes, Eq. 15 can be generalized as follows

$$l_V^+(t) = \prod_{v \in V} f_v^+(t) \cdot g_v^+(t+1). \tag{16}$$

Equation 16 exhibits the probability of nodes becoming infected by the rumor at time t and their subsequent influence at $t+1$. This formulation enables the encapsulation of the dynamic propagation of a rumor in a dynamic evolving network. Thus, based on Eqs. 15 and 16, we introduce two functions $\sigma^+(.)$.

$$\sigma^+(K^+) = l_{K^+}^+(t) \qquad = \prod_{v \in K^+} f_v^+(t) \cdot g_v^+(t+1). \tag{17}$$

$\sigma^+(A)$ estimates the likelihood of the infection of an individuals and adopting a supporting opinion. Thus, the objective is to identify the set of nodes K that exert the most significant influence on the dissemination of the rumor. Subsequently, the solution is formulated as follows

$$\min\left[\sigma^{+}(K)\right] \tag{18}$$

To enhance the nodes' selection persistence, and inspired by the work of Wijayanto [6], we opt the optimal set of nodes from the reachable nodes set within the minimum vertex cover (MVC) of the graph. Since the network structure is dynamic, the MVC must be recalculated periodically.

Multi-turn AGPUX. Applying a preemptive strategy could be efficient in late rumor detection cases, however, exploiting the whole budget in a single preventive turn cannot capture the dynamic flow of rumors. Hence, AGPUX assumes an adaptive RIM approach applying a multi-turns process. Beyond the rumor detection time, each timestamp, if the available budget is greater than the cardinality of the reachable nodes set **RN**, all nodes within **RN** will be blocked temporary depending on their respective blocking thresholds. Subsequently, the RIM procedure progresses to the next step. Upon the cardinality of the reachable nodes set surpassing the remaining budget, the best ranked nodes of RN are selected using the proposed objective function, and the RIM process reaches its final turn.

```
Data: G = (V, E), Rumor_Parametre, List_spreaders, T_det, k
Result: SN_List
while k ≠ 0 do
    if time ≥ T_det then
        SN(time) = emptyList()
        tempo_list = emptyList()
        for u ∈ G do
            if u ∈ neighbors(List_spreaders) and u ∈ MVC(G, time) and not
            u ∈ SN(timee) then
                add u to tempo_list
            end
        end
        if |tempo_list| ≥ k then
            for i = 1..K do
                v = arg max_{u∈tempo_list} [I(u)]
                add v SN(time)
            end
        end
        else

        end
        k = k − |tempo_list|
        SN(time) < − tempo_list
    end
    Block_tempo_tolerance(SN(time))
end
```

Algorithm 1: The AGPUX RIM algorithm

Proposed AGPUX Algorithm. The illustrated pseudocode Algorithm 1 outlines the implementation steps for the proposed Rumor Influence Minimization (RIM) approach in dynamic Online Social Networks (OSNs). It starts with essential inputs: dynamic network, node-specific rumor parameters, initial spreaders, rumor detection time, and RIM budget. After rumor detection, AGPUX intervenes, identifying the reachable nodes set (RN) as neighbors of spreaders within the network's Minimum Vertex Cover. If RN's cardinality is less than the budget, temporary blocking based on node thresholds is applied. The algorithm deducts RN's cardinality from the budget for the next step. Otherwise, i.e. the RN exceeds the budget, we select optimal k nodes via the objective function for temporary blocking. The AGPUX process ends when the budget is null.

5 Experimental Results

To validate the effectiveness of the proposed approach, experiments were conducted across diverse minimization budgets and rumor detection times. This section presents and discusses the outcomes of comparative simulations, subjected to statistical analysis of the different approaches' efficiency in minimizing the rumor influence on the two dynamic networks listed in Table 1. The GNN algorithm [17] was employed for link prediction, assuming a predetermined probability for each link's removal from the network.

Table 1. Employed Datasets Descriptions

Name	Layers	Type	Nodes	Links	$<k>$	ASPL	Clustering degree
Facebook	1	undirected	4039	88234	21.84	2.0	0.655
Twitter	1	directed	76245	1667885	12.875	3.12	0.0378

The algorithms selected for comparison include:

1. **NP:** It denotes the natural propagation without applying RIM any algorithm,
2. **BNSW:** the blocking nodes approach of Wijayanto [6], which involves blocking nodes within the MVC network with the highest degree.
3. **TCScen:** The truth campaign strategy is based on the centrality metric to select the anti-rumor propagators,
4. **DRIMUX:** Dynamic rumor influence minimization with user experience of [15],
5. **BNLScen:** Blocking node-link strategy b based on the centrality degree.
6. **BLS:** Blocking link strategy based on the betweenness metric and
7. **AGPUX:** the proposed approach.

At this stage, we present the results of simulating the spread of rumors across the two aforementioned datasets. In the first experiment, we focus on tracking the infection rate, the number of spreaders, the rumor-supporting rate, and the rumor popularity. We simulate the rumor propagation through the Facebook network under normal circumstances and after applying three different Rumor Influence Minimization (RIM) strategies, along with the proposed approach. Given the $5th$ timestamp as the rumor detection instant and 8% is the minimization budget. As illustrated in Fig. 2a and Fig. 2b, both the infection rate and the rumor-supporting rate exhibit an exponential evolution trend in the five different experiments, reaching varying final rates. Notably, the proposed approach minimizes the two aforementioned metrics more effectively compared to the previous RIM algorithms. Additionally, Fig. 2c and Fig. 2d emphasize the evolution of rumor popularity and the set of rumor spreaders, respectively. These proportional metrics experience rapid increases until reaching a peak, after which they decrease, ultimately becoming null beyond the $40th$ time step.

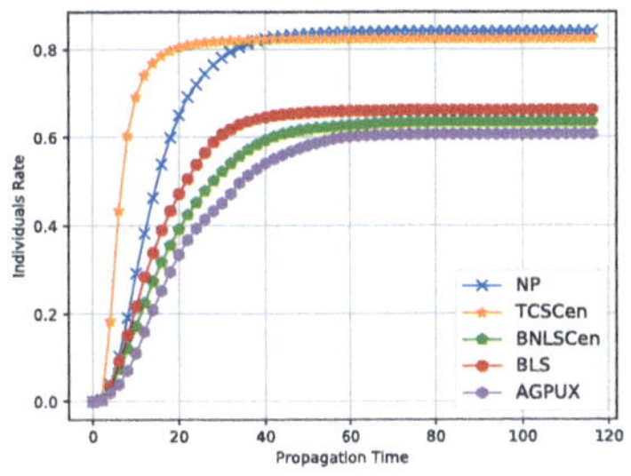

(a) Infected individuals rate in Facebook

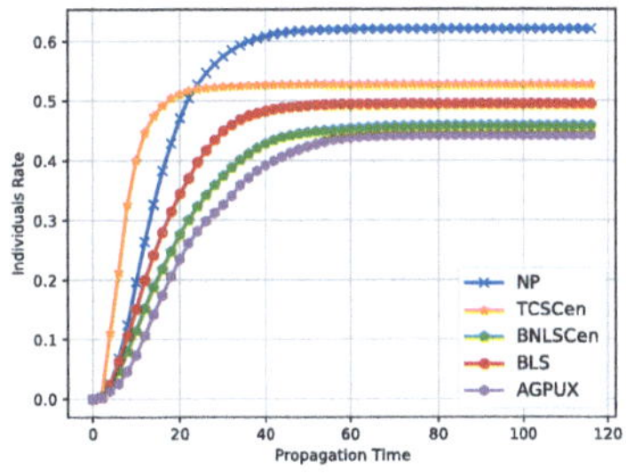

(b) Rumor supporting rate in Facebook

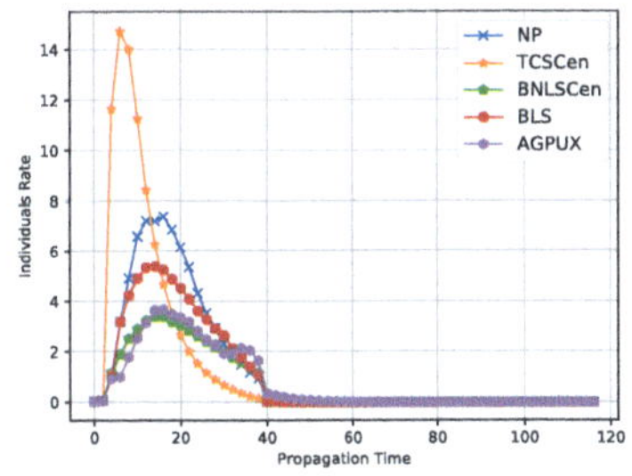

(c) Rumor popularity evolution in Facebook

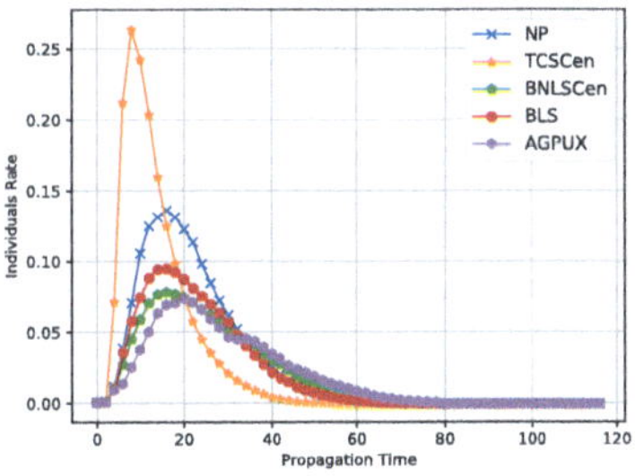

(d) Rumor spreaders rate in Facebook

Fig. 2. Evolution of the rumor tracking metrics during its propagation after applying different RIM algorithms

Subsequently, we conducted multiple simulations using the Human Individual and Social Behavior (HISB) model, setting the initial infection percentage to 0.1% and 0.01% for Facebook and Twitter network simulations respectively. Systematically, we conducted the simulation results in varied scenarios, i.e. different detection times and minimization budgets. Regarding the opted RIM efficiency evaluation metric; the final rumor-supporting ratio, which reflects the rumor influence on the individuals' set. Table 2 depicts the influence of the rumor after applying the RIM algorithms to the simulation process. Overall, the blocking-based strategies exhibit higher efficiency compared to clarification-based methods. AGPUX yielded the lowest supporting ratio at $t_detection = 1$ for different minimization budgets on either Facebook or Twitter.

Table 2. Comparison Of Proposed Approach with previous RIM Strategies

		Budget (k)									
	Networks	Facebook					Twitter				
Tdet	Algorithms	1%	5%	10%	15%	20%	1%	5%	10%	15%	20%
1	NP	0,6431	0,6431	0,6431	0,6431	0,6431	0,4126	0,4126	0,4126	0,4126	0,4126
	BNSW	0,5899	0,5354	0,4718	0,3818	0,2718	0,3674	0,2669	0,0981	0,0091	0,0052
	TCSCen	0,6386	0,5623	0,4707	0,3829	0,2931	0,3901	0,3253	0,2417	0,1612	0,0814
	DRIMUX	0,6111	0,5684	0,5555	0,5383	0,4905	0,3992	0,3562	0,3096	0,2708	0,2305
	BNLSCen	0,5826	0,5038	0,4159	0,3201	0,2208	0,3712	0,1997	**0,0159**	**0,0012**	**0,0001**
	BLS	0,4914	0,4292	0,3065	0,1864	0,2359	0,3284	0,2681	0,1087	0,0301	0,0036
	AGPUX	**0,4143**	**0,3661**	**0,1949**	**0,0844**	**0,0031**	**0,0917**	**0,0648**	0,0525	0,0032	0,0042
5	BNSW	0,5823	0,5225	0,5029	0,4361	0,3728	0,3668	0,2417	0,1245	0,0103	0,0081
	TCSCen	0,6135	0,573	0,4741	0,3894	0,3018	0,3938	0,3253	0,2449	0,1607	0,0824
	DRIMUX	0,6372	0,5729	0,5659	0,5419	0,5113	0,40097	0,3587	0,3201	0,2791	0,2334
	BNLSCen	0,576	0,506	0,4442	0,3646	0,3291	0,3688	0,2483	0,0641	0,0154	0,0039
	BLS	**0,5467**	**0,4381**	**0,2913**	**0,2511**	0,2144	0,3646	0,2694	0,143	0,0209	0,0056
	AGPUX	0,5587	0,5066	0,3981	0,2642	**0,1778**	**0,2751**	**0,2313**	**0,1029**	**0,0047**	**0,0029**
10	BNSW	0,5916	0,5711	0,5158	0,4723	0,4136	0,3741	0,2497	0,1421	0,0324	0,0154
	TCSCen	0,6281	0,5805	0,4967	0,4334	0,3517	0,3941	0,3292	0,2455	0,1707	0,0879
	DRIMUX	0,6394	0,6043	0,5756	0,5551	0,5273	0,4082	0,3664	0,3337	0,2802	0,2305
	BNLSCen	0,5894	0,5417	0,4567	0,4294	0,3788	0,3678	0,2689	**0,116**	**0,028**	0,0097
	BLS	**0,5164**	**0,4828**	**0,3404**	0,3157	0,2158	0,3303	**0,2173**	0,1449	0,0551	**0,0118**
	AGPUX	0,5236	0,5035	0,4135	**0,2805**	**0,1987**	**0,3075**	0,2792	0,1707	0,1271	0,042
15	BNSW	0,5284	0,5832	0,5253	0,4893	0,3921	0,3881	0,2997	0,1732	0,0898	0,071
	TCSCen	0,6226	0,5905	0,5316	0,4397	0,3911	0,4018	0,3364	0,2523	0,174	0,09744
	DRIMUX	0,6413	0,6372	0,5915	0,5862	0,5306	0,4108	0,3804	0,3404	0,2926	0,2472
	BNLSCen	0,5825	0,5299	0,5071	0,4794	0,3487	**0,3734**	0,2789	**0,1132**	**0,049**	0,0521
	BLS	**0,475**	**0,4668**	**0,4378**	**0,3558**	**0,2908**	0,3897	**0,2745**	0,1571	0,0728	**0,0301**
	AGPUX	0,5254	0,5011	0,4491	0,4207	0,3004	0,3807	0,3355	0,2253	0,1418	0,092

For the RIM in an undirected dynamic network (i.e. Facebook in our case), the proposed approach and BLS were the most efficient methods, followed by $BNLScen$, $TCScen$, and then $BNSW$, where $DRIMUX$ exhibits the lowest minimization efficiency in this dynamic network, attributed to its high computing time. Whereas, for the directed dynamic graph of Twitter, the best outcomes were achieved by AGPUX and $BNLScen$. Here we notice that blocking the nodes is the most efficient approach in the directed graphs, while, in the undirected, the blocking links surpass the other methods besides the proposed approach. Furthermore, $BNSW$ performs better in scalable networks, which is Twitter in this case, owing to its large scale compared to the used Facebook graph. The truth campaign method is more efficient in early rumor detection due to its notable dependency on the time detection constraint, while blocking-based approaches, including AGPUX, show lower sensitivity to this constraint. However, Blocking-based methods exhibit a high dependency on the minimization budget. For instance, at $t_detection = 1$, the efficiency of $BNLscen$ and BLS witnesses a remarkable degradation between the highest and the lowest budgets. Furthermore, AGPUX overcomes the other algorithms' efficiency for limited minimization budgets; however, it was surpassed by BLS or $BNLScen$ in scenarios involving late rumor detection and high budgets.

Table 3 demonstrates the outcome of the statistical study conducted on the executed simulations, evaluating each method at each rumor detection instance: the best case, the worst case, and the average minimization ratio.

Table 3. Critical study of RIM methods comparison results

Method	Network		worst / Best		Detection time Tdet				Minimization budget K			
	F	Twit	W	B	1	5	10	15	1	5	10	
BNSW	12.45%	24.59%	11.28%	38.94%	28.51%	31.66%	28.51%	26.08%	5.43%	11.91%	20.86%	28.77%
TCSCen	12.39%	16.78%	4.84%	34.06%	21.64%	25.80%	23.41%	21.62%	1.75%	7.50%	15.82%	23.77%
DRIMUX	5.52%	9.02%	1.70%	16.74%	11.86%	12.86%	11.37%	9.31%	0.93%	4.73%	7.88%	10.98%
BNSCen	15.02%	26.53%	9.72%	41.74%	33.09%	35.37%	31.38%	29.46%	5.14%	14.32%	26.12%	31.70%
BLS	22.39%	25.13%	**15.31%**	40.81%	36.13%	41.00%	**39.72%**	**34.92%**	9.75%	17.21%	**28.66%**	**36.69%**
AGPUX	**23.11%**	**26.58**	9.74%	**52.42%**	**49.99%**	**41.34%**	36.48%	28.59%	**14.32%**	**17.93%**	27.70%	36.20%

Table 3 compares the supporting ratio difference between normal propagation and after applying each RIM method. AGPUX shows the best minimization ratio for $t_{\text{det}} = 1, 5, 10$ in Facebook simulations, however, it was surpassed by BLS at $t_{\text{det}} = 15$. Where it performs worse than BLS in all cases except for $t_{\text{det}} = 1$. For Twitter, the best-case scenario balances between AGPUX, $BNLScen$, BLS, and $BNSW$. Nevertheless, AGPUX had the highest ratio of the worst cases at $t_{det} = 1, 5, 10$. Furthermore, the best average minimization in Facebook balance between AGPUX and BLS, while it was yielded by AGPUX in early rumor detection, and both $BNLScen$ and BLS in the other cases.

6 Conclusions

In this work, the rumor influence minimization problem has been investigated on dynamic social networks. First, we introduce a dynamic network representation to imitate the evolving pattern of real social networks. Afterward, we designed a multi-turn adaptive approach to minimize the rumor-supporting rate evolution, commencing with selecting the optimal set of reachable nodes that belong to the network's MVC whose removal decreases the most the supporting rate. The proposed approach is based on a greedy algorithm using a heuristic that can model the behavioral and structural factors besides the opinion of the rumor spreaders. The conducted experiments on Facebook and Twitter graphs proved that AGPUX surpasses in general the most relevant RIM method in the literature. Thus, it presents a higher efficiency in early rumor detection cases even with limited RIM budget. Nevertheless, our proposed approach couldn't outperform the blocking-based approaches in some scenarios. Despite this limitation, AGPUX is a feasible RIM method unlike the other blocking-based methods because of the user experience consideration.

References

1. Hosni, A.I.E., Li, K., Ahmad, S.: Analysis of the impact of online social networks addiction on the propagation of rumors. Phys. A: Stat. Mech. Appl. **542**, 123456 (2019)
2. Yao, Q., Shi, R., Zhou, C., Wang, P., Guo, L.: Topic-aware social influence minimization, pp. 139–140, May 2015
3. Shi, Q., Wang, C., Ye, D., Chen, J., Feng, Y., Chen, C.: Adaptive influence blocking: minimizing the negative spread by observation-based policies, pp. 1502–1513, April 2019
4. Song, C., Hsu, W., Lee, M.: Temporal influence blocking: minimizing the effect of misinformation in social networks, pp. 847–858, April 2017
5. Tong, G., Wu, W., Du, D.-Z.: Distributed rumor blocking with multiple positive cascades. IEEE Trans. Comput. Soc. Syst. **5**(2), 468–480 (2018)
6. Wijayanto, A.W., Murata, T.: Effective and scalable methods for graph protection strategies against epidemics on dynamic networks. Appl. Netw. Sci. **4**(1), 18 (2019)
7. Zhan, J., Rafalski, T., Stashkevich, G., Verenich, E.: Vaccination allocation in large dynamic networks. J. Big Data **4**, 1–17 (2017)
8. Tanınmış, K., Aras, N., Altinel, K., Güney, E.: Minimizing the misinformation spread in social networks. IISE Trans. **52**, 1–37 (2019)
9. Dey, P., Roy, S.: Centrality based information blocking and influence minimization in online social network, pp. 1–6, December 2017
10. Yan, R., Li, Y., Weili, W., Li, D., Wang, Y.: Rumor blocking through online link deletion on social networks. ACM Trans. Knowl. Discov. Data **13**, 1–26 (2019)
11. Hosni, A.I.E., Li, K., Ding, C., Ahmad, S.: Least cost rumor influence minimization in multiplex social networks, December 2018
12. He, X., Song, G., Chen, W., Jiang, Q.: Influence blocking maximization in social networks under the competitive linear threshold model technical report. In: Proceedings of the 12th SIAM International Conference on Data Mining, SDM 2012, October 2011
13. Hosni, A.I.E.: Rumor influence minimization in online social networks and its application on cyber defense, June 2020
14. Hosni, A.I.E., Li, K.: Minimizing the influence of rumors during breaking news events in online social networks. Knowl.-Based Syst. **193**, 105452 (2019)
15. Wang, B., Chen, G., Fu, L., Song, L., Wang, X.: DRIMUX: dynamic rumor influence minimization with user experience in social networks. IEEE Trans. Knowl. Data Eng. **29**(10), 2168–2181 (2017)
16. Hosni, A.I.E., Li, K., Ahmad, S.: HISBmodel: a rumor diffusion model based on human individual and social behaviors in online social networks. In: Cheng, L., Leung, A., Ozawa, S. (eds.) Neural Information Processing. ICONIP 2018. LNCS, vol. 11302, pp. 14–27. Springer, Cham (2018). https://doi.org/10.1007/978-3-030-04179-3_2
17. Zhang, M., Chen, Y.: Link prediction based on graph neural networks. In: NeurIPS, p. 17, February 2018

Optimization and Operational Research

Solving LEDs Placement Problem in Indoor VLC System Using an Efficient Coronavirus Herd Immunity Optimizer

Abdelbaki Benayad[1(✉)], Amel Boustil[1], Yassine Meraihi[2], Selma Yahia[2], and Sylia Mekhmoukh[2]

[1] LIMOSE Laboratory, University of M'Hamed Bougara, Boumerdes, Algeria
{a.benayad,a.boustil}@univ-boumerdes.dz

[2] LIST Laboratory, University of M'Hamed Bougara, Boumerdes, Algeria
{y.meraihi,s.yahia,s.mekhmoukh}@univ-boumerdes.dz

Abstract. Visible Light Communication (VLC) is a new emerging technology. It uses Light Emitting Diodes (LEDs) as transmitters and Photo-Detectors (PDs) as receivers to provide both communication and illumination for users. In indoor VLC systems, the deployment of numerous LEDs to serve numerous users is a complex problem. Because this is an NP-Hard problem, approximation methods, particularly meta-heuristics, are appropriate for solving it. In this paper, we introduce a novel algorithm called ECHIO (Enhanced Chaos-based Herd Immunity Optimizer) that integrates elements from the Improved Herd Immunity Optimizer (CHIO), incorporating the concept of chaotic maps and the Opposition-Based Learning (OBL) mechanism. The chaotic map is adopted to increase the chaotic stochastic behavior of the CHIO. Moreover, the Opposition-Based Learning (OBL) mechanism is applied to enhance the convergence speed of CHIO and explore the search space effectively. Finally, The effectiveness of the proposed ECHIO is tested on several scenarios under different settings, taking into account the throughput and the user coverage metrics. Simulation results demonstrate the accuracy and superiority of our proposed ECHIO algorithm in finding optimal LEDs positions when compared with the standard CHIO, Firefly algorithm (FA), Particle Swarm Optimization (PSO), Marine Predators algorithm (MPA), Manta Ray Foraging Optimization (MRFO), Whale Optimization Algorithm (WOA), Bat algorithm (BAT), Grey Wolf Optimizer (GWO), and Genetic algorithm (GA).

Keywords: Visible Light Communication · LEDs Placement Problem · Coronavirus Herd Immunity Optimize · Chaotic map · Opposition-Based Learning

B. Djamaa et al. (Eds.): CSA 2024, LNNS 1145, pp. 375–385, 2025.
https://doi.org/10.1007/978-3-031-71848-9_34

1 Introduction

In recent years, the Light Communication (LC) technology is gaining more interest due to its enormous bandwidth available in the unregulated optical spectrum. Visible Light Communication (VLC) stands as a technology that offers a viable option or addition to conventional radio communication. It finds practical utility in various indoor scenarios, including tasks like positioning [1], home networking setups, and communication within environments where the use of radio frequency (RF) radiation is restricted, such as medical facilities and aircraft cabins. This technology is based on Light-Emitting Diodes (LEDs), which are used as transmitters to send data and illumination simultaneously, thus achieving dual functionality. The positioning of LEDs within indoor VLC communication systems is a critical issue. Therefore, it is essential to carefully evaluate the best placement of LEDs to guarantee sufficient coverage, optimal throughput and appropriate illumination. The placement of LEDs is recognized as an NP-hard problem [2], and as a result, meta-heuristics, which are intelligent techniques, offer effective solutions for addressing it. Several research studies have introduced meta-heuristic-based approaches to tackle the problem of LED placement in indoor VLC communication systems. Rui et al. [3] uses PSO algorithm to solving the LEDs deployment problem in an indoor VLC communication system. The discussion encompassed the influence of the average outage area rate. Shangsheng Wen et al. [4] tackled the challenge of LED placement within an indoor VLC system, considering positioning accuracy, using a Modified Artificial Fish Swarm Algorithm (MAFSA). Kumawat et al. [5] used the Whale Optimization Algorithm (WOA) to address the challenge of positioning LED panels within an indoor VLC system, while taking into account received power and SNR parameters. In recent years, numerous of metaheuristic algorithms inspired by the behavior of viruses in nature have emerged. Among them, CHIO [6] has proven to be effective, due to its dynamic and adaptable control parameters. It has gained a huge interest and has been applied in numerous application fields such as feature selection [7], Vehicle routing problem [8], Scheduling problem [9], Travelling Salesman Problem (TSP) [10], and so on. The diverse applications of CHIO showcased its effectiveness, inspiring us to employ it in tackling the challenge of optimizing LED placement. However, to enhance the CHIO algorithm, Naderipour et al. [11] applied a nonlinearly decreasing inertia weight strategy (NDIWS) to improve the performance of CHIO in global and local discovery. Regarding our work, we employed two techniques. One involves utilizing a chaos mechanism to generate initial values, thereby boosting the algorithm's convergence speed. The other entails applying the opposition-based learning method to strike a balance between exploration and exploitation capabilities, helping the algorithm escape local optimal solutions. We assess the performance of our improved ECHIO algorithm considering user coverage and throughput parameters in a conference room with dimensions of 10 m × 10 m × 3 m.

The rest of this paper is organized as follows. In Sect. 2, the formulation of the LED placement problem in indoor VLC communication systems is presented. A description of CHIO, the chaotic map concept, and the OBL mechanism are given in Sect. 3. Section 4 describes the ECHIO algorithm for solving the LED placement problem. The outcomes of the simulations are elaborated upon in Sect. 5, followed by a conclusion and our future work.

2 LEDs Placement Problem Formulation

2.1 System Model

In this work, we focus on a common indoor VLC system. (V) operating in an empty conference room with dimensions of $D \times$ W $\times H$, as shown in Fig. 1. The room is equipped with multiple Lighting Emitting Diodes (LEDs) which act as access points, and multiple receiving users that are randomly distributed within the room.

- L is the set of N LEDs: $L = \{L_1, L_2, ..., L_N\}$ to be placed on the ceiling at locations (x_i, y_i, z_i), where $i \in (1, 2, 3, ...N)$.
- U is the set of M users $U = \{U_1, U_2, ..., U_M\}$, each user is equipped with a photo-detector (PD) used as a wireless receiver.

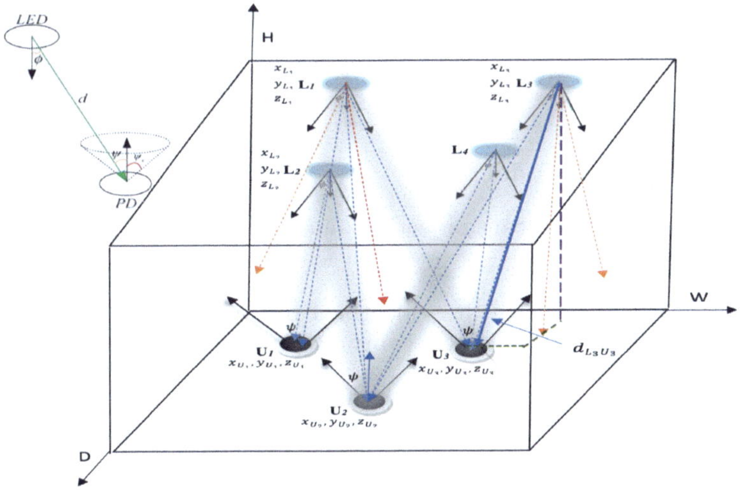

Fig. 1. The Chanel and VLC Room Model

The LEDs can be regarded as a source of Lambertian radiation. Following the Lambertian radiation model, Eq. 1 can be used to describe the channel gain between i^{th} LED and j^{th} user [2].

$$H(0)_{ij} = \begin{cases} \frac{(m+1)A}{2\pi d_{ij}^2} \cos^m(\phi_i) T_S(\psi_j) g(\psi_j) \cos(\psi_j), & 0 \le \psi_j \le \psi_c \\ 0, & \psi_j > \psi_c \end{cases} \tag{1}$$

In this context, the Lambertian order m is calculated using the formula $m = -\log 2/\log\left(\cos\left(\phi_{1/2}\right)\right)$, where $\phi_{1/2}$ denotes the half power angle of the LED. The variables A represent the area of the photodiode, and d_{ij} indicates the distance between the ith LED and the jth photodiode.

Additionally, ϕ_i represents the radiation angle, while ψ_j corresponds to the incident angle. The terms $T_s(\psi_j)$ and $g(\psi_j)$ denote the optical filter's gain and the optical concentrator's gain, respectively, with $g(\psi_j)$ being determined as follows:

$$g(\psi_j) = \begin{cases} \frac{n^2}{\sin^2(\psi_c)}, 0 \leq \psi_j \leq \psi_c \\ 0, \psi_j > \psi_c \end{cases} \tag{2}$$

where n is the optical concentrator's refractive index. Assuming Pt_i is the transmit power of the ith LED, the reception power of the jth PD can be stated as follows:

$$Pr_{ij} = Pt_i \times H(0)_{ij} \tag{3}$$

From [12], the end-to-end SNR can be expressed as:

$$SNR_{ij} = \frac{(R.Pr_{ij})^2}{\sigma_t^2} \tag{4}$$

where R is the PD responsivity and σ_t^2 is the variance noise total.

We note that each user U_j can be associated with a single LED (i.e., L_i) based on the highest received power. In the following, We provide our problem's mathematical model.

2.2 Mathematical Model

In our research, two significant objectives must be optimized, network coverage and throughput.

- The network coverage, which is identified based on each user's coverage, is the field of diffusion of the power transmitting from a set of LEDs. It is given by the following formula:

$$Cov(V) = \sum_{j=1}^{M} max_{i \in \{1,\ldots N\}}(Cov_{U_j}^{L_i}) \tag{5}$$

Here, $Cov_{U_j}^{L_i}$ represents the coverage of the user U_j by the LED L_i. If the SNR_{ij} value is greater than the threshold level (P^{th}) of SNR, the user U_j is considered to be covered by the LED L_i, as shown in equation (6).

$$Cov_{U_j}^{L_i} \begin{cases} 1 & if\ SNR_{ij} > P^{th} \\ 0 & Otherwise \end{cases} \tag{6}$$

– The network's overall throughput depends on the individual user's throughput, which has the potential to be extremely high, sometimes even reaching several gigabits per second. It is represented as follows:

$$Tr(V) = \sum_{j=1}^{M} max_{i\in\{1,...N\}}(Tr_{U_j}^{L_i}) \quad (7)$$

Here, $Tr_{U_j}^{L_i}$ reflects the throughput of user U_j covered by LED L_i, which might be indicated as [13]:

$$Tr_{U_j}^{L_i} = B \times log2(1 + SNR_{ij}) \quad (8)$$

where B represents the bandwidth, and SNR_{ij} is defined by Eq. 4.

Therefore, our objective function is considered as a weight-aggregation multi-objective function, aiming to maximize the total coverage and throughput of our VLC system while respecting and satisfying some constraints. This may be written using the linear weighted sum approach, as shown below.

$$\textbf{Maximize } f = (\lambda)(\frac{Cov(V)}{M}) + (1-\lambda)(\frac{Tr(V)}{M \times Tr_{max}}) \quad (9)$$

$$\textbf{S.t.} \begin{cases} SNR_{ij} \geq P^{th} \\ 0 \leq x_i \leq W \qquad i \in \{1,...N\}, j \in \{1,...M\} \\ 0 \leq y_i \leq D \end{cases} \quad (10)$$

where $Cov(V)$ and $Tr(V)$ represent two objective functions as shown in Eq. 5 and Eq. 7 respectively. λ serves as a weighting coefficient that regulates the significance of the associated function and can assume any value within the range of [0,1]. P^{th} denotes the threshold value of SNR, and D, W indicate the dimensions of the room, with D representing the length and W representing the width.

3 Preliminaries

3.1 Coronavirus Herd Immunity Optimizer (CHIO)

The coronavirus herd immunity optimizer (CHIO) is a novel nature-inspired optimization algorithm introduced by Al-Betar [6] in 2021. CHIO draws inspiration from both the herd immunity and social distancing strategies. It formulates the optimization process by modeling susceptible, infected, and immune populations. CHIO comprises four primary steps, detailed as follows:

Step 1. At the start of the CHIO algorithm, the control parameters are set up, which encompass the initial number of primary infected cases (C_0), the basic reproduction rate (BR_r), and the maximum age for infected cases (Max_{age}).

Step 2. During this stage, the population undergoes an evolution in response to the virus. The enhancement of X_i^j is influenced by the basic reproduction rate (BR_r) and is governed by three specified rules, as detailed below:

$$X_i^j(t+1) \leftarrow \begin{cases} C\left(X_i^j(t)\right) & r < \frac{1}{3} \times BR_r \\ N\left(X_i^j(t)\right) & r < \frac{2}{3} \times BR_r \\ R\left(X_i^j(t)\right) & r < BR_r \end{cases} \tag{11}$$

$C(X_i^j(t))$, $N(X_i^j(t))$ and $R(X_i^j(t))$ correspond to the infected, susceptible, and immune cases, respectively. These cases are defined as follows:

$$C\left(x_i^j(t)\right) = x_i^j(t) + r \times \left(x_i^j(t) - x_i^c(t)\right) \tag{12}$$

$$N\left(x_i^j(t)\right) = x_i^j(t) + r \times \left(x_i^j(t) - x_i^m(t)\right) \tag{13}$$

$$R\left(x_i^j(t)\right) = x_i^j(t) + r \times \left(x_i^j(t) - x_i^v(t)\right) \tag{14}$$

where r is a random number between 0 and 1. $X_i^c(t)$ is a randomly generated value taken from the infected case. The random value $X_i^m(t)$ is selected from the susceptible case, while $X_i^C(t)$ is derived from the best immuned case.

Step 3. To update the population of herd immunity, the $X_i^j(t+1)$ is incorporated into the population if it exhibits a superior fitness value compared to $X_i^j(t)$. Otherwise, if $S_i = 1$, the age vector will be incremented.

Step 4. In cases of fatality, when the current infection level S_i reaches 1 and does not show improvement for the specified number of iterations as indicated by the Max-Age parameter, that case will be classified as deceased. Furthermore, both Max_{age} and S_i will be set to zero to introduce diversity into the existing population and prevent it from converging to a local optimum. Subsequently, X_i will be randomly generated within the search space.

3.2 Chaotic Map and Opposition-Based Learning

The chaos method offers a deterministic approach for studying dynamic and non-linear systems, featuring non-repetitive and stochastic ergodicity. Integrating chaotic maps into meta-heuristics accelerates convergence, improves search capabilities, and prevents entrapment in local optima [14]. One such chaotic map, the Sine map, is defined as follows:

$$SM_{k+1} = \frac{ac}{4} \sin\left(\pi SM_k\right), 0 \leq SM_0 \leq 1 \tag{15}$$

Here, SM_k represents the chaotic map value at the k-th iteration, constrained within [0, 1]. Opposition-Based Learning (OBL), introduced by Tizhoosh in 2005

[15], enhances convergence rates and search space exploration in various metaheuristic optimization algorithms. The opposite of a real number $Pmo \in [lb, ub]$ in a multidimensional search space is determined using the equation:

$$P\bar{m}oi = ubi + lb_i - Pmo_i, \quad i = 1, 2, \ldots, d \tag{16}$$

Here, $P\bar{m}o \in \mathbb{R}^d$ is the opposite vector derived from the real vector $P \in \mathbb{R}^d$.

4 The Proposed ECHIO Algorithm for Solving the LEDs Placement Problem

This section delves into the implementation details of our proposed algorithm ECHIO. First, we have replaced the use of a random values with the sine map values. This modification empowers ECHIO to exercise greater control over the parameter used for updating solutions and broadens the range of available solutions. Second, in order to enhance convergence speed and facilitate effective exploration of the search space, we have integrated the OBL strategy. The implementation of the ECHIO algorithm for the LEDs placement problem encompasses six stages, Fig. 2 illustrates the Flowchart of CHIO and ECHIO.

- **Initialization** In this phase, the parameters of ECHIO and the VLC system environment are initialized. The solutions are initialized in a random manner. The parameter r initialized using the sine map, which governs the solution evolution in ECHIO, to $r_0 = 0.7$.
- **OBL Strategy section** After the initialization step and at each iteration, the OBL strategy is applied to the solutions $x_{i..n}$. Using Eq. 16 to determine the opposite solutions $\overline{x_{i..n}}$. Afterwards, both $x_{i..n}$ and $\overline{x_{i..n}}$ are evaluated using the fitness function f. According to the fitness value, only the n best solutions $\mathrm{x}^{i \ldots n}$ from $\overline{x_{i..n}} \cup x_{i..n}$ are selected for the evolution phase. This step represents the first improvements to ECHIO.
- **Solution evolution and chaotic integration** In this step, the chaos value of the parameter r is determined based on Eq. 17, This step represents the second improvements to ECHIO. Here the r_i is the chaotic map value of parameter r at the $i - th$ iteration.
- **Solution update** In this step, ECHIO updates the solutions $\mathrm{X}_{i..n}$ according to CHIO.
- **Termination** The last three steps 3 are repeated until the stopping condition is met.

$$r_{i+1} = \frac{ac}{4} \sin(\pi r_i) \tag{17}$$

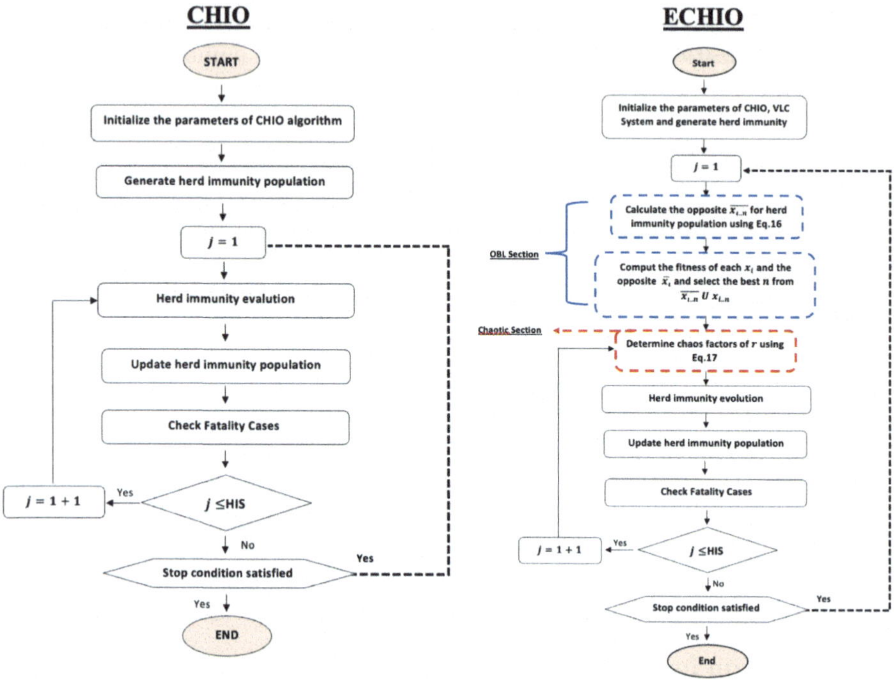

Fig. 2. Flowchart of CHIO and ECHIO

5 Simulation Results

In this section, we evaluate and analyze the efficiency of our proposed ECHIO algorithm in resolving the optimal LED placement problem. We conduct a comparative analysis between ECHIO and the original CHIO, and eight other optimization algorithms. All simulations are run on a Core i5 2.5 GHz-CPU machine and Matlab software is used to code all the algorithms. We consider an empty conference room with dimensions of 10 m × 10 m × 3 m, i.e., without any objects hindering line-of-sight (LOS) conditions. A typical scheme of such room is depicted in Fig. 1. The efficiency of the ECHIO algorithm is validated based on various scenarios with different numbers of LEDs (from 1 to 10) and different numbers of users (from 5 to 40). The assessment is carried out taking into account three metrics such as coverage, mean throughput per user, and fitness value. The total iteration count has been established as 1000. Each result represents the average outcome derived from 30 runs. Under these conditions, we proceed to analyze the effects of varying the number of LEDs and varying the number of users.

Table 1. Coverage, mean throughput, and fitness under various values of number of LEDs and Users

number of LEDs	1	2	3	4	5	6	7	8	9	10
Coverage (%)										
*ECHIO	**26.66**	**43.33**	50	**60**	73.33	**86.66**	**86.66**	**93.33**	**100**	**100**
CHIO	23.33	26.66	40	46.66	50	56.66	63.33	73.33	80	83.33
FA	23.33	26.66	40	46.66	50	56.66	63.33	73.33	80	83.33
PSO	20	36.666	**53.33**	56.66	57.33	70	83.33	86.66	93.33	96.66
MPA	23.33	40	43.33	50	70	76.66	80	90	96.66	**100**
MRFO	**26.66**	30	43.33	53.33	66.66	73.33	76.66	80	90	93.33
WOA	23.33	30	40	43.33	60	66.66	78.33	80	83.33	86.66
BAT	20	26.66	40	46.66	50	53.33	56.66	60	70	73.33
GWO	**26.66**	33.33	43.33	56.66	**76.66**	83.33	83.33	90	90	93.33
GA	23.33	36.66	50	56.66	70	70	**86.66**	90	96.66	**100**
Mean throughput per user (Mbps)										
*CHIO	16.66	**27.302**	31.60	**38.10**	46.43	**55.22**	55.57	**60.23**	**63.64**	64.67
CHIO	14.38	16.81	26.08	29.40	31.32	36.25	40.55	45.32	50.86	52.79
FA	14.38	16.81	26.08	29.40	31.32	36.25	40.55	45.32	50.86	52.79
PSO	12.48	22.99	**33.99**	35.83	36.86	45.58	52.90	56.67	62.04	64.84
MPA	14.71	25.72	28.35	32.09	44.15	50.07	51.36	57.96	62.38	65.12
MRFO	16.14	18.78	27.35	34.32	42.41	46.80	49.63	51.93	59.53	60.96
WOA	15.13	19.10	26.36	28.11	39.82	42.57	50.97	51.73	54.12	56.20
BAT	12.78	16.49	24.28	29.38	30.68	32.50	36.02	38.41	43.21	45.53
GWO	**17.00**	21.94	27.77	35.25	**48.71**	53.69	52.92	57.53	59.41	61.20
GA	14.36	22.51	32.40	35.60	43.42	44.10	**56.36**	57.31	62.59	**65.65**
Fitness										
*ECHIO	0.240	**0.392**	0.453	**0.545**	0.665	**0.788**	0.791	**0.854**	**0.909**	0.916
CHIO	0.209	0.241	0.367	0.422	0.451	0.516	0.577	0.658	0.727	0.756
FA	0.209	0.241	0.367	0.422	0.451	0.516	0.577	0.658	0.727	0.756
PSO	0.180	0.331	**0.485**	0.514	0.520	0.643	0.757	0.798	0.866	0.900
MPA	0.211	0.365	0.399	0.456	0.634	0.705	0.730	0.823	0.885	0.919
MRFO	0.237	0.27	0.374	0.478	0.588	0.648	0.690	0.724	0.833	0.859
WOA	0.214	0.282	0.356	0.410	0.512	0.567	0.698	0.703	0.765	0.795
BAT	0.182	0.239	0.356	0.422	0.447	0.475	0.515	0.547	0.628	0.659
GWO	**0.242**	0.307	0.395	0.510	**0.697**	0.762	0.757	0.820	0.832	0.860
GA	0.209	0.321	0.458	0.512	0.629	0.633	**0.796**	0.819	0.886	**0.922**

Number of users	5	10	15	20	25	30	35	40
Coverage (%)								
*ECHIO	**100**	**91.66**	**89.33**	73	70	**66.66**	**63.33**	55
CHIO	**100**	90	80	70	64	60	56.66	45
FA	**100**	80	70	60	52	50	42.5	42.85
PSO	**100**	86.66	80	70	68	60	57.14	52.5
MPA	**100**	90	80	**75**	**72**	65.71	60	54.28
MRFO	**100**	90	86.66	70	68	**66.66**	59.45	**57.5**
WOA	**100**	80	66.66	65	60	56.66	57.14	54
BAT	**100**	70	66.66	60	52	46.66	45.71	42.5
GWO	**100**	90	86.66	70	68	63.33	55	48.88
GA	**100**	90	86.66	73	70	**66.66**	61.71	55.66
Mean throughput per user (Mbps)								
*ECHIO	**69.39**	58.67	**58.25**	46.30	45.78	42.38	**40.93**	35.99
CHIO	68.71	59.17	51.19	45.49	41.18	38.51	36.70	29.56
FA	65.12	51.02	43.61	37.54	32.72	30.23	27.64	26.45
PSO	68.11	57.01	52.87	45.43	44.15	38.94	36.50	33.79
MPA	69.33	57.75	52.49	**49.17**	**46.52**	41.41	38.18	34.97
MRFO	69.35	59.01	56.73	45.67	43.79	42.56	37.74	**37.19**
WOA	66.95	49.59	45.02	41.74	37.25	36.54	35.98	34.60
BAT	66.49	44.57	42.90	37.01	33.98	29.81	28.53	27.30
GWO	68.11	**60.20**	55.44	45.38	44.33	40.61	34.89	31.54
GA	69.34	59.34	56.10	46.62	45.41	**42.64**	39.44	36.12
Fitness								
*ECHIO	**0.947**	0.836	**0.825**	0.671	0.644	0.606	**0.580**	0.506
CHIO	0.942	0.831	0.729	0.642	0.580	0.547	0.519	0.415
FA	0.919	0.728	0.630	0.541	0.470	0.444	0.390	0.384
PSO	0.938	0.800	0.740	0.642	0.624	0.550	0.520	0.480
MPA	0.946	0.821	0.737	**0.691**	**0.659**	0.595	0.545	0.496
MRFO	0.946	0.829	0.798	0.644	0.622	0.607	0.540	**0.526**
WOA	0.945	0.719	0.623	0.593	0.539	0.518	0.517	0.492
BAT	0.928	0.636	0.609	0.538	0.478	0.425	0.412	0.388
GWO	0.938	**0.837**	0.790	0.642	0.625	0.578	0.499	0.447
GA	0.939	0.833	0.804	0.673	0.641	**0.661**	0.569	0.511

5.1 Discussion and Analysis

Here, we investigate the impact of varying the number of LEDs from 1 to 10, while maintaining a fixed number of 30 users. Additionally, we analyze also the impact of increasing the number of users from 5 to 40 while keeping the number of LEDs fixed at 5, focusing on coverage, mean throughput per user, and fitness metrics. The findings are summarized in Table 1 and Fig. 3.

As the number of LEDs increases, coverage expands, eventually encompassing almost all users. This improvement occurs because more light sources are available, increasing the chances for users to receive light and enhancing the coverage metric. Notably, ECHIO consistently outperforms other algorithms in achieving coverage in most cases. Mean throughput per user increases with the number of LEDs, as more light sources lead to improved SNR and higher throughput. Furthermore, the fitness metric increases with the number of LEDs, demonstrating the efficiency of ECHIO in achieving coverage and mean throughput per user compared to other algorithms in most cases. Meanwhile, when the number of users in the room increases, the coverage metric decreases. This decrease occurs because additional users may not receive light from the deployed LEDs, resulting in reduced coverage. When varying the number of users, mean throughput per user decreases with more users in the room. This decline is due to LEDs being

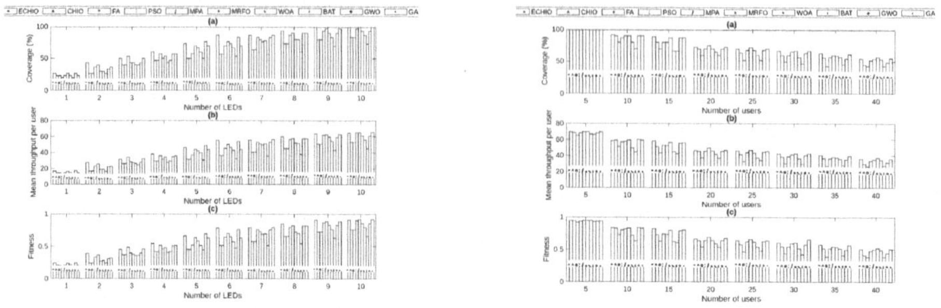

Fig. 3. Coverage (a), Throughput (b) and Fitness (c) under various numbers of LEDs and Users

placed farther away from users, resulting in lower SNR and reduced throughput per user. The fitness metric decreases as the number of users increases across all algorithms. ECHIO consistently outperforms other algorithms in terms of coverage and mean throughput per user in most cases.

6 Conclusion

In this paper, we introduce ECHIO, a novel algorithm combining chaotic mapping and Opposition-Based Learning (OBL) to optimize LED placement in indoor VLC systems. ECHIO enhances CHIO diversity with chaotic mapping and balances exploration-exploitation with OBL. Evaluation in a 10 m × 10 m × 3 m indoor space shows ECHIO outperforms CHIO, FA, PSO, MPA, MRFO, WOA, BAT, GWO, and GA methods in achieving optimal LED layouts. Future work will focus on multi-objective challenges and other optimization algorithms.

References

1. Botsinis, P., et al.: Quantum-assisted indoor localization for uplink mm-wave and downlink visible light communication systems. IEEE Access **5**, 23327–23351 (2017)
2. Benayad, A., Boustil, A., Meraihi, Y., Mirjalili, S., Yahia, S., Taleb, S.M.: An enhanced whale optimization algorithm with opposition-based learning for LEDs placement in indoor VLC systems. In: Handbook of Whale Optimization Algorithm, pp. 279–289. Elsevier (2024)
3. Guan, R., Wang, J.-Y., Wen, Y.-P., Wang, J.-B., Chen, M.: PSO-based led deployment optimization for visible light communications. In: 2013 International Conference on Wireless Communications and Signal Processing, pp. 1–6 (2013)
4. Wen, S., Cai, X., Guan, W.P., Jiang, J., Chen, B., Huang, M.: High-precision indoor three-dimensional positioning system based on visible light communication using modified artificial fish swarm algorithm. Opt. Eng. **57**(10), 1 (2018)
5. Kumawat, I.R., Nanda, S., Maddila, R.: Positioning LED panel for uniform illuminance in indoor VLC system using whale optimization, pp. 131–139, January 2018

6. Al-Betar, M., Alyasseri, Z., Awadallah, M., Doush, I.: Coronavirus herd immunity optimizer (CHIO). Neural Comput. Appl. **33**, 05 (2021)
7. Alweshah, M., Alkhalaileh, S., Al-Betar, M., Bakar, A.A.: Coronavirus herd immunity optimizer with greedy crossover for feature selection in medical diagnosis. Knowl.-Based Syst. **235**, 107629 (2022)
8. Dalbah, L.M., Al-Betar, M.A., Awadallah, M.A., Zitar, R.A.: A modified coronavirus herd immunity optimizer for capacitated vehicle routing problem. J. King Saud Univ. - Comput. Inf. Sci. **34**(8, Part A), 4782–4795 (2022)
9. Makhadmeh, S.N., et al.: A modified coronavirus herd immunity optimizer for the power scheduling problem. Mathematics **10**(3), 315 (2022)
10. Dalbah, L.M., Al-Betar, M.A., Awadallah, M.A., Zitar, R.A.: A coronavirus herd immunity optimization (CHIO) for travelling salesman problem. In: Khanna, A., Gupta, D., Bhattacharyya, S., Hassanien, A.E., Anand, S., Jaiswal, A. (eds.) International Conference on Innovative Computing and Communications, pp. 717–729. Springer, Cham (2022). https://doi.org/10.1007/978-981-16-3071-2_58
11. Naderipour, A., Abdullah, A., Marzbali, M.H., Nowdeh, S.A.: An improved coronavirus herd immunity optimizer algorithm for network reconfiguration based on fuzzy multi-criteria approach. Expert Syst. Appl. **187**, 115914 (2022)
12. Yahia, S., Meraihi, Y., Ramdane-Cherif, A., Benmessaoud, A., Eldeeb, H.B.: Performance evaluation of vehicular visible light communication based on angle-oriented receiver. Comput. Commun. **191**, 500–509 (2022)
13. Eroglu, Y.S., Sahin, A., Guvenc, I., Pala, N, Yuksel, M.: Multi-element transmitter design and performance evaluation for visible light communication. In: 2015 IEEE Globecom Workshops, pp. 1–6. IEEE (2015)
14. Meraihi, Y., Acheli, D., Ramdane-Cherif, A.: QoS multicast routing for wireless mesh network based on a modified binary bat algorithm. Neural Comput. Appl. **31**(7), 3057–3073 (2019)
15. Tizhoosh, H.R.: Opposition-based learning: a new scheme for machine intelligence. In: International Conference on Computational Intelligence for Modelling, Control and Automation and International Conference on Intelligent Agents, Web Technologies and Internet Commerce (CIMCA-IAWTIC 2006), vol. 1, pp. 695–701. IEEE (2005)

Assuming Carbon Tax in a Simultaneous Production and Maintenance Planning Optimization

Hind Abdi[1](✉) and Ali Berrichi[2]

[1] RECITS Laboratory, Faculty of Mathematics, USTHB University, BP 32 El-Alia, Bab-Ezzouar, Algiers, Algeria
habdi@usthb.dz

[2] LIMOSE Laboratory, The Department of Computer Science, UMBB University, Boumerdes, Algeria
ali.berrichi@univ-boumerdes.dz

Abstract. In this paper, we address the challenge of simultaneous production and maintenance planning under carbon emission (CE) regulations, aiming to minimize the combined costs of production, setup, inventory, maintenance, and CE. The study focuses on the multi-item capacitated lot-sizing problem, incorporating carbon emission considerations to ensure compliance with environmental regulations and promote sustainable manufacturing practices. We develop a mixed-integer non-linear program to model the problem and subsequently perform linearization techniques to enable its efficient solution. The linearized model is solved using the CPLEX solver, demonstrating a significant reduction in carbon emissions when faced with high carbon taxes. The results highlight the importance of integrating maintenance and production planning with carbon emission constraints, emphasizing the potential for achieving both economic and environmental benefits in manufacturing operations. The paper contributes to the ongoing efforts to develop sustainable and responsible strategies for production and maintenance planning in the context of increasingly stringent carbon emission regulations.

Keywords: Carbon Tax · Lot-sizing · Condition-based Maintenance

1 Introduction

Minimizing breakdowns and machine downtime, as well as ensuring reliability, are essential aspects of production planning, which require simultaneous consideration of maintenance activities. Addressing maintenance and production plans in a combined manner leads to reduced costs compared to planning them separately, as these processes are closely interconnected. Over the past few decades, extensive literature has been built on joint planning at strategic, tactical, or operational levels. In this paper, we focus on tactical decisions with a finite time horizon, divided into multiple periods. The time horizon is sufficiently broad,

B. Djamaa et al. (Eds.): CSA 2024, LNNS 1145, pp. 386–395, 2025.
https://doi.org/10.1007/978-3-031-71848-9_35

spanning weeks, months, or even longer. Production planning problems at this level are commonly known as lot-sizing problems.

Historically, the lot-sizing problem has been widely addressed in the literature, covering aspects such as setup, limited capacity, back-order, multi-level, multi-component, and inventory. For those interested, several review papers are available [5,6,11,16]. Recently, integrating maintenance scheduling into the lot-sizing model has emerged as a challenging research area. An intriguing state-of-the-art review of joint maintenance planning and lot-sizing in manufacturing systems can be found in [20], with another review in [17]. To closely monitor the state of the system, we will integrate a condition-based maintenance (CBM) policy into the production system planning. This maintenance approach takes into account the actual state of the machine, rather than systematically planned preventive maintenance. Recent papers that consider CBM with lot-sizing problems include [7,8,13,14,19,22,23].

Addressing greenhouse gas emissions has become a critical concern to prevent further acceleration of global warming. High fossil fuel consumption, resulting from population growth and the Industrial Revolution, has significantly impacted manufacturing systems. Consequently, it is crucial to incorporate carbon emission constraints into mid-term production system planning. Energy consumption leads to the emission of greenhouse gases, as discussed in several review papers [1,15,18,21,24].

Various papers have considered carbon taxes in lot-sizing using different approaches [2,9,12]. The carbon tax policy aims to regulate carbon emissions by imposing a tax on simultaneous production and maintenance planning.

In prior research, CE was consistently treated as a constant value across all periods, irrespective of the system's conditions. Nevertheless, in practice, this assumption does not always hold, especially when dealing with machine degradation over time caused by production quantity, increased energy used, or elevated production speeds. In the study of [10], the authors examine the numerical assessment of carbon emissions in a lathe utilized for turning operations. Their findings indicate a significant correlation between carbon emissions and processing duration, the emissions increased by 93% within $800(s)$. The researchers concluded that longer processing times result in higher carbon emissions due to increased energy consumption. In this paper, we consider that a degradable system might result in an increase in carbon emissions as the machine ages, a factor that has not been considered in prior research. This study aims to integrate condition-based maintenance into production planning while minimizing total costs due to production, maintenance, stock, set-up and CE. The principal contribution of this work involves developing a Mixed integer linear program to model the Multi-item Multi-period capacitated lot-sizing with CE tax consideration and CBM scheduling. Another contribution is to assume that the machine's longer processing times result in higher carbon emissions.

2 Mathematical Formulation

We consider a manufacturing system consisting of one machine that produces N different items within a finite time horizon, divided into several discrete periods $t \in T$ with fixed lengths. The challenge is to determine the lot size to produce in each period t to meet the demand d_{it} while simultaneously identifying the periods for maintenance actions. These one are scheduled depending on the available capacity to produce in each period. When capacity is deemed too low CBM takes place. The goal is to minimize breakdowns and interruptions while maximizing production. Figure 1illustrates the problem and demonstrates variations in CE.

2.1 Assumptions

The assumptions outlined in our mathematical model are as follows:

- The demand for each item is variable but deterministic across the time horizon.
- The maintenance policy restores the system to an as-good-as-new state after completion.
- Maintenance planned at period t takes place at the beginning of the period.
- The machine capacity is limited and decreases exponentially over time at a rate of $0 \leq \lambda \leq 1$.
- A tax is imposed on carbon emissions due to production, set-up, and inventory processes.
- The machine's CE increases over time.

2.2 Model Parameters

We use the following parameters in our model:

f_{it}: Production costs for item i during time t.
h_{it}: Holding costs for item i during time t.
s_{it}: Set up costs for item i during time t.
m_t: Maintenance cost during time t.
c_{max}: Maximum capacity of the machine.
d_{it}: Demand of item i during time t.
e: Fixed carbon emissions amount per unit of production.
u: Production upper bound for each item.
r_{it}: Carbon emissions emitted by holding a unit of item i in period t.
p_{it}: Carbon emissions emitted by setting up item i in period t.
tax: Carbon unit price.
λ: Capacity reduction coefficient due to machine deterioration.
α: Carbon emission increase rate.
UB: Upper Bound used for constraint linearization.
L: A range extending from 0 to $\lfloor log(u) \rfloor$.

2.3 Decision Variables

The decision variables in our model are:

x_{it}: Quantity of item i produced during time t.
y_{it}: Binary variable equal to 1 if item i is produced during time t and 0 otherwise.
I_{it}: Inventory value of item i at the end of period t.
z_t: Binary variable equal to 1 if maintenance is performed in period t and 0 otherwise.
c_t: Available capacity in period t.
ce_{it}: Variable carbon emission due to item i production in period t.
Q_i: Replaces the product of binary and continuous variables $Q_i = ce_{it}(1-z_t)$.
l_{itk}: Binary variable used for constraint linearization.
w_{itk}: Replaces the product of binary and continuous variables $w_{itk} = ce_{it}l_{itk}$.
X_{it}: Carbon emissions emitted by producing item i in period t.

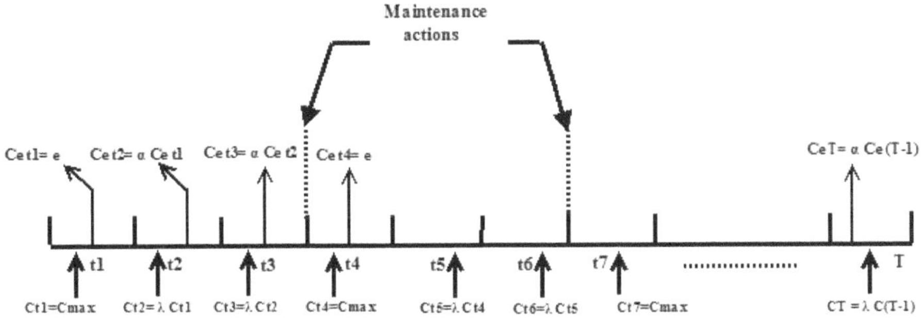

Fig. 1. System state illustration

2.4 Mathematical Model Description

The objective function aims to minimize the total costs associated with production, inventory, set-up, carbon emissions, and maintenance.

$$\min \sum_{t \in T \in I} \sum_{t \in T} (f_{it}x_{it} + h_{it}I_{it} + s_{it}y_{it} + tax(ce_{it}x_{it} + r_{it}I_{it} + p_{it}y_{it})) + \sum_{t \in T} m_t z_t \quad (1)$$

The product of the two continuous decision variables x_{it} and ce_{it} makes the objective function non-linear. We proceed by using substitute variables, we start by replacing the product of the two continuous variables with X_{it}. Then, the following linearization constraints are added to the model according to [3] to ensure model linearization:

$$x_{it} = \sum_{k=0}^{\lfloor log(u) \rfloor} 2^k l_{itk} \quad \forall t \in T, i \in N \quad (2)$$

$$X_{it} = \sum_{k=0}^{\lfloor log(u) \rfloor} 2^k w_{itk} \quad \forall i \in N, t \in T \tag{3}$$

$$w_{itk} \leq ce_{it}, \quad \forall i \in I, \forall t \in T, \forall k \in 1,2,3 \tag{4}$$

$$w_{itk} \leq UBl_{itk}, \quad \forall i \in I, \forall t \in T, \forall k \in 1,2,3 \tag{5}$$

$$w_{itk} \geq ce_{it} - UB(1 - l_{itk}), \quad \forall i \in I, \forall t \in T, \forall k \in 1,2,3 \tag{6}$$

Since we expect a variable carbon emission that increases over time, the following constraint expresses the variation based on the machine state.:

$$ce_{it} = \alpha ce_{i(t-1)}(1 - z_t) + ez_t \quad \forall t \in T, i \in N \tag{7}$$

The nonlinearity of this constraint arises from the interaction between binary (z_t) and continuous variables (ce_{it}). To linearize the constraint, incorporate the following additional constraint based on the approach presented by [4]:

$$ce_{it} = \alpha Q_{it} + ez_t \quad \forall t \in T, i \in N \tag{8}$$

$$ce_{it} \geq e \quad \forall i \in Nt \in T \tag{9}$$

$$Q_i \leq ce_{it}, \quad \forall i \in I, \forall t \in T \tag{10}$$

$$Q_i \leq UB(1 - z_t), \quad \forall i \in I, \forall t \in T \tag{11}$$

$$Q_i \geq ce_{it} - UBz_t, \quad \forall i \in I, \forall t \in T \tag{12}$$

$$0 \leq x_{it} \leq u, \quad \forall i \in I, \forall t \in T \tag{13}$$

Equation 13 expresses the production upper bound for each item and period.

The available capacity in each period is determined by whether or not maintenance is conducted. If the capacity is determined to be too low, a maintenance action is performed [19] and restores the system to its nominal state $c_t = c_{max}$:

$$c_t \leq c_{max} \tag{14}$$

$$c_t \leq \lambda c_{t-1} + c_{max} z_t \quad \forall t \geq 2 \tag{15}$$

The classical constraints of the capacitated lot-sizing problem are added and described below:

$$I_{it} = I_{i,t-1} + x_{it} - d_{it}, \quad \forall i \in I, \forall t \in T \tag{16}$$

$$\sum_{i \in I} x_{it} \leq c_t, \quad \forall t \in T \tag{17}$$

$$x_{it} \leq (\sum_{s \geq t, s \in T} d_{is}) y_{it} \quad \forall i \in I, \quad t \in T \tag{18}$$

$$x_{it}, I_{it}, c_t, ce_{it}, w_{itk}, Q_{it}, X_{it} \geq 0 \quad \forall i \in N, \quad t \in T, k \in L \tag{19}$$

$$y_{it}, z_t, l_{itk} \in 0,1, \quad \forall i \in I, \forall t \in T, \forall k \in L \tag{20}$$

The constraint 16 maintains inventory balance across all items and periods. The machine capacity constraint is expressed by 17, and demand satisfaction by 18. Finally, 19 describes the non-negative constraints, whereas 20 conveys the binary constraints.

The problem is modelled by a Mixed integer non-linear program. The aim is to determine the optimal production and maintenance decisions while considering taxes on carbon emissions. The objective is to minimize total costs associated with production, inventory, set-up, maintenance and CE. The Mixed Integer Linear model can be solved using the CPLEX solver.

3 Results and Discussion

In this section, we evaluate the performance of the joint production and maintenance planning model. To accomplish this, we use the benchmark from [2] and assume a single machine with a single item to produce within a time horizon divided into 15 periods. The demand is variable and follows a uniform distribution within the interval [20, 70]. The carbon tax rate is selected in the interval [0, 1.5]. Table 1 displayed the numerical values of the different parameters.

Table 1. Parameter settings

f_{it}	s_{it}	h_{it}	m_t	e	p_{it}	r_{it}	λ	α
04	60	01	30	02	20	02	0.8	1.2

Since [2] assumed an uncapacitated lot-sizing system without maintenance consideration, we set a sufficient capacity cap to satisfy the demand $C_{max} = 84$. The results presented in Fig. 2 are the outcomes of running ten demand series using IBM ILOG CPLEX. The choice of CPLEX is supported by its simplicity and effectiveness in solving this type of program in a fair amount of time.

First, we analyze the impact of carbon tax rates on the total cost, production, inventory, and maintenance decisions. As the carbon tax rate increases, we observe a general trend of rising total costs, which include production, inventory, setup, maintenance, and carbon emissions costs. This is due to the added expense of managing carbon emissions and the need to adjust production, inventory, and maintenance strategies accordingly.

The production and inventory levels tend to fluctuate based on the carbon tax rate, with higher carbon taxes incentivizing the manufacturing system to adopt more environmentally friendly practices. When the carbon tax is low or zero, the production and inventory levels are primarily driven by production, setup, and holding costs. However, as the carbon tax rate increases, we observe a shift towards more sustainable production and inventory levels that minimize carbon emissions. Consequently, we can observe decreased production quantities and inventory levels for higher carbon tax rates, leading to lower carbon emissions.

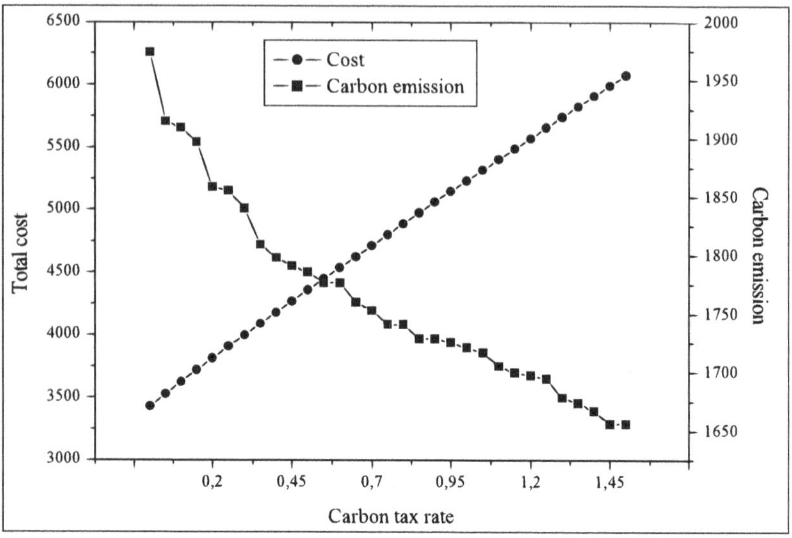

Fig. 2. Costs and CE variation according to tax rate

Additionally, we examine the influence of the carbon tax rate on maintenance decisions. As the carbon tax rate increases, the model tends to schedule maintenance more frequently to ensure the machine operates at its optimal state, minimizing carbon emissions. This results in an increased maintenance cost; however, the overall environmental impact is reduced due to lower carbon emissions.

We also investigate the relationship between the time horizon and the optimal decisions. As the time horizon lengthens, the model can better optimize production, inventory, and maintenance decisions to minimize total costs and carbon emissions. The model can take advantage of longer planning periods to balance production and inventory levels, schedule maintenance more effectively, and mitigate the impact of carbon taxes.

In summary, the joint production and maintenance planning model demonstrates its ability to provide optimal decisions for production, inventory, and maintenance, considering the carbon tax rate and time horizon (Fig. 3). The model highlights the importance of integrating carbon emissions and maintenance policies into the decision-making process. Furthermore, it offers valuable insights into how varying carbon tax rates and time horizons can impact overall costs, production levels, inventory levels, and maintenance decisions in manufacturing systems.

The Cplex solver was able to solve small instances until twelve items and twenty-four periods (12,24), beyond this instance no results are provided. For large instances an approach method is necessary.

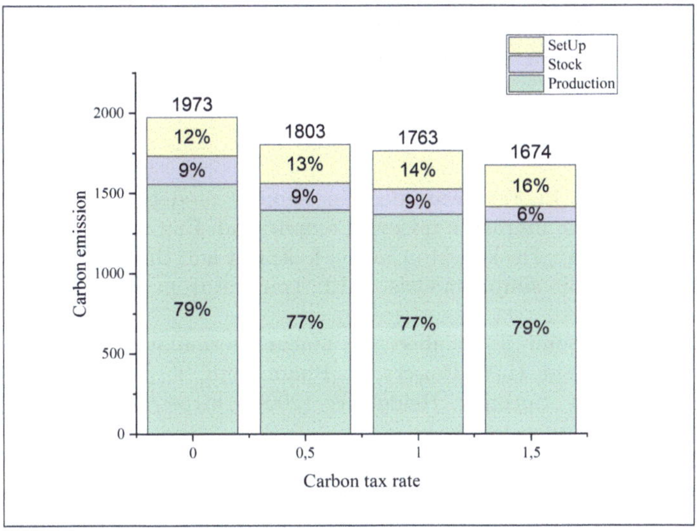

Fig. 3. Decision variables sensitivity over tax variation

4 Conclusion

This paper presented a joint production and maintenance planning model that incorporates carbon emissions costs and carbon tax rates. The primary goal was to provide a holistic approach to decision-making in manufacturing systems by considering not only the traditional cost factors, such as production, inventory, and maintenance but also the environmental impact due to carbon emissions. The model was evaluated using a benchmark from the literature, and its performance was analyzed under various carbon tax rates and time horizons.

The results demonstrated the model's effectiveness in providing optimal decisions for production, inventory, and maintenance while taking into account the carbon tax rate and time horizon. As the carbon tax rate increased, the model successfully adapted production, inventory, and maintenance strategies to minimize total costs and reduce carbon emissions. Furthermore, a longer time horizon allowed for better optimization of decisions, balancing production and inventory levels, scheduling maintenance more effectively, and mitigating the impact of carbon taxes.

In conclusion, this study highlights the importance of integrating carbon emissions and maintenance policies into the decision-making process for manufacturing systems. By adopting a joint production and maintenance planning model that considers environmental impacts, organizations can make more sustainable decisions, ultimately contributing to the global effort of reducing carbon emissions and mitigating climate change. Future research could extend this model by incorporating additional environmental factors, exploring the impact of different demand patterns, or studying multi-level and multi-machine systems to provide a more comprehensive understanding of sustainable manufacturing practices.

References

1. Bänsch, K., et al.: Energy-aware decision support models in production environments: a systematic literature review. Comput. Ind. Eng. **159**, 107456 (2021)
2. Benjaafar, S., Li, Y., Daskin, M.: Carbon footprint and the management of supply chains: insights from simple models. IEEE Trans. Autom. Sci. Eng. **10**(1), 99–116 (2012)
3. Billionnet, A., Elloumi, S., Lambert, A.: Linear reformulations of integer quadratic programs. In: Le Thi, H.A., Bouvry, P., Pham Dinh, T. (eds.) MCO 2008. CCIS, vol. 14, pp. 43–51. Springer, Heidelberg (2008). https://doi.org/10.1007/978-3-540-87477-5_5
4. Bisschop, J.: AIMMS optimization modeling. Lulu.com (2006)
5. Brahimi, M., Boukhalfa, K., Moussaoui, A.: Deep learning for tomato diseases: classification and symptoms visualization. Appl. Artif. Intell. **31**(4), 299–315 (2017)
6. Buschkühl, L., Sahling, F., Helber, S., Tempelmeier, H.: Dynamic capacitated lot-sizing problems: a classification and review of solution approaches. OR Spectrum **32**, 231–261 (2010)
7. Cheng, G.Q., Zhou, B.H., Li, L.: Integrated production, quality control and condition-based maintenance for imperfect production systems. Reliab. Eng. Syst. Saf. **175**, 251–264 (2018)
8. Darendeliler, A., Claeys, D., Khatab, A., Aghezzaf, E.H.: Joint optimization of dynamic lot-sizing and condition-based maintenance. In: 9th International Conference on Operations Research and Enterprise Systems, pp. 151–158. SCITEPRESS–Science and Technology Publications (2020)
9. Heck, M., Schmidt, G.: Lot-size planning with non-linear cost functions supporting environmental sustainability. Int. J. Green Comput. (IJGC) **1**(1), 34–39 (2010)
10. Jiang, Z., Gao, D., Lu, Y., Kong, L., Shang, Z.: Quantitative analysis of carbon emissions in precision turning processes and industrial case study. Int. J. Precis. Eng. Manuf.-Green Technol. **8**, 205–216 (2021)
11. Karimi, B., Ghomi, S.F., Wilson, J.: The capacitated lot sizing problem: a review of models and algorithms. Omega **31**(5), 365–378 (2003)
12. Kian, R.: A production planning problem with lost sales and nonlinear convex production cost function under carbon emission restrictions. J. Energy Manage. Technol. **4**(2), 1–13 (2020)
13. Lemoine, D., Castanier, B.: Tactical planning and predictive maintenance: towards an integrated model based on ε-reliability. In: Dolgui, A., Bernard, A., Lemoine, D., von Cieminski, G., Romero, D. (eds.) APMS 2021. IAICT, vol. 630, pp. 92–101. Springer, Cham (2021). https://doi.org/10.1007/978-3-030-85874-2_10
14. Peng, H., van Houtum, G.J.: Joint optimization of condition-based maintenance and production lot-sizing. Eur. J. Oper. Res. **253**(1), 94–107 (2016)
15. Qasim, M., Wong, K.Y., Saufi, M.S.R.M.: Production planning approaches: a review from green perspective. Environ. Sci. Pollut. Res. **30**, 90024–90049 (2023)
16. Quadt, D., Kuhn, H.: Capacitated lot-sizing with extensions: a review. 4OR **6**(1), 61–83 (2008)

17. de Souza, N.L.S., Mendes, L.G., Rovaris, E.S., Frazzon, E.M., Braghirolli, L.F.: Integrated production and maintenance planning: a systematic literature review. In: Rossit, D.A., Tohmé, F., Mejía Delgadillo, G. (eds.) ICPR-Americas 2020. CCIS, vol. 1407, pp. 342–356. Springer, Cham (2021). https://doi.org/10.1007/978-3-030-76307-7_26
18. Sabtu, M.I., Hishamuddin, H., Saibani, N., Ab Rahman, M.N.: A review of environmental assessment and carbon management for integrated supply chain models. EVERGREEN Joint J. Novel Carbon Resour. Sci. Green Asia Strategy **8**(3), 628–641 (2021)
19. Shamsaei, F., Van Vyve, M.: Solving integrated production and condition-based maintenance planning problems by MIP modeling. Flex. Serv. Manuf. J. **29**(2), 184–202 (2017)
20. de Souza, N.L.S., Mendes, L.G., Rovaris, E.S., Frazzon, E.M., Braghirolli, L.F.: Integrated production and maintenance planning: a systematic literature review. In: Rossit, D.A., Tohme, F., Mejia Delgadillo, G. (eds.) Production Research: 10th International Conference of Production Research-Americas, ICPR-Americas 2020, Bahía Blanca, Argentina, 9–11 December 2020, Revised Selected Papers, Part I, pp. 342–356. Springer, Cham (2021). https://doi.org/10.1007/978-3-030-76307-7_26
21. Suzanne, E., Absi, N., Borodin, V.: Towards circular economy in production planning: challenges and opportunities. Eur. J. Oper. Res. **287**(1), 168–190 (2020)
22. Wang, L., Lu, Z., Han, X.: Joint optimal production planning and proactive maintenance policy for a system subject to degradation. J. Qual. Maintenance Eng. **25**(2), 236–252 (2019)
23. Zhang, N., Tian, S., Liu, B., Zhang, J.: Joint optimization of production lot-sizing and condition-based maintenance in an imperfect production process with dependent indicators. Qual. Technol. Quant. Manage. **894**, 511–527 (2022)
24. Zhou, X., Wei, X., Lin, J., Tian, X., Lev, B., Wang, S.: Supply chain management under carbon taxes: a review and bibliometric analysis. Omega **98**, 102295 (2021)

Space-Time Clustering of Seismicity in Algeria

Oussama Sohaib Mazari[1(✉)], Abderrazak Sebaa[1,2], and Francisco Martínez-Álvarez[3]

[1] LIMED Laboratory, Faculty of Exact Sciences, University of Bejaia, 06000 Bejaia, Algeria
oussamasohaib.mazari@univ-bejaia.dz

[2] Laboratoire LITAN, École supérieure en Sciences et Technologies de l'Informatique et du Numérique. RN 75, Amizour, 06300 Bejaia, Algeria
sebaa@estin.dz

[3] Data Science and Big Data Lab, Pablo de Olavide University, 41013 Seville, Spain
fmaralv@upo.es

Abstract. Each year, earthquakes pose a significant threat to human life, attributed to their sudden and unpredictable nature. Over time, a heightened awareness of this phenomenon has driven increased attention from researchers and experts. This paper seeks to demonstrate the applicability of the k-means algorithm to seismic data, focusing on the identification of seismic zones in Algeria. Initially, we conducted a comprehensive review of existing literature on clustering seismic data, revealing an unexplored niche in the context of Algeria's seismicity. Subsequently, we introduce our dataset comprising 5876 seismic events. A detailed explanation of the k-means algorithm is provided, with a breakdown of each parameter. Visualization of our findings, including determining the optimal value for k using Elbow and Silhouette scores, is presented and thoroughly discussed. In conclusion, we identify and delineate the seismic zones in Algeria, highlighting the four most critical regions encapsulating these zones. This study contributes to a better understanding of seismic patterns in Algeria, potentially aiding in the development of more effective earthquake preparedness and mitigation strategies.

Keywords: Seismic zones · Algerian Seismicity · *k*-means clustering · Earthquakes

1 Introduction

Algeria stands out as one of the African countries most susceptible to seismic activities. Situated within the geographical coordinates of 19°-38.5° N and 9.5° W- 12.5° E, it lies on the convergent plate boundary of the African and Eurasian tectonic plates. The north-central region experiences notably intense seismic activities, contrasting with the milder occurrences in the east and west extremes.

B. Djamaa et al. (Eds.): CSA 2024, LNNS 1145, pp. 396–405, 2025.
https://doi.org/10.1007/978-3-031-71848-9_36

Conversely, the southern part of the country experiences minimal seismic activity. Given Algeria's strategic location, a comprehensive global analysis and study of its earthquake characteristics are strongly recommended. Seismic zoning for a confined area poses a well-recognized challenge, as tracking earthquake occurrences and analyzing features of similar areas is crucial [24].

Several studies attempted to study Algeria's seismicity, focusing on specific regions to study their seismic characteristic, including [2,5], and [13]. For instance, [13] examined the Constantine region, [2] investigated the Oran region, and [5] studied the Algiers region. However, none has attempted to partition seismic events across Algeria.

Various clustering algorithms, such as k-means [9], DBScan [4] and Hierarchical clustering [10] are well-established in unsupervised learning. Advances in sensors and data storage tools have elevated the field, enhancing information extraction [3]. Geological data, particularly earthquake events, holds paramount importance for scientists, providing valuable insights into phenomena like tsunamis and volcanic eruptions [12,18].Recently, clustering has become a pivotal tool for identifying patterns and structures within seismic zones. This study specifically applies the K-Means algorithm to analyze seismic zones in Algeria. Widely used for clustering and partitioning data, the K-Means algorithm is instrumental in categorizing seismic events. Silhouette and Elbow methods assist in determining the optimal number of clusters, providing a robust approach to identifying patterns.

This paper investigates whether Algerian seismic episodes and seismicity occur in specific regions and explores their temporal correlation. The objective is to discover and understand distinct activity zones within Algeria's seismogenic structures. The clustering algorithm k-means is utilized to cluster earthquakes, extracting seismic zones based on the temporal parameters of events captured in Algeria. The remainder of the paper is organized as follows: Sect. 2 provides a review of related work; Sect. 3 outlines the data processing methodology, including the clustering algorithm and time clustering; Sect. 4 presents and discusses the results; and finally, Sect. 5 concludes the paper.

2 Related Work

The clustering of seismic zones is a crucial aspect of research for scientists studying seismic activities. Numerous studies worldwide have delved into extracting seismic zones in various regions. For instance,[26] presented a statistical approach for cluster analysis, distinguishing between two separate earthquake populations: clustered and non-clustered. [1] undertook the clustering of seismic zones in Iran, while [22] employed the k-means algorithm to identify seismic zones with shallow events in the Aegean Area. Additionally, [7] used the k-means algorithm to detect seismic zones in India. In Indonesia, [15] applied the k-means algorithm to cluster seismic events successfully, underscoring the algorithm's efficacy in identifying seismic zones. Notably, several studies have explored Space-Time clustering, as evidenced by works like those of [16,19]. Subsequently, [23] devised a density-based clustering technique specifically tailored for evaluating the three-dimensional spatial patterns of short-term seismic events, with a particular focus on mining activities. Together, these investigations underscore the importance of clustering in comprehending seismic activity and its potential utility in hazard assessment. In contrast, there has been no comparable research conducted in Algeria to extract seismic zones, despite the existence of numerous seismic catalogs. This notable gap in the literature motivated our undertaking of this study.

3 Clustering

Our approach is achieved in four steps, namely data preparation, determining the optimal number of clusters, clustering, eliminating unnecessary clusters and merging the closest clusters. We detail such steps in the following subsections:

3.1 Data Sources

The data used for clustering is obtained from the homogenized catalog [11], from the year 1960 to 2020 and an extended catalog to fill the gap between 2020 and 2023 from ISC. In sum, there are 5876 events where the magnitude is unified for all the events which is moment magnitude "Mw" the best magnitude to be used in such studies depicted in Fig. 1.

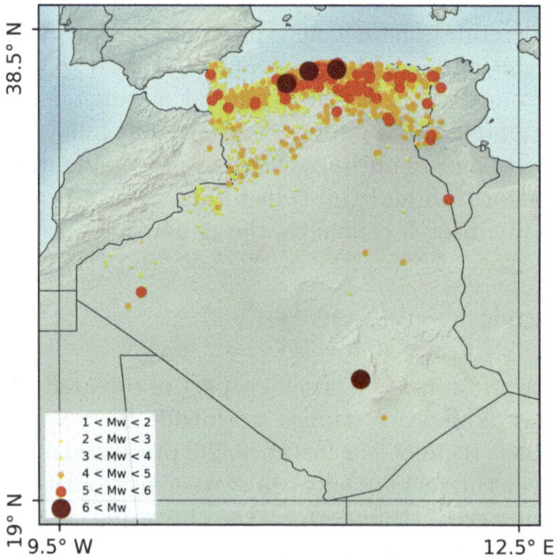

Fig. 1. All the events occurred in the catalog mapped with their magnitude for the whole area of Algeria.

3.2 Space Clustering

The process of identifying seismic zones entails selecting the noteworthy events from our catalog. To establish the lower threshold for earthquake magnitude that piques our interest, we refer to findings from previous earthquake studies. According to [1], earthquakes with a magnitude of 5 or greater are considered significant and warrant in-depth examination. It is crucial to acknowledge that Iranian earthquakes generally occur at greater depths compared to those in Algeria, where they typically manifest between depths of 10 km to 20 km. Consequently, we deem earthquakes with a magnitude of 4 or higher as potentially hazardous, especially in urban areas characterized by higher population density.

k-Means Algorithm. The k-means algorithm is widely recognized as one of the most prominent clustering techniques categorized under unsupervised learning algorithms in the field of machine learning. It is employed to delineate clusters within a dataset based on the similarity between its elements. The parameter 'k' denotes the number of clusters, while similarity is determined by the distance between individual elements. Various distance metrics, such as Euclidean Distance, Manhattan Distance, and Minkowski Distance, as discussed by [20], can be utilized to measure the dissimilarity between events. Once the optimal 'k' value is determined, we implemented the k-means algorithm in the seismic catalog, as outlined in this step [25]. k-means algorithm can be formulated with the Formula 1. Where, D is the distance between two instances. When c_i is the number of data points in i cluster and c is the number of clusters, the center of the cluster can be calculated each time by the Formula 2 based on [20].

$$J = \sum_{i=1}^{c} \sum_{j=1}^{c_i} D(x_i, v_j)^2 \qquad (1) \qquad\qquad v_j = (1/c_i) \sum_{j=1}^{c_i} x_i \qquad (2)$$

Determining the Optimal Number of Clusters. To create clusters, the process consists of certain parameters which are a minimum magnitude which makes it considerable, k parameters which is the number of clusters, and the distance metric. There are many methods for k-value selection. The most effective methods are Elbow and Silhouette according to [8,17,21]. Therefore, we used these methods to get the optimal k value. In the first step, we applied the Elbow algorithm to the seismic catalog with the Sum of Squared Errors (SSE) measurement as in Formula 3, which is also called cluster inertia. In addition, we also applied the Silhouette algorithm to select the best k values for k-means clustering. Finally, the best choice of distance metric for k-means is the Minkowski detailed in Formula 4, where each event is defined with its latitude and longitude.

$$SSE = \sum_{i=1}^{n} (d)^2 \qquad (3) \qquad\qquad D(x, y) = \left(\sum_{i=1}^{n} |x_i - y_i|^p \right)^{1/p} \qquad (4)$$

Where d is the distance between the cluster center and its elements [14]. And, D is the distance between x and y in n dimensions and the order of the norm p [17].

Determining the Seismic Regions. Extracting the seismic regions in Algeria requires some processing. First, we have to cluster the seismic zones with the k means algorithm, and it is preferred to raise the number of clusters as much as possible so we can distinguish them from each other. Then, we should remove the ones with a low density of events. Finally, we gather the lasting clusters in regions. Thus, we created 70 clusters, removed the ones that had less than 20 events, and grouped the rest in groups (regions).

3.3 Time Clustering

Clustering extends beyond spatial dimensions and applies to the temporal aspect, allowing events to be grouped based on time. Time clustering, a pivotal step in geostatistical studies, enhances data accessibility. The primary data source for this analysis is a catalog detailed in the data section, encompassing 5876 events, with a focus on the parameters of date and magnitude [6]. Time clustering involves grouping data into clusters based on temporal criteria. In our context, this process entails categorizing earthquakes into multiple clusters, each linked to a specific date. Initially, earthquakes are clustered according to year seasons across various magnitude ranges. Subsequently, events are grouped into 12 monthly clusters, each further divided based on magnitude ranges. This approach aims to discern periods when seismic activity is more likely, providing insights into temporal patterns and earthquake occurrences.

4 Result and Discussion

Applying the k-means algorithm to data has yielded many valuable results, which will be discussed in detail later. Additionally, Fig. 2 illustrates Elbow diagram, indicating the optimal k value as 18, Similarly, Fig. 3 displays the silhouette score diagram, with the chosen k value of 28. Seismicity clustering of Algeria based on a k value of 18 is depicted in Fig. 4, while the seismicity clustering of Algeria based on a k value of 28 is presented in Fig. 5.

To focus on extracting zones with a concentration of earthquakes, we increased the number of clusters to 70 and excluded those with fewer than 20 events, resulting in a final mapping of 17 clusters shown in Fig. 6. These clusters, outlined in Fig. 6, are categorized into four major regions: Oran area with two clusters, Chlef region with four clusters, Algiers region the largest with seven clusters, and Bejaia region with four clusters detailed in Table 1.

Notably, these regions are identified as the most vulnerable to seismic activity, primarily situated along the north coast of Algeria, flanking both sides of the Tellian and Saharan Atlas Mountain ranges. Furthermore, the time clustering of earthquakes by seasons is presented as a bar chart in Fig. 7. It illustrates that Earthquake events with a magnitude superior to 5 occur in the winter and fall seasons, while Earthquake events with a magnitude inferior to 5 occur in the spring and summer seasons. Figure 8 complements this by providing a bar chart of time clustering based on months. It illustrates that Earthquake events with

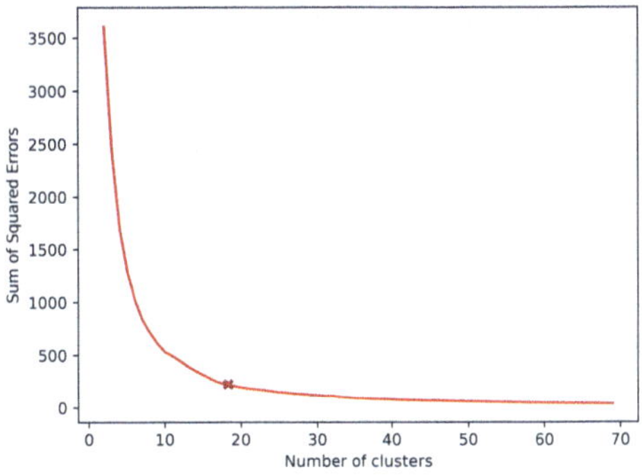

Fig. 2. Elbow diagram with 18 clusters.

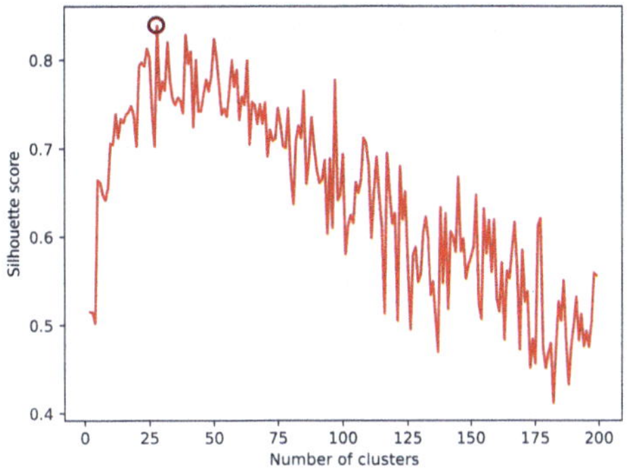

Fig. 3. Silhouette diagram with 28 clusters.

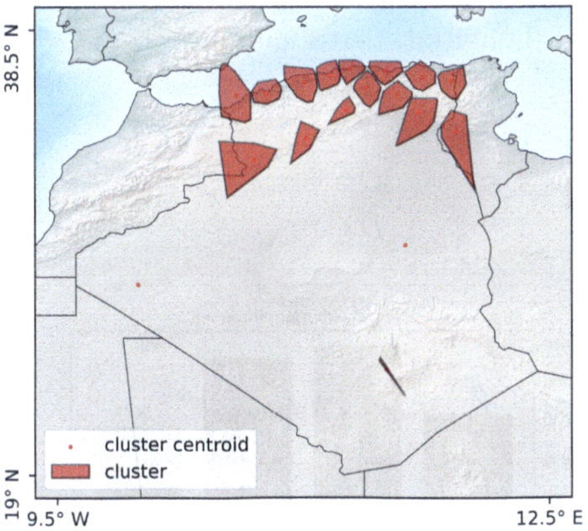

Fig. 4. The clusters of the seismic events for $k = 18$ and $Mw > 4$.

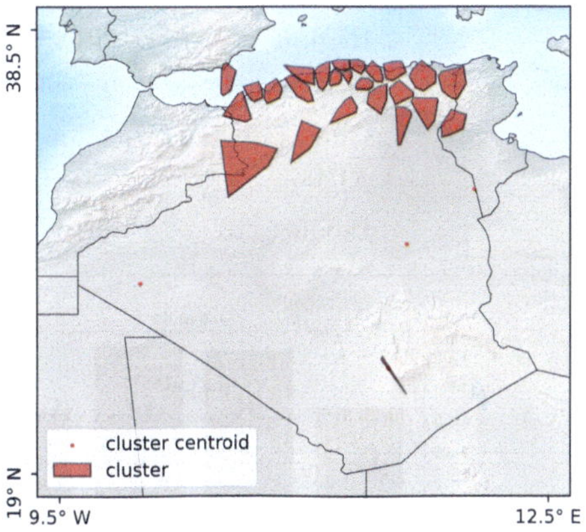

Fig. 5. The clusters of the seismic events for $k = 28$ and $Mw > 4$.

a magnitude superior to 5 occur in the May and October months, while Earthquake events with a magnitude inferior to 5 occur in May. The results of time clustering reveal that the seasons exhibit relatively similar seismic activity when considering all magnitude ranges. However, as the magnitude increases, there is a noticeable rise in the number of events during winter and fall. Conversely, with higher magnitudes, there is a decrease in the number of events during summer, the hottest season.

In contrast, the clustering analysis by month provides more detailed insights. May and October emerge as the most seismic months in Algeria, each accounting for 14% of large earthquakes. On the other hand, the months preceding May and October, such as March, April, June, July, August, and September,

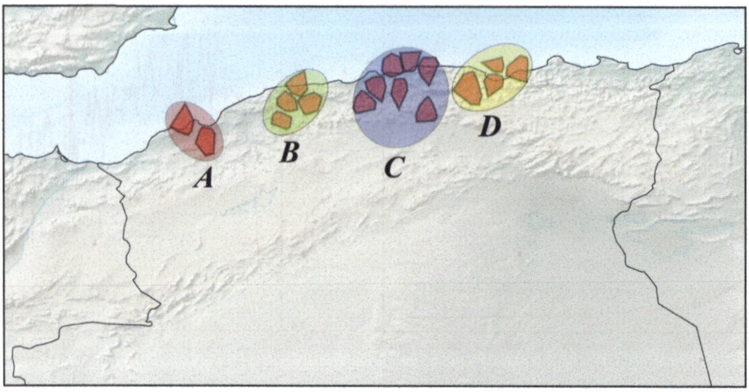

Fig. 6. The four major seismic regions.

Table 1. The four major seismic regions

Region	Clusters	Latitude	Longitude	Letter on map
Oran	2	35.8	−0.3	*A*
Chlef (El Asnam)	4	35.2	1.5	*B*
Algiers	7	36.7	3	*C*
Bejaia	4	36.6	5	*D*

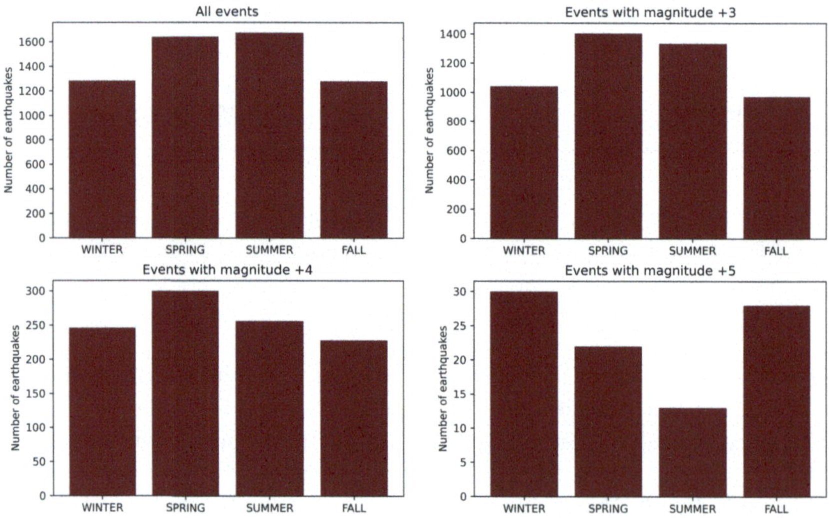

Fig. 7. Clustering the events according to seasons.

exhibit lower seismicity. Consequently, the concluding months of spring and summer stand out as the most seismic, collectively representing a third of all significant seismic events. Additionally, the hot season is characterized as the

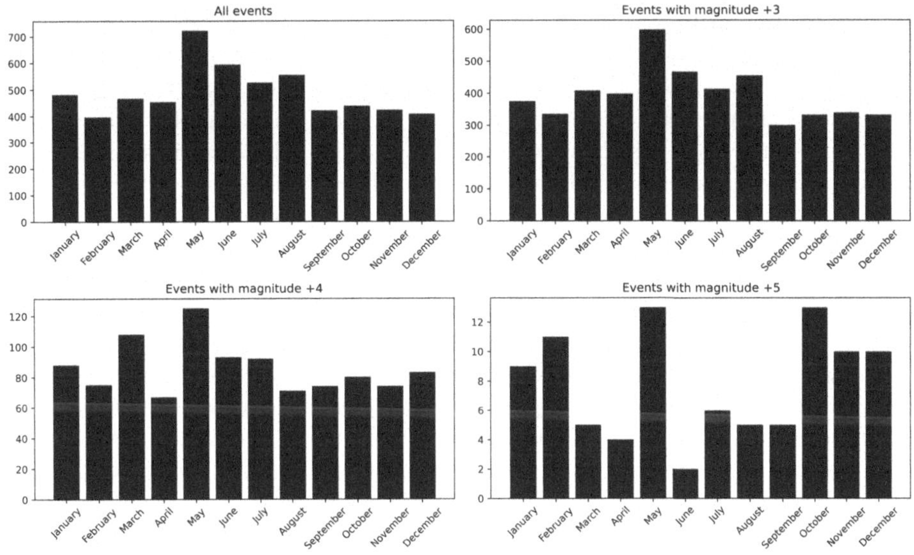

Fig. 8. Clustering the events according to Months.

least seismic period for large events. Additionally, conducting comparative analyses with seismic data from different regions presents an exciting avenue for future research. By juxtaposing seismic data from diverse geographic locations, researchers can elucidate commonalities and differences in seismic activity patterns, thereby enriching our understanding of regional seismic dynamics. Such comparative analyses could reveal underlying factors influencing seismic risk, including geological characteristics, tectonic plate boundaries, and urbanization patterns.

Although the insights drawn from our research, it is critical to acknowledge numerous limitations inherent in our technique. Firstly, the efficacy of the k-means algorithm may be modified by the starting centroids chosen and the distance measure used. Additionally, while our analysis focused on seismic events exceeding a magnitude threshold of 4 degrees, variations in data quality and reporting may have affected the completeness of the seismic catalog. Moreover, our study employed the k-means algorithm to identify seismic zones within Algeria, leveraging both spatial and temporal clustering techniques. However, future research endeavors could benefit from exploring alternative clustering algorithms to enhance the robustness and accuracy of seismic zone delineation. Algorithms such as hierarchical clustering, density-based clustering (e.g., DBSCAN), or spectral clustering may provide additional insights and allow for a more complete understanding of seismic activity patterns.

5 Conclusions

In this study, we employ both spatial and temporal clustering through the k-means algorithm to extract seismic zones in Algeria, utilizing the date parameter to cluster events based on seasons and months. The data is sourced from a homogenized catalog. Initially, we mapped all events from the catalog to gain a comprehensive understanding, revealing that the majority of events are concentrated in the northern region of Algeria. Furthermore, we selectively retained meaningful events with a magnitude exceeding 4°. For the k-means algorithm, we opted for the Minkowski metric to measure distances and determined the optimal 'k' using both the Elbow and Silhouette methods, resulting in values of 18 and 28, respectively, as illustrated in Figs. 4 and 5. Subsequently, we identified the most seismic zones, organized into regions detailed in Table 1, with two major cities, Algiers and Oran, among them. It's noteworthy that only significant seismic events were considered. In contrast, for time clustering, Fig. 7 shows clustering according to seasons, and Fig. 8 shows the clustering according to months, where May and October are the most seismic ones for large events which are before and after the lowest seismic season Summer. In summary, the four extracted regions will know significant events in the future side to side with the May and October months, so it is extremely important to take into consideration the structure of the building to avoid any human and material losses. However, additional study is needed to solve the stated shortcomings and improve the prediction capacities of seismic hazard assessments.

References

1. Ansari, A., Noorzad, A., Zafarani, H.: Clustering analysis of the seismic catalog of iran. Comput. Geosciences **35**(3), 475–486 (2009)
2. Bouhadad, Y., Laouami, N.: Earthquake hazard assessment in the oran region (northwest algeria). Nat. Hazards **26**(3), 227–243 (2002)
3. Celebi, M.E., Aydin, K.: Unsupervised learning algorithms, vol. 9. Springer (2016)
4. Ester, M., Kriegel, H.P., Sander, J., Xu, X., et al.: A density-based algorithm for discovering clusters in large spatial databases with noise. In: kdd. vol. 96, pp. 226–231 (1996)
5. Harbi, A., et al.: Seismicity, seismic input and site effects in the Sahel-Algiers region (North Algeria). Soil Dyn. Earthq. Eng. **27**(5), 427–447 (2007)
6. Howarth, J.D., et al.: Spatiotemporal clustering of great earthquakes on a transform fault controlled by geometry. Nat. Geosci. **14**(5), 314–320 (2021)
7. Kamat, R., Kamath, R.: Earthquake cluster analysis: K-means approach. J. Chem. Pharm. Sci. **10**(1), 250–253 (2017)
8. Kodinariya, T.M., Makwana, P.R.: Review on determining number of cluster in k-means clustering. Int. J. **1**(6), 90–95 (2013)
9. Likas, A., Vlassis, N., Verbeek, J.J.: The global k-means clustering algorithm. Pattern Recogn. **36**(2), 451–461 (2003)
10. Maimon, O., Rokach, L.: Data mining and knowledge discovery handbook, vol. 2. Springer (2005). https://doi.org/10.1007/b107408

11. Mazari, O.S., Sebaa, A., Amaro-Mellado, J.L., Martínez-Álvarez, F.: Creating a homogenized earthquake catalog for Algeria and mapping the main seismic parameters using a geographic information system. J. Afr. Earth Sc. **201**, 104895 (2023)
12. Mimura, N., Yasuhara, K., Kawagoe, S., Yokoki, H., Kazama, S.: Damage from the great east japan earthquake and tsunami-a quick report. Mitig. Adapt. Strat. Glob. Change **16**(7), 803–818 (2011)
13. Mouloud, H., Badreddine, S.: Probabilistic seismic hazard assessment in the constantine region, northeast of algeria. Arab. J. Geosci. **10**, 1–20 (2017)
14. Nainggolan, R., Perangin-angin, R., Simarmata, E., Tarigan, A.F.: Improved the performance of the k-means cluster using the sum of squared error (SSE) optimized by using the elbow method. J. Phys. Conf. Ser. **1361**, 012015 (2019). IOP Publishing
15. Novianti, P., Setyorini, D., Rafflesia, U.: K-means cluster analysis in earthquake epicenter clustering. Int. J. Adv. Intell. Informatics **3**(2), 81–89 (2017)
16. Santoyo, M.A., Singh, S.K., Mikumo, T., Ordaz, M.: Space-time clustering of large thrust earthquakes along the Mexican subduction zone: an evidence of source stress interaction. Bull. Seismol. Soc. Am. **95**(5), 1856–1864 (2005)
17. Saputra, D.M., Saputra, D., Oswari, L.D.: Effect of distance metrics in determining k-value in k-means clustering using elbow and silhouette method. In: Sriwijaya International Conference on Information Technology and Its Applications (SICONIAN, 341–346. Indonesia: Atlantis Press (2020)
18. Segall, P.: Earthquake and Volcano Deformation. Princeton University Press (2010)
19. Shearer, P.M.: Space-time clustering of seismicity in California and the distance dependence of earthquake triggering. J. Geophysical Res. Solid Earth **117**(B10) (2012)
20. Singh, A., Yadav, A., Rana, A.: K-means with three different distance metrics. Int. J. Comput. Appl. **67**(10) (2013)
21. Syakur, M., Khotimah, B., Rochman, E., Satoto, B.D.: Integration k-means clustering method and elbow method for identification of the best customer profile cluster. In: IOP Conference Series: Materials Science and Engineering, vol. 336, pp. 012017. IOP Publishing (2018)
22. Weatherill, G., Burton, P.W.: Delineation of shallow seismic source zones using k-means cluster analysis, with application to the Aegean region. Geophys. J. Int. **176**(2), 565–588 (2009)
23. Woodward, K., Wesseloo, J., Potvin, Y.: A spatially focused clustering methodology for mining seismicity. Eng. Geol. **232**, 104–113 (2018)
24. Yang, J., Cheng, C., Song, C., Shen, S., Zhang, T., Ning, L.: Spatial-temporal distribution characteristics of global seismic clusters and associated spatial factors. chinese geogr. sci. **29**, 614–625 (2019)
25. Yuan, R.: An improved k-means clustering algorithm for global earthquake catalogs and earthquake magnitude prediction. J. Seismolog. **25**, 1005–1020 (2021)
26. Zaliapin, I., Gabrielov, A., Keilis-Borok, V., Wong, H.: Clustering analysis of seismicity and aftershock identification. Phys. Rev. Lett. **101**(1), 018501 (2008)

Matrix Factorization for Cloud Service Recommendation Based on Social Trust

Fatma Zohra Lebib[1(✉)], Ichrak Djebrit[2], and Khadidja Mahmoudi[2]

[1] Research Center for Scientific and Technical Information, CERIST, Algiers, Algeria
zmatouk@cerist.dz
[2] University of Science and Technology Houari Boumediene, USTHB, Bab Ezzouar, Algeria

Abstract. Recommending trustworthy cloud services is essential to establishing credibility and ensuring better user decision-making based on their specific needs. Traditional recommendation approaches based on collaborative filtering face many challenges, including cold start problem, data sparsity and low accuracy and reliability of recommendations. We propose a social trust-based recommendation approach using matrix factorization which improves the recommendation accuracy and alleviates the problems of traditional recommendation systems. First, users' trust level is inferred from their interactions on social networks. Then, the social trust model is integrated with matrix factorization to generate reliable recommendations for users. Experiments conducted on two public datasets Epinions and WSdream demonstrate that social trust significantly improves recommendation accuracy in terms of error minimization compared to state-of-the-art recommender systems that do not take trust into account.

Keywords: cloud service · trust-based recommendation system · social trust · matrix factorization

1 Introduction

In cloud computing, resources are provided as a service (Infrastructure-as-a-Service, Platform-as-a-Service and Software-as-a-Service). These resources are provided on-demand via the Internet and on a pay-as-you-go basis [19]. Large IT companies like Amazon, IBM, Microsoft, and Google are massively supporting cloud computing and offering different types of cloud services with different capabilities. The large number of functionally equivalent cloud services presents users with a major challenge in selecting the most appropriate services for their requirements. Quality of Service (QoS) attributes represent non-functional performance aspects for measuring performance, reliability and quality of service, such as throughput, response time, reliability, availability, etc. QoS provides a basis for selecting cloud services from a set of functionally equivalent candidates [19].

B. Djamaa et al. (Eds.): CSA 2024, LNNS 1145, pp. 406–415, 2025.
https://doi.org/10.1007/978-3-031-71848-9_37

The rapid growth in the number of existing resources has led to an increased need for an efficient and intelligent system [7,9] to help the user choose which resources to use according to their needs [8]. Service recommendation systems based on user experience, especially those relying on QoS values observed by the user, have been proposed to solve these problems, which generally adopt similarity-based collaborative filtering techniques [16]. These approaches tend to identify similar users and recommend to the target user the services preferred by the most similar users (nearest neighbors). However, users similar to the target user are not necessarily trustworthy. The results recommended by these systems are therefore probably not reliable. Trust-based recommendation has emerged as a promising avenue to improve rating prediction. Leveraging additional information provided by a trust network among users helps solve the cold start problem and provide more personalized and reliable suggestions. Trust is a fundamental factor in the context of cloud services [1], as users rely on these services to store, process, and manage their data. The guarantees mentioned in formal contracts between users and cloud providers in SLAs (Service Level Agreements) are not sufficient to establish trust between consumers and cloud providers due to their clauses which can be vague and inconsistent [3]. In this work, we propose to integrate social trust into the recommendation of cloud services using Matrix Factorization (MF) [5]. MF-based techniques have shown their effectiveness to improve the precision of the rating prediction under high sparsity [17]. However, traditional MF methods only use information from user feedback to predict missing QoS values, without considering trust (user or service side). Trust can be represented as an additional layer, introducing a level of confidence or influence based on the trustworthiness inferred from social networks. Cloud user reviews have been considered an important source of information to evaluate trust in the cloud [3,11,12]. Today, social networks have become another important source for analyzing trust relationships between users [13,15]. This work aims to exploit relationships between users on social networks in recommending cloud services to guarantee trust and credibility in the cloud.

In the remainder of the article, we find in Sect. 2 a summary of the documentary research on service recommendation. Section 3 describes the proposed approach. Section 4 presents the results of the experiments carried out on real datasets. Section 5 presents the conclusion of the article.

2 Related Work

Faced with the proliferation of cloud services offering similar functionalities, the community of service recommendation researchers is starting to take much more interest in the cloud domain. They first focused more on historical QoS ratings to generate recommendations [19]. QoS attributes alone are not a sufficient criterion to generate more accurate recommendations in the cloud environment. Different users, across diverse locations and devices, may perceive different QoS values for the same attribute due to Internet conditions (bandwidth variations, geographical distance). Additionally, the presence of malicious ratings can negatively impact the prediction accuracy. The recommendation should filter out malicious users who might inject unfair ratings. Trust has therefore become another key factor to take into account in recommending cloud services, trust-based approaches have then been proposed [6,11,18]. These approaches typically use collaborative filtering [16] to select the trustworthy neighbors needed to generate predictions for the active user. However, collaborative filtering-based approaches face data sparsity issues that involve generating inaccurate predictions when user and service data is missing. Recommendation approaches based on matrix factorization technique are proposed to overcome the sparsity challenge [4,5,10,17]. Matrix factorization has proven to be a powerful technique for capturing latent factors and making accurate predictions in recommendations based on collaborative filtering. We propose to integrate social trust into matrix factorization to create a recommendation system that not only provides precise and relevant cloud service recommendations but also ensures trustworthiness in the recommendations. By considering both user-service interactions and social trust, the system aligns more closely with users' preferences and social contexts, ultimately leading to a positive user experience and satisfaction.

3 Proposed Recommendation Approach

A recommendation approach based on social trust using matrix factorization, called CSRec-STMF (Cloud Service Recommendation using Social Trust with Matrix Factorization) is proposed. Before presenting the CSRec-STMF process, a formal problem description is given below.

3.1 Problem Definition

Consider n users, such that $S = \{u_1, ...u_n\}$ and m cloud services, $C = \{c_1, ...c_m\}$. $Q = \{q_1, ..., q_k\}$ is a set of k QoS attributes (e.g. throughput, bandwidth, etc.). The QoS value aggregated over all attributes is calculated and normalized in the interval [0 1] (see Sect. 2). Let R be the rating matrix, $R = [R_{u,c}(n \times m)]$, $R_{u,c}$ designates the score of QoS given by user u to cloud service c, which represents the aggregated and nominalized QoS value given in Eq. 2. Let T be the trust matrix, such that: $T = [Tu, v(n \times n)]$, $T_{u,v}$ is the trust level from u to v. We formalize the recommendation problem as follows:

Input : R, T, user u ($u \in S$) and cloud service c ($c \in C$), the value of $R_{u,c}$ is unknown

Output : the predicted rating value $\hat{R}_{u,c}$

3.2 Combination of QoS Criteria

The QoS attributes that characterize a cloud service can have different dimensions and value ranges. Therefore, the QoS value for each attribute is calculated and normalized in the interval [0, 1] according to the two categories: "cost" and "benefit" [2] (Eq. 1).

$$\bar{Q}^i_{u,c} = \begin{cases} \dfrac{Q^i_{u,c} - Q^i_{min}}{Q^i_{max} - Q^i_{min}}, & Q^i_{u,c} \text{ is "benefit"} \\ \dfrac{Q^i_{max} - Q^i_{u,c}}{Q^i_{max} - Q^i_{min}}, & Q^i_{u,c} \text{ is "cost"} \end{cases} \quad (1)$$

where, $Q^i_{u,c}$ is the QoS value of attribute i given by user u to service cloud c. Q^i_{min} and Q^i_{max} are the minimum value and maximum value of attribute i, respectively (considering all defined QoS values) and $\bar{Q}^i_{u,c}$ is the normalized QoS value of the i th QoS attribute. Each QoS attribute affects the QoS value with a certain weight. Equation 2 calculates the aggregated QoS value (denoted $Q_{u,c}$) by integrating the QoS attributes with their weights.

$$R_{u,c} = Q_{u,c} = \sum_{i=1}^{k} w_i \times \bar{Q}^i_{u,c} \quad (2)$$

where $w_i \in [0, 1]$, $\sum_{i=1}^{k} w_i = 1$ is the weight assigned to i-th QoS attribute.

3.3 Description of CSRec-STMF : Cloud Service Recommendation Using Social Trust with Matrix Factorization

CSRec-STMF combines social trust and matrix factorization technique to recommend to a given user a list of trusted cloud services that meet their QoS requirements (availability, throughput, response time, etc.). Figure 1 presents the general architecture of CSRec-STMF. The recommendation process of CSRec-STMF is described in Algorithm 1. The steps of the algorithm will be described in detail below.

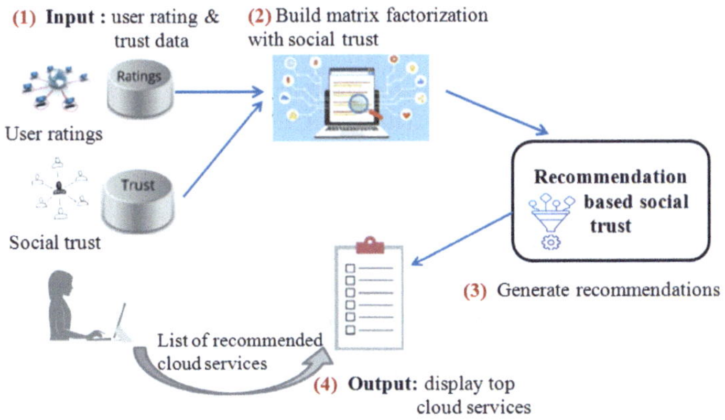

Fig. 1. General architecture of CSRec-STMF.

Algorithm 1. CSRec-STMF Algorithm

Input: user u, user rating, social trust relationships
Output: Top k cloud services
begin
1:Build the rating matrix R from the QoS values given by users to cloud services
2: Build the social trust matrix T from the user trust network
3: Use the classical matrix factorization on R to construct the two latent feature matrices of users and cloud services
4: Integrate the trust matrix T into the users' latent features matrix
5: Build the learning model
6: Generate the predictions (the missing QoS values)
7: Order predictions from maximum value to minimum value
8: Display the top-k cloud services to user u
end

Preprocessing. This step consists of constructing the two rating and trust matrices.

(a) Rating matrix. First, user rating data on cloud services is transformed into a user-cloud service matrix named R. In R, rows represent users and columns represent services. The cell value is the rating score given by a user to a given cloud service, as shown in the Fig. 2a.

(b) Trust matrix. In a trust network, nodes represent users and edges represent trust links between each pair of users. Trust is classified into two categories: explicit trust and implicit trust. In explicit trust, the user explicitly specifies their trusted users while implicit trust is inferred from social network relationships (friendships, user activities on the network, user interactions, etc.). Social media data is analyzed to construct the user's trust matrix. we consider that the existence of a social relationship (friendship or interaction) from u to v means that "u trusts v". In the constructed trust matrix called T (Fig. 2b), the rows and columns represent the users. For example, u and v two users from S, the cell $T_{u,v}$ designates the trust value that user u has towards user v, such that :

$$T_{u,v} = \begin{cases} 1 \text{ if u trusts v} \\ 0 \text{ otherwise} \end{cases}$$

Trust can be direct or indirect. Direct trust means trust obtained through the direct interaction of two users connected directly with a link in the trust network. Indirect trust is inferred by exploiting the different paths that may exist between the two users on the trust network.

Suppose the cloud users are the nodes of the social network, then S represents the set of nodes in trust network. The total number of nodes is equal to n, $n = \mid S \mid$. If user u interacts with v, an edge connecting them is created and the length of the edge represents the direct trust level equal to 1. When no direct path exists between two nodes, the trust propagation in the trust network can be

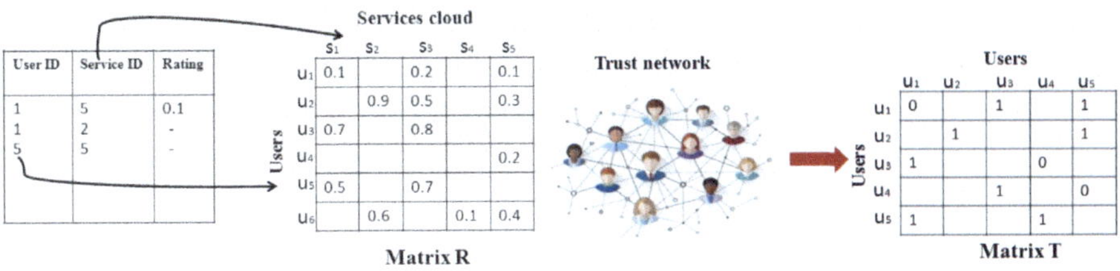

(a) Construction of the rating matrix. (b) Construction of the social trust matrix.

Fig. 2. Pre-processing step.

used to estimate the degree of indirect trust between the two nodes. We propose to use transitivity which is a key property of trust in order to propagate trust in the trust network. The trust between two nodes not directly connected can then be deduced as follows: Let three users u_1, u_2 and u_5 from S such that: $T_{u_5,u_1} = 1$ and $T_{u_1,u_3} = 1$ then $T_{u_5,u_3} = 1$. This implies that user 3 is a direct neighbor of user 5.

Construction of the Social Trust Matrix Factorization. In CSRec-STMF, matrix factorization technique is used to learn the latent features of users and cloud services and predict unknown ratings using these latent features.

Let $U \in R^{K\times n}$ and $V \in R^{K\times m}$ be latent feature matrices of the user and cloud service, respectively. The vectors U_u and V_c represent K-dimensions vectors of user u and cloud service c, respectively. The objective of matrix factorization is to learn the latent features and use them for recommendation, for this the conditional probability formula of observed ratings is used, defined as follows [14]:

$$p(R\backslash U, V, \sigma_R^2) = \prod_{u=1}^{n}\prod_{c=1}^{m}[\mathcal{N}(R_{u,c}\backslash g(U_u^T V_c), \sigma_r^2]^{I_{u,c}^R} \tag{3}$$

where $\mathcal{N}(x\backslash\mu, \sigma^2)$ is the Gaussian distribution with mean μ and variance σ^2, and $I_{u,c}^R$ is the indicator function that is equal to 1 if u rated service c and equal to 0 otherwise. The function $g(x) = 1/(1 + e^{-x})$, which bounds the range of $U_u^T V_c$ within [0,1].

In this step, we integrate social trust into the matrix factorization. From the trust matrix T, each user u has a N_u set of direct trust neighbors (the trusted users). Let u and v be two users, such that $u, v \in n$, if $T_{u,v} = 1$ then user v is a direct neighbor of user u. User u is affected by his/her direct trust neighbors N_u, that is, the latent feature vector of u depends on the latent feature vectors of all his/her direct trust neighbors v, $v \in N_u$, as shown in Eq. 4.

$$\hat{U} = \frac{\sum_{v\in N_u} T_{u,v} U_v}{\sum_{v\in N_u} T_{u,v}} = \frac{\sum_{v\in N_u} T_{u,v} U_v}{| N_u |} \tag{4}$$

where $\hat{U}$ is the estimated latent feature vector of u that depends on user vectors of his/her direct trust neighbors. Equation 4 shows that the estimated user's latent feature vector is the weighted average of the latent feature vectors of his/her direct trust neighbors. The estimated feature vector of user u can be derived as follows:

$$\begin{pmatrix} \hat{U}_{u,1} \\ \hat{U}_{u,2} \\ \dots \\ \hat{U}_{u,K} \end{pmatrix} = \begin{pmatrix} U_{1,1} & U_{2,1} & \dots & U_{N,1} \\ U_{1,2} & U_{2,2} & \dots & U_{N,2} \\ \dots & \dots & \dots & \dots \\ U_{1,K} & U_{2,K} & \dots & U_{N,K} \end{pmatrix} \begin{pmatrix} T_{u,1} \\ T_{u,2} \\ \dots \\ T_{u,N} \end{pmatrix} \tag{5}$$

Adding trust to the matrix factorization does not change Eq. 3 of the conditional probability of observed rantings of cloud services but only that of latent user features. Two factors are considered for the user latent features : the zero Gaussian distribution to avoid overfitting, and the conditional distribution of user latent features given the latent features of his/her direct trust neighbors, defined in the equation below.

$$\begin{gathered} p(U \backslash T, \sigma_U^2, \sigma_T^2) \propto p(U \backslash \sigma_U^2) \times p(U \backslash T, \sigma_T^2) = \\ \prod_{u=1}^{N} \mathcal{N}(U_u \backslash 0, \sigma_U^2 \mathbf{I}) \times \prod_{u=1}^{N} \mathcal{N}(U_u \backslash \sum_{v \in N_u} T_{u,v} U_v, \sigma_T^2 \mathbf{I}) \end{gathered} \tag{6}$$

The Fig. 3 shows the graphical model of the proposed recommendation approach based on social trust.

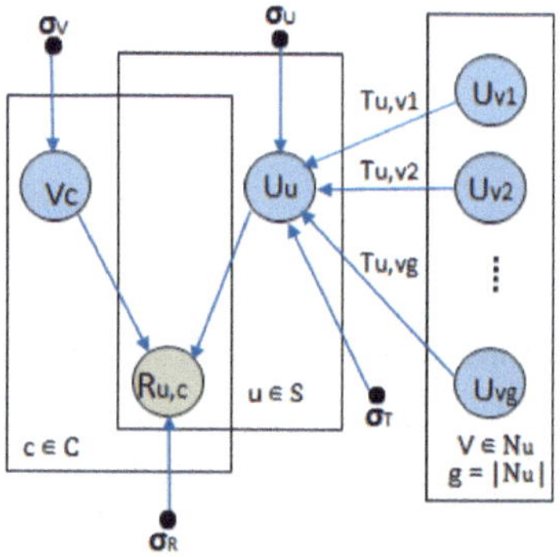

Fig. 3. Graphical model of social trust integration in matrix factorization.

Construction of Learning Matrices for Prediction. The learning matrices are used to find the right factorizations for the latent feature matrices of users and cloud services. For this, gradient descent is used on U_u and V_c for all users and all cloud services to find a local minimum of the objective function (noted L[1]), using Eq. 7 and Eq. 8 respectively [4].

$$\frac{\partial L}{\partial U_u} = \sum_{c=1}^{m} I_{u,c}^{R} V_c g'(U_u^T V_c)(g(U_u^T V_c) - R_{u,c}) + \lambda_U U_u$$
$$+\lambda_T(U_u - \sum_{v \in N_u} T_{u,v} U_v)) - \lambda_T \sum_{v \backslash u \in N_v} T_{u,v}(U_v - \sum_{w \in N_v} T_{v,w} U_w) \quad (7)$$

$$\frac{\partial L}{\partial V_c} = \sum_{u=1}^{n} I_{u,c}^{R} U_v g'(U_u^T V_c)(g(U_u^T V_c) - R_{u,c}) + \lambda_V V_c \quad (8)$$

where, $\lambda_U = \frac{\sigma_R^2}{\sigma_U^2}$, $\lambda_V = \frac{\sigma_R^2}{\sigma_V^2}$, $\lambda_T = \frac{\sigma_R^2}{\sigma_T^2}$, $g'(x) = \frac{e^{-x}}{(1+e^{-x})^2}$

The model built is used for prediction using the following formula:

$$\hat{R}_{u,c} = \sum_{v \in N_u} \hat{U} V_c \quad (9)$$

Cloud services are ranked based on their predicted QoS values from largest to smallest, and then the top-k cloud services are recommended to the user.

4 Experiments

4.1 Datasets

Due to the lack of real data on the cloud service, the web service dataset WS-DREAM[2] is used. WS-DREAM describes the real QoS of the 5825 web service provided by 339 users, including response time and throughput. For trust data, the social network Epinion[3] is used, which describes 582613 trust relationships between 163651 users. The dataset is divided into a training set used to train the models, a validation set used to choose the best model, and a test set used for experiments.

4.2 Evaluation Metrics

We evaluate the performance of the proposed approach on the basis of error minimization (i.e., the ability to provide correct results), using Mean Absolute Error

[1] The objective is to minimize the difference between the observed ratings in the original matrix R and the predicted ratings obtained from the factorized matrices.

[2] https://wsdream.github.io/.

[3] The Epinions dataset is constructed from a who-trust-whom online social network of a general consumer review site Epinions.com.

(MAE) and Squared Error average (RMSE). Smaller MAE and RMSE values mean higher prediction accuracy. MAE and RMSE are calculated as follows.

$$MAE = \frac{1}{|R_{pred}|} \sum_{u,c \in R_{pred}} |(R_{u,c} - \hat{R}_{u,c})| \quad RMSE = \sqrt{\frac{\sum_{u,c \in R_{pred}} (R_{u,c} - \hat{R}_{u,c})^2}{|R_{pred}|}} \tag{10}$$

where $\bar{R}_{u,c}$ and $R_{u,c}$ denote respectively the predicted and real rating values for the pair (u,c) and R_{pred} is the set of all predicted values.

4.3 Experiment Results

The performance of CSRec-STMF is compared with the state-of-the-art item-based collaborative filtering approach, denoted item-item CF. Performance results on test data consisting of 66 users, 4500 services and 1553 trust relationships are shown in Table 1, where the best MAE and RMSE values are in boldface. The results show that the proposed approach outperforms the item-based CF approach by providing a low error rate. We conclude that integrating social trust into the recommendation process helps improve recommendation accuracy, thereby ensuring credibility and reliability between cloud users and providers. CSRec-STMF learns the feature vector of each user using the feature vectors of his direct (trusted) neighbors in the social network. One limitation of the proposed approach is the lack of direct trust neighbors of a given user which can lead to low prediction quality, which is due to the sparsity of the trust matrix. A potential solution is to study the propagation of trust in the network by taking into account the distance and quality of trust links between users.

Table 1. Evaluation results of the proposed approach

Algorithm/Metric	MAE	RMSE
CSRec-STMF (K=5)	**0.5367** (+17%)	**0.7031** (+15.71%)
item-item CF	0.6467	0.8341

5 Conclusion

This paper proposes CSRec-STMF, a novel recommendation approach that integrates social links into matrix factorization to improve the accuracy and reliability of collaborative filtering-based cloud service recommendations. Experiments carried out on two public datasets Wsdream and Epinions show that CSRec-STMF outperforms state-of-the-art item-based collaborative filtering in accuracy and error minimization. It would be interesting to compare CSRec-STMF with trust-based recommendation approaches that ignore social links on real-world datasets from cloud services using other metrics.

References

1. Armbrust, M., et al.: A view of cloud computing. Commun. ACM **53**, 50–58 (2010)
2. Chen, M., Ma, Y.: A hybrid approach to web service recommendation based on QoS-aware rating and ranking. CoRR arxiv preprint arxiv:abs/1501.04298 (2015)
3. Habib, S., Ries, S., Mühlhäuser, M.: Towards a trust management system for cloud computing. In: TrustCom, pp. 933–939 (2011)
4. Jamali, M., Ester, M.: A matrix factorization technique with trust propagation for recommendation in social networks. In: ACM Conference on Recommender Systems (2010)
5. Koren, Y., Bell, R., Volinsky, C.: Matrix factorization techniques for recommender systems. Computer **42**(8), 30–37 (2009)
6. Kuang, L., et al.: A personalized QoS prediction approach for cps service recommendation based on reputation and location-aware collaborative filtering. Sensors (Basel) **18**(5) (2018)
7. Lebib, F.Z., Drias, H., Mellah, H.: Selection of information sources using a genetic algorithm. In: WorldCIST, pp. 60–70 (2017)
8. Lebib, F.Z., Mellah, H., Meziane, A.: Knowledge discovery from log data analysis in a multi-source search system based on deep cleaning. In: WEBIST, pp. 257–264 (2019)
9. Lebib, F.Z., Mellah, H., Meziane, A.: A multi-dimensional source selection based on topic modelling. J. Inf. Sci. Eng. **38**(3), 619–644 (2022)
10. Li, S., Wen, J., Luo, F., Ranzi, G.: Time-aware QoS prediction for cloud service recommendation based on matrix factorization. IEEE Access **6**, 77716–77724 (2018)
11. Liu, J., Chen, Y.: A personalized clustering-based and reliable trustaware QoS prediction approach for cloud service recommendation in cloud manufacturing. Knowl.-Based Syst **174**, 43–56 (2019)
12. Mahmud, K., Usman, M.: Trust establishment and estimation in cloud services: a systematic literature review. J. Netw. Syst. Manage. **27**, 489–540 (2019)
13. Mayadunna, H., Rupasinghe, L.: A trust evaluation model for online social networks. In: 2018 NITC, pp. 1–6 (2018)
14. Mnih, A., Salakhutdinov, R.R.: Probabilistic matrix factorization. In: Advances in Neural Information Processing Systems, vol. 20 (2007)
15. Rahangdale, R., Thakar, U.: A user action based approach to determine trustworthiness among users in social network. In: International conference on I-SMAC, pp. 148–152 (2017)
16. Schafer, J.B., Frankowski, D., Herlocker, J., Sen, S.: Collaborative filtering recommender systems. In: The Adaptive Web: Methods and Strategies of Web Personalization, pp. 291–324 (2007)
17. Sharifi, Z., Rezghi, M., Nasiri, M.: A new algorithm for solving data sparsity problem based-on non negative matrix factorization in recommender systems. In: ICCKE, pp. 56–61 (2014)
18. Yin, C., Wang, J., Park, J.H.: An improved recommendation algorithm for big data cloud service based on the trust in sociology. Neurocomputing **256**, 49–55 (2017)
19. Zheng, X., Xu, L.D., Chai, S.: QoS recommendation in cloud services. IEEE Access **5**, 5171–5177 (2017)

Author Index

B. Djamaa et al. (Eds.): CSA 2024, LNNS 1145, pp. 417–418, 2025.
https://doi.org/10.1007/978-3-031-71848-9

SPRINGER NATURE

GPSR Compliance

The European Union's (EU) General Product Safety Regulation (GPSR) is a set of rules that requires consumer products to be safe and our obligations to ensure this.

If you have any concerns about our products, you can contact us on ProductSafety@springernature.com

In case Publisher is established outside the EU, the EU authorized representative is:

Springer Nature Customer Service Center GmbH
Europaplatz 3
69115 Heidelberg, Germany